MW00700090

Beverage Basics

Understanding and Appreciating Wine, Beer, and Spirits

Robert W. Small, PhD, and Michelle Couturier

with Michael Godfrey on Beer

WILEY

JOHN WILEY & SONS, INC.

This book is printed on acid-free paper. ∞

Copyright © 2011 by John Wiley & Sons, Inc. All rights reserved

Published by John Wiley & Sons, Inc., Hoboken, New Jersey

Published simultaneously in Canada

No part of this publication may be reproduced, stored in a retrieval system, or transmitted in any form or by any means, electronic, mechanical, photocopying, recording, scanning, or otherwise, except as permitted under Section 107 or 108 of the 1976 United States Copyright Act, without either the prior written permission of the Publisher, or authorization through payment of the appropriate per-copy fee to the Copyright Clearance Center, Inc., 222 Rosewood Drive, Danvers, MA 01923, (978) 750-8400, fax (978) 646-8600, or on the web at www.copyright.com.

Requests to the Publisher for permission should be addressed to the Permissions Department, John Wiley & Sons, Inc., 111 River Street, Hoboken, NJ 07030, (201) 748-6011, fax (201) 748-6008, or online at http://www.wiley.com/go/permissions.

Limit of Liability/Disclaimer of Warranty: While the publisher and author have used their best efforts in preparing this book, they make no representations or warranties with respect to the accuracy or completeness of the contents of this book and specifically disclaim any implied warranties of merchantability or fitness for a particular purpose. No warranty may be created or extended by sales representatives or written sales materials. The advice and strategies contained herein may not be suitable for your situation. You should consult with a professional where appropriate. Neither the publisher nor author shall be liable for any loss of profit or any other commercial damages, including but not limited to special, incidental, consequential, or other damages.

For general information on our other products and services or for technical support, please contact our Customer Care Department within the United States at (800) 762-2974, outside the United States at (317) 572-3993 or fax (317) 572-4002.

Wiley also publishes its books in a variety of electronic formats. Some content that appears in print may not be available in electronic books. For more information about Wiley products, visit our web site at www.wiley.com.

Library of Congress Cataloging-in-Publication Data:

Small, Robert W., 1947-

Beverage basics: understanding and appreciating wine, beer, and spirits / Robert W. Small and Michelle Couturier ; with Michael Godfrey.

p. cm.

Includes index.

ISBN 978-0-470-13883-0 (cloth); ISBN 978-1-118-15714-5 (ebk); ISBN 978-1-118-15715-2 (ebk); ISBN 978-1-118-15716-9 (ebk)

1. Alcoholic beverages. I. Couturier, Michelle, 1947- II. Godfrey, Michael, 1959- III. Title.

TP505.S63 2011

641.2'1--dc22

2010016917

Printed in the United States of America

10 9 8 7 6 5 4 3

Contents

Acknowledgments

We would like to thank the many individuals who provided support on the research for this book and assisted in reviewing the chapters.

A special thanks to John Swetnam, who read the manuscript several times. He offered invaluable advice and, more importantly, encouraged us at every step. Also, thanks to Christine McKnight, our editor at Wiley, who kept us on track.

The following reviewed various chapters and provided tremendous insight: Professor Robert Palmer, Bob's colleague at Cal Poly Pomona, was instrumental in making sure the legal chapter was factually correct, and Dr. Michael Apstein did the same with the chapter on alcohol and health. Dr. Margie Jones of Cal Poly provided reviews and guidance on a number of chapters. Daryl Groom, Mike Stutler, Peter Sichel, Jim Trezise, Dan Berger, Gary Eberle, Jon McPherson, Don Galleano, Harry McWatters, Tim McDonald, Sharron McCarthy, Rebecca Chappa, Robin Kelley O'Connor, and Darrell Corti all provided consultation, and Nelson Barber, James A. Chandler, Robert A. Green, Brian Hay, Wally Rande, Ann Littlefield, Coke Roth, John Hailman, Fred Brander, and Michael Wray reviewed the manuscript at various stages.

Alex Puchner, Owen Williams, Peter D'Souza, Lowell and Lois Godfrey, Jennyfer and Justin Godfrey all provided support on the beer chapter.

The following helped arrange winery visits around the world: Rory Callahan of Wine and Food Associates; Steve Metzler of Classical Wines; Sandrine Dupouy of Boutique Wines; Leann Lindsey of Young's Market; David Schneiderman of Southern Wines and Spirits; Maureen Tripp of Chateau Ste. Michelle; Roberto and Catarina Gerometta; Elizabeth Koenig of Banfi Vintners; Fabrizio Vignolini; Lia Teo; Alain Junguenet of Wines of France; Patrick Terrail; Lucio Gomiero; Sammie Daniels; Roger Voss of Wine Enthusiast; Chris Braun of Ciatti Group; Paul Wagner of Balzac Communications; Niccolo Lorimer; Nick Goldschmidt; Pancho Campo; Javier Arauz, and Wendy Vallester of The Wine Academy of Spain; Charlie and Arden Small; James Healy of Dog Point Winery; Kirk Wille of Loosen Bros. USA; Mary Ellen Cole of The Los Angeles International Wine Competition; Fred Dame of Icon Estates; and Michael Jordan.

The many hosted winery visits included:

Neil McCallum, Dry River; Cloudy Bay Vineyards; James and Wendy Healy, Dog Point Vineyard; Rudi Bauer, Quartz Reef; Steve Smith, Craggy Range; Michael Brajkovich, Kumeu River; Nick Mills, Rippon Vineyard & Winery; Alwyn Corban, Ngatarawa; John Hancock, Trinity Hill; Felton Road; Neudorf Vineyards; Jon McPherson, South Coast Winery; Joe Hart, Hart Winery; Ted Bennett, Navarro Vineyards; Tim McDonald, Gallo; Grady Wann, Quivira; Ernst Loosen, Dr. Loosen Estate; Carlo and Ludovica Fabbri, Savignola Paolina; John Mariani, Castello Banfi; Marzia Morganti Tempestini and Costanza Lorenzet, Carpene Malvolti; Carolina Alvino, Tenuta La Cipressaia; Pacenti Franco, Canalicchio; Dennis Lepore, Bastianich; Roberto Damonte, Malvirà; Francesco Spadafora, Spadafora; Paolo Librandi, Librandi; Lisa Sapienza, Benanti; Francesco Planeta, Planeta; Lluigi Peroni, La

Fiorita; Paola and Paula Bartolommei, Caprili; Filomena Ruppi D'Angelo, D'Angelo; Teresa and Angelo D'Uva, D'Uva; Dr. Libero Rillo, Fontana Vecchia; Francesca Festa, Feudi di San Gregorio; Gianluca Grasso, Elio Grasso; Aldo Vaira, G.D. Vajra; Valentina Marku, Feudi della Medusa; Lucio Gomiero, Vignalta; Lorenzo Borletti, Dominio di Bagnoli; Riccardo Baracchi, Baracchi; Trabalzini Dott. Enrico; Vecchia Cantina di Montepuciano; Villa Sparina; Valter Scarbolo, Scarbolo; Marco and Alessandra Sbernadori, Palazzo Vecchio; Antonio Caggiano, Cantine Antonio Caggiano; Stojan Ššurek, Ššurek Wines; Valerija Simcic, Simcic; Catherine et Didier Champalou, Champalou; Jean-Bernard Larrieu, Clos Lapeyre; Xavier Pierre, Cave Irouléguy; Domaine Cauhape, Trimbach; Jean-Luc Soty, Pascal Jolivet; Christian and Fabien Moreau, Domaine Christian Moreau Pere et Fils; Mario Rios, Louis Latour; Anthony Ravat and Olivier Masmondet, Maison Louis Jadot; Jean (Johnny) Hugel, Hugel et Fils; Jean-Jacques Vincent, Château-Fuissé; William Chevalier, Georges Duboeuf; Bernard Burgaud, Domaine Bernard Burgaud; Fabrice Gripa, Domaine Bernard Gripa; Robert Moulin, Domaine Moulin-Tacussel; Karine and Jean-Marc Diffonty, Vignobles Diffonty; Pierre Pastre, Château Fortia; Olivier Hillaire, Domaine Relagnes; Pascal LaFond, Domaine LaFond; Bosquet Des Papes, Isabelle Coustal; Marc Lurton, Château Reynier; Château Sainte Eulalie; Luis Patrao, Herdade de Esporao; Ana Isabelle, Ramos Pinto; Jorge Moreira, Quinta de la Rosa; Tomas Roquette, Quinto de Crasto; Joana Mesquita, Amorim; Amy Hopkinson, Bodegas Cenit; Pablo Rodriguez Lopez, Castro Ventosa; Friedrich Schatz, Bodega F. Schatz; Jose Manuel, Ramos-Paul; Ruano and Pilar Martinez-Mejias Laffitte, Ramos-Paul; Peter de Trolio, Bodega Hidalgo; Jose Gomez Ariza, Grupo Osborne; Miguel Castejon-Aguado and Carmen Rodriguez, Bodegas Castejon; Marc Kolling, Jean Leon; Eva Inglada, Enante; Martin Abell, Pirineos; Marcos Yllera, Grupo Yllera; Xosé Carlos Morell Gonzáles, Lugar de Fillaboa; Joan and Josep Huguet i Gusi, Can Feixes; Luis Valentín, Bodeguera de Valenciso; Felicia Skira, Codorníu; Torres Winery; Ysios Winery; Elena Adell, Juan Alcorta; Pablo Blanco, Martin Códax; Ana Quintela Suarez, Pazo de Señoráns; Javier Vieitez Castro and Jose Antonio Lopez, Lusco do Mino; Juan Gil de Aranjo, Palacio de Fefiñanes; Joan Cusiné Carol, Parés Baltà; René Barbier, Clos Mogador; Jordi Fernàndez Davi, Gratavinum; Carles Escolar Cunillé, Celler El Masroig; Mercedes Lopez de Heredia, R. Lopez de Heredia Viña Tondonia; Jean François Hébrard, Montecastro; Tinto Pesquera; Candado de Haza; Eulogio Calle Jr., Bodegas Naia; Mario Rico and Pedro González Mittelbrunn, Dominio de Tares; Angel Suarez Cicente, Lagar de Cervera; Raquel Pérez Cuevas, Ontanon; Marifé Blanco, Bodegas Julian Chivite; Zavier Flores Benítez, Freixenet; and Raimon Olivella, Masia Olivera.

We apologize for not including any individuals who provided advice or consultation or who so graciously hosted winery visits, making this book possible. Many thanks to all of you.

Foreword by Darrell Corti

When you open *Beverage Basics*, you will be entering a fascinating world of beverages that has been with mankind from the beginning of recorded history. The scope of this work is very ambitious. Yet, there is no other way of learning about alcoholic beverages than to simply set out on the voyage.

The very confusing and sometimes contradictory world of official regulations is complex; yet it must be learned by any hospitality professional. In fact, this might be the most important part of *Beverage Basics*. Retailing of alcohol and serving alcoholic beverages in the hospitality industry are both highly regulated. The regulations must be known before even attempting to understand the products of the industry.

I like that the authors have organized the Wine section according to grape varietal. Currently in the United States, this is how we tend to look at wine—through its major varietal. It also neatly pigeonholes the varieties and facilitates the understanding of each wine. It is systematical, and we do like systems these days.

Although the amount of information presented is enormous, it is written with a modicum of technical jargon. Foreign words and technical terms are nicely explained and easy to understand, even if you do not speak that particular language. Every industry and trade has its own special vocabulary—here you learn that of alcoholic beverages.

The essentials are all here, including information on the basic elements of wine, beer, and spirits: grapes and grains. Varied as the finished products are, wine, beer, and spirits all begin in nature and are transformed by the work of human hands, palates, and ingenuity. Welcome to this new world of exciting information. *Beverage Basics* is about much more than the history, production, and enjoyment of alcoholic beverages. It is also about the world of human beings who have enjoyed these drinks for centuries.

Preface

A jug of wine with lunch after a morning's work in the field. A pint of Guinness following a rugby match. A Champagne toast to celebrate the New Year. Irish whiskey at a wake. From routine activities to rituals and special occasions, wine, beer, and spirits have been an integral part of life since early civilizations first learned how to ferment fruits and grains to create alcoholic beverages. Today, alcoholic beverages continue to play an important role in business activities, family gatherings, social interactions, and even religious ceremonies around the world. Some of life's greatest moments—marriages, graduations, baptisms or brises, business successes, birthdays and anniversaries, personal achievements, and even death—often involve drinking wine, beer, or spirits as recognition of the passage of a significant event.

Learning about alcoholic beverages is a starting point on a lifelong journey that will help you grow professionally and personally while providing a tremendous amount of pleasure along the way. In this book, we will provide you with information to expand your awareness and knowledge of alcoholic beverages and show you how to use this knowledge to enhance your business and social relationships. The ability to choose an alcoholic beverage and pair it with the appropriate food, have an understanding of alcoholic beverage etiquette, and knowledge about the responsible use of alcoholic beverages are all essential to fully enjoying wine, beer, and spirits in social settings, as well as to anyone working in the beverage or food-service industry. Gaining an understanding of alcoholic beverages can be an educational experience that includes learning about cultures, history, languages, geography, and agriculture. The experience should develop your senses of sight, smell, taste, and touch. It should help you meet new people and discover new worlds.

Wine as we know it today came from Europe, where it was largely identified by its region of origin and not by the type of grapes that were used to produce the wine. In keeping with the established patterns, the broad subject of wine is typically taught by country.

Our approach, by contrast, focuses on the major grape varieties grown throughout the world, and also includes information on lesser-known varieties. As we discuss each variety, we integrate information on the countries or regions where that grape grows, as well as specifics on grape growing, winemaking techniques, and food-and-wine pairings for each variety.

Our goal is to create an educated consumer, one who knows about the benefits and enjoys the attributes of wine, beer, and spirits while fully understanding that there are responsibilities that go along with drinking alcoholic beverages. We expect that you should and will seek other sources for additional information about particular topics. There are many exceptional books on wine, beer, and spirits, and some helpful Web sites, which we have included in the Bibliography (see page 453). Keep in mind that wine, beer, and spirits and issues related to their consumption have changed throughout the ages in response to new environments, new technology, social trends, and personal taste. New and interesting viewpoints and information on the subject are always available. We urge you to make alcoholic beverage education a lifelong process.

Let the journey begin...

Introduction to Alcoholic Beverages

Alcoholic beverages have been made and consumed since before time was recorded for myriad reasons. Throughout history, alcoholic beverages have been popular for their taste, for their ability to make people feel happy and forget their troubles, and even because they were safer to drink than water.

In the last few centuries, alcoholic beverages have become more diverse, with a range of wine, beer, and spirits available. Of course, the prevalence of alcohol throughout the world has also raised legal, health, and other concerns, including debates over taxation, drinking and life expectancy, and how best to consume and enjoy each type of alcoholic beverage.

WINE

Wine is a beverage with the potential to provide great pleasure, and one that has been influenced by centuries of history, places, and people. When the right combination of grapevine selection, growing conditions, and winemaking occurs, excellent wines can be produced.

Wine is not a complex beverage. Most wines are made from a single product, grapes. Both of the ingredients required for the "recipe" to make wine—sugar and yeast—are found within the grape or adhering to its surface. The first wine probably made itself by following the natural process of rotting. From this simple beginning thousands of years ago, wine has now become a beverage with innumerable variations that is produced worldwide. The grape species *VITIS VINIFERA* is responsible for a majority of the world's wine production. This grape species has the capacity to produce a broad range of wines with widely different colors, aromas, and taste sensations. There are tens of thousands of different wines produced from more than 2,500 different wine-grape varieties, but most commercial wine is made from less than 200 varieties, and 90 percent of commercial wine produced comes from fewer than 30 different varieties.

(handwritten margin note:) Yeast + Sugar Alcohol CO_2 Energy

Marketing Wine

In spite of the wealth of grape varieties and wines produced from them, in the United States it sometimes seems that Chardonnay, Cabernet Sauvignon, and Merlot are virtually the only varieties produced. This is one indication of how marketing has come to play an important role in the world of wine. In most years more wine is produced worldwide than is consumed. This makes the role of marketing and selling the wine a driving force in the wine business. In large wine-producing companies, it is sometimes the marketing arm that determines the style of wine produced, not the winemaker. In some markets, particularly the U.S. market, consumers want a product that always tastes the same. Variation or change is difficult for many consumers to accept. The result is wine that, like soft drinks, can be a homogeneous, even boring, product. The wine meets a narrow flavor profile that tends to favor consumer preferences for sweet beverages. Yet there are thousands of winemakers throughout the world that are making interesting, complex, and distinctive wines.

Learning to Appreciate Wine

For many people the world of wine is daunting—full of obscure language, complicated geography, subtle climate differences, elaborate techniques, and most of all, expert opinions that make it seem remote and inaccessible. The simple act of tasting wine can, in the wrong hands, become an elaborate ritual, a contest in which your knowledge and taste are on trial. This book provides you with an accessible guide to the world of wine so that you can learn to appreciate this most ancient beverage.

Tasting many different kinds of wines is part of the process of learning to appreciate wine.

TOM ZASADZINSKI/CAL POLY POMONA

We will explore the world of wines by grape **VARIETY** rather than by country of origin and will guide you through the basic art and science of winemaking to help you understand how wine travels from the vineyard to the bottle to your glass. The seven primary grape varieties covered in this text are the international varietals that are most commonly available in the United States, the ones you are most likely to see on the grocery store or wine shop shelf. These include Chardonnay, Sauvignon Blanc, Riesling, Cabernet Sauvignon, Merlot, Pinot Noir, and Syrah. Viticultural and vinicultural characteristics, such as ripeness, **ACIDITY**, level of oak, and so on, vary widely from one varietal to the next, giving each of these grape varieties its own unique profile. Why is Chardonnay the most popular and the most manipulated wine? Why is Cabernet Sauvignon perceived to be best for aging? What are the factors that make Sauvignon Blanc a great food wine? Armed with a knowledge of each grape variety's characteristics and production techniques, you will be able to answer these questions and choose wines from the wine list or shelf with greater confidence.

(handwritten note:) NEW WORLD VS OLD WORLD

Johnny Hugel of Hugel & Fils in the tasting room of the family winery in Alsace, France

In addition to these international grape varieties, less well-known grapes such as Grenache, Viognier, Malbec, Gewürztraminer, and Chenin Blanc, not to mention the indigenous varieties of Italy, Spain, and France, should also become part of your knowledge base. These varieties, which are often more popular elsewhere in the world or are used as blending grapes in conjunction with better-known varieties, can result in expressive wines that provide a broad range of profiles and work with an array of different foods.

Stylistically, there are some major differences between **OLD WORLD** wines and **NEW WORLD** wines. Learning to recognize these differences is an important part of learning to understand wines. Old World wine refers to wine produced in Europe, where wines have been made for hundreds of years. Old World winemaking tends to result in wines with less fruit, more acid, moderate to no oak, and less alcohol. New World wines generally refer to wines from countries with less winemaking experience. New World winemaking tends to result in wines with a riper, fruit-forward style, less acid, more alcohol, and increased use of new oak. An awareness of the history, grape varieties, and substantially different **WINE STYLES** of various regions, and particularly of key winemaking countries, such as France, Italy, and Spain, are essential to developing a general understanding of wine.

Stylistic wines are created using a particular and distinct processing method. The most common stylistic wines are sparkling wines, fortified wines, and dessert wines. Less popular than table wines, they are nonetheless an important segment of the wine industry and are the perfect wines for the right place, time, and food.

Sample Different Wines

Jean Hugel, the patriarch of Hugel et Fils, a wine-producing family for more than 350 years in Alsace, France, provided excellent advice on learning to appreciate wine. He suggested that before reading anything about wine or listening to reviews or advice, you need to taste a different wine every day for a year. After sampling 365 different wines, you will know what you prefer and will be ready to form your own opinions about wines without depending on an expert to tell you what you should like. Of course, you need to try 365 different wines, not 365 Chardonnays and Cabernets, but as many different **VARIETALS**, styles, and blends as possible from different countries, regions, and producers.

After this exercise, you will have sampled only a tiny percentage of the world's wines but probably will have experienced more different wines than most people will try during their lifetime. You will also come to understand what we think is one of the greatest aspects of enjoying wine, learning about and appreciating the history and people of different cultures and countries. Wine can be a way of taking a life-fulfilling trip around the world without leaving home. And the experience may encourage you to travel to see for yourself wine in its own culture.

A Word of Caution

Be cautious of the wine media. The press can keep you informed of new or different wines in the marketplace or suggest wineries you might want to visit. It can also help you continue to expand your wine knowledge. But many critics also emphasize scoring wines on a 100-point scale. It is easy to fall into the trap of buying only the wines that receive high scores.

Do not only buy wines that receive high scores. Many wonderful wines that do not fit the profile of a particular evaluator are available. By being your own evaluator, you will learn to identify wines suitable to your individual taste that complement the foods you prefer.

BEER

Beer is the second broad category used to classify alcoholic beverages. Like wine, beer is a fermented beverage. Unlike wine, the sugar required for FERMEN-TATION in beer comes from grain rather than fruit. In one sense, the brewer has a tougher job than the winemaker. The grape is, in effect, wine-to-be, with no additional ingredients required. The brewer must add water and yeast to the grain in order to have the ingredients necessary for beer. And for contemporary beers, a brewer must include a fourth ingredient, HOPS, to create the finished product.

water (primary ingredient)
malted grain
hops
yeast

In the world of beer, the broadest way to classify styles is usually based on the fermentation method. TOP-FERMENTING beers include classic ALE styles as well as most wheat beers. The other broad category is the BOTTOM-FERMENTING beers, or LAGERS. In addition, there are a few styles that are hybrids of both fermentation methods.

A brewpub is a good place to sample a range of beer styles.
OWEN WILLIAMS

Barley is the primary grain for most beer styles. Wheat beers use wheat as the sugar source for fermentation, although almost all wheat beers contain barley. Other grains, like rice, oats, and corn, may be used to supplement the sugars derived from barley. A few styles even contain fruit, but fruit is generally an addition to the beer rather than a primary source of fermentable sugar.

Beer Styles

The vast majority of beers are based on the historic styles from present-day Belgium, the Czech Republic, Germany, Ireland, and the United Kingdom. Any style of beer can be brewed anywhere in the world; however, a conscientious brewer recognizes the beer's historic lineage. Whenever you see a term with a "dash style" appended—for example, Pilsner-style—on a beer label, it implies that the beer is based on a traditional beer from another city, region, or country (in this case, classic Pilsner from the Czech Republic). Yet, with tens of thousands of beers available around the globe, even this nod to a beer's pedigree is no guarantee that the beer is representative of the original style.

TOP — ALE
BOTTOM — LAGER

As mentioned earlier, a winemaker aims to make the best possible wine from the grapes he or she is given each year. This year's Zinfandel may be as good as last year's Zinfandel, but it is likely there will be at least subtle differences between the two bottles because grape quality varies from year to year. By comparison, annual quality variations in grain are minimal. Therefore, a brewer's goal is to combine art, science, and common sense to make an ale or lager that tastes the same, year after year after year. With good-tasting water, relatively stable grain quality, specific yeast strains, and hops, the brewer creates a recipe that works for a particular beer style. The better the brewer, the more likely he or she can reproduce the same beer, batch after batch.

With wine, one winemaker's Chardonnay will taste different from another's Chardonnay. Likewise with beer, one brewer's pale ale will have its distinctive characteristics and will taste somewhat different—or significantly different— from another brewer's pale ale. Using water of varying mineral composition, different combinations of grains, the optional use of **ROASTED GRAINS** and crystallized grains, dozens of hops varieties, and numerous yeast strains, the number of variations available to the brewer is virtually limitless.

With the numerous possible combinations of ingredients, a brewer can develop a new style. It may become popular, and others may copy the style. If enough commercial brewers make the style, it might become a new classification that is formally recognized by the brewing world. This is one of the reasons tasting, learning, and studying beer is such a great lifelong learning experience—things change.

Selecting Your Preferred Style

With enough variety to taste a different beer every day for your entire life and not run out of options, where do you begin? Maybe begin with Jean Hugel's advice, modified for this beverage. Try 365 different beers from different regions, different producers, and different styles before making decisions about your favorite styles or favorite beer brands. Part of the enjoyment is to ferret out a preferred beer for each of your favorite foods, for every season of the year, and for any trip you take. Even when you cannot travel, beer can be a tour of the globe in a glass.

DISTILLED SPIRITS

The third classification of alcoholic beverages is **DISTILLED SPIRITS**. Spirits have long enjoyed popularity because they are relatively easy to produce, are easy to transport and store, and keep forever. Plus, they are available in a wide range of categories, styles, and flavor profiles.

All spirits are derived from plant materials, usually grains or fruits. To make a spirit of any kind, the distiller must start with a fermented beverage made from these raw ingredients. The alcohol in the fermented liquid is heated in a still, which separates the water in the fermented beverage from its alcohol and concentrates it. When **DISTILLATION** is complete, spirits are clear and generally

GRAINS
AGAVE
POTATOES
GRAPES

have high alcohol content. At this point, the distiller manipulates the spirit by any combination of redistilling, cutting with water, blending, aging, coloring, and adding flavors to create the desired beverage.

The base, type of yeast, water source, and flavorings that are used determine the final category and style of spirit that is created. Grain-based spirits encompass a wide range of beverages from whiskey, which is primarily made from barley, to gin and vodka. Rum uses sugar cane as its base, and tequila is made from agave. Fruit is the base for brandies, including Cognac and Armagnac from grapes, Calvados from apples, and Kirsch from cherries. Even the leftover bits from wine production, the pomace, can be used to make spirits such as grappa.

Liqueurs, the most diverse category of spirits, are made from distilled alcohol, usually in a pure form, which is sweetened and flavored with fruits, herbs, and other ingredients. They are frequently colored using natural products, for example, mint, which produces a green liquid. Finally, bitters are made from stems, roots, and other parts of plants that are less sweet than the fruits. These are considered digestive aids.

Learning to Appreciate Spirits

Spirits are a diverse category of alcoholic beverages, and are extremely versatile in terms of how they can be served. Because spirits can be derived from such a broad spectrum of plant materials and because alcohol is so easily manipulated during and after distilling, there is a great deal of variation in style from one spirit to the next. In addition, because alcohol itself is neutral in taste, it has the ability to take on the characteristic flavors of a wide range of added ingredients.

Once a distilled spirit is sold to a consumer, the possibilities for combining it with other alcoholic or nonalcoholic beverages abound. This versatility means that in the hands of a skilled bartender or consumer, the art of drink mixing can result in myriad new drink styles. Spirits can be served cold or hot, alone, with water or ice, or combined with almost any other flavoring or ingredient. Most often, they are cut with other beverages to dilute their intensity.

As with wine and beer, spirits are subject to consumer whims. For years, cocktails were the rage and lent social occasions an air of sophistication. Then they fell out of favor. Single **BARREL** and small batch bourbons, tequila, and scotch took the limelight, and sweet mixed drinks gained a following. Today there is a resurgence of interest in the art of the cocktail, and "classic" drinks are becoming popular once again.

You can learn to be a connoisseur of distilled beverages by exploring different styles and brands as well as selecting from hundreds of cocktails. Over time your palate will discern subtle differences in premium bourbons, tequilas, gins, brandies, and so on. You will also learn to appreciate the unique tastes of those spirits that are blended or combined with flavorings and sweeteners. As you will discover, there is a spirit to suit every occasion, mood, and time of day.

Cocktails have become popular among men and women of all ages.
WILFRED WONG

Alcohol and the Law

Governments worldwide have established laws to control the sale and consumption of alcoholic beverages. In this book, however, we will focus on the U.S. laws that have the greatest impact on American consumers and business operators.

PROHIBITION

PROHIBITION was a flamboyant era of bootleg whiskey, speakeasies, and gangsters with their glamorous molls—at least as portrayed in the movies. In reality Prohibition was sometimes violent, had devastating social and economic impacts, and left a lasting imprint on the American psyche and on the laws governing alcoholic beverages long after it was repealed.

The seeds of Prohibition were sown by the Temperance Movement. Although **TEMPERANCE** has had a long history in the United States, it took hold in the 1800s in response to social issues, including the large number of saloons selling cheap liquor. Supporters of temperance were driven by the belief that drinking alcoholic beverages was a sin that threatened the family and fostered drunkenness and debauchery. They blamed drink for causing increased crime, violence, and moral depravity.

In the mid-1800s the movement encouraged moderation. However, as the century came to a close, temperance groups led by the Woman's Christian Temperance Union, the Anti-Saloon League, and Protestant churches called for a total ban on alcohol and demanded that saloons be eliminated. Their efforts paid off at the state level: by 1919, nine states prohibited all alcohol consumption and 31 others gave local communities the option to be dry.

In 1919 the 18th Amendment to the U.S. Constitution, forbidding the manufacture, sale, or transportation of "intoxicating liquor" nationally, was adopted. Congress soon enacted the National Prohibition Act, better known as the Volstead Act, to enforce the amendment. The act defined intoxicating liquor as any beverage containing more than 0.5 percent alcohol. It also allowed exceptions for alcohol used for religious or medicinal purposes and authorized the production of a small amount of alcoholic beverage for home use.

Problems Develop

Prohibition was a dismal failure. People who drank did not stop drinking; they just found ways to get around the law. The average law-abiding citizen began buying bootleg alcohol or making it at home. Prohibition was a bonanza for organized crime, which established illicit stills, distilleries, and fermenters and replaced saloons with speakeasies. Smuggling thrived, especially from Canada. Because it was relatively fast and easy to produce, strong liquor displaced beer and wine as the drink of choice.

Prohibition was weakly enforced because the government did not have the resources to adequately deal with those who were illegally producing alcoholic beverages. At the same time that the federal government was expending money to enforce the law, it was losing revenue because it could no longer tax alcoholic beverages. States lost an important source of income as well.

Prohibition damaged the alcoholic beverage industry by forcing the closure of legitimate wineries, breweries, and distilleries. The wine industry was particularly hard hit because of the costly investment required in vineyards. Prohibition forced owners to dig up their vines and often go out of business, and it took years for the wine industry to replant the lost vineyards.

Prohibition Repealed

As the 1920s neared an end, societal forces were moving toward greater acceptance of personal choice and responsibility. This attitude, coupled with the other problems attributed to Prohibition, resulted in passage of the 21st Amendment in 1933, which repealed the 18th Amendment and ended Prohibition.

Federal agents closed establishments that sold alcoholic beverages during Prohibition.
PRINTSANDPHOTOS.COM

Fearing a backlash from those opposed to its role in Prohibition, the federal government was cautious about overly regulating alcohol. Individual states, the District of Columbia, and U.S. territories were given enormous power over alcoholic beverage regulation. This created independent jurisdictions that determined laws within their borders. More importantly, each state could regulate both interstate and intrastate distribution. State regulation of the alcoholic beverage industry has had a significant long-term impact on the industry, particularly the distribution and sales chain.

FEDERAL, STATE, AND LOCAL LAWS

Governments regulate the alcoholic beverage industry for two reasons: to bring in revenue through taxes and to control consumption. Alcoholic beverage laws determine who controls the various components of production, distribution, and sales and how much revenue each government entity receives.

Alcohol is the most highly taxed product in the United States, and everyone wants a piece of this lucrative pie. Federal, state, and in some cases, local entities levy excise taxes in addition to collecting license and permit fees.

Taxes on alcoholic beverages, often called sin taxes (see Sin Tax, at left), serve the dual purposes of raising government income while attempting to limit alcohol use. Excessive alcohol taxes are justified as a way to recover the costs of irresponsible drinking, such as expenses related to health care for alcoholics. Anti-drinking factions also use federal, state, and local laws to control drinking, for example, by encouraging the passage of blue laws, which prohibit alcohol sales on Sunday (see Blue Laws, page 17). With hundreds of inconsistent and arbitrary regulations, it is almost impossible to make generalizations about alcohol law. But here is what is consistent throughout the United States:

1 You cannot sell alcohol to a minor, defined by law as a person under 21 years of age.
2 You are driving under the influence if your blood alcohol level is 0.08 percent or higher.
3 You are taxed for buying alcoholic beverages.

SIN TAX

A sin tax is a state tax on a product or activity that is considered to be a vice or a sin. Sin taxes were first used by the Puritans to discourage such sinful behaviors as drinking, gambling, and smoking. To this day, they are levied on alcohol and tobacco in an effort to curtail their consumption based on religious or moral grounds. Those in favor of this form of taxation argue that sin taxes discourage drinking by increasing the cost of alcohol. However, raising taxes does not necessarily change behavior, and in many instances, alcoholic beverage consumption increases in spite of higher taxes.

Federal Law

Federal law has regulated alcoholic beverages in the United States since the country was founded, when the first Congress approved taxing imported liquor as a way to provide revenue. Although the laws and the agencies assigned to oversee those laws have changed in the intervening decades, the federal government continues to regulate and tax alcoholic beverages.

As of 2003, the Alcohol and Tobacco Tax and Trade Bureau (TTB) [formerly the Bureau of Alcohol, Tobacco, and Firearms (BATF)], a unit in the Department of the Treasury, is specifically charged with collecting excise taxes and protecting consumers by ensuring that businesses comply with labeling, advertising, and marketing regulations. In addition to these functions, the federal government regulates winery, brewery, and distillery operations and oversees importing.

The federal government determines taxes based on the type of alcoholic beverage, the percent of alcohol in the beverage, and the container size. Distilled spirits are taxed at the highest rate, followed by wine, then beer. Table 2.1 provides examples of the federal taxes that are imposed on wine, beer, and spirits.

Wineries, breweries, and distilleries must comply with federal regulations and also with state statutes, which may be more restrictive than federal law.

Sometimes federal law preempts state law. For example, the Government Warning Label that is required on beverage bottles prohibits states from imposing similar laws.

Importing

Importing alcoholic beverages into the United States for commercial purposes is very complex. Foreign producers must comply with the TTB, the U.S. Customs and Border Protection (CBP), and the Food and Drug Administration (FDA), as well as the regulations in all the states where their products will be sold.

All imported beverages are charged customs duty, which is based on the type of beverage, the alcohol content, the container size, and the value of the beverage. States may also levy taxes on imported beverages. An importer's permit and a certificate of label approval are required before a product can enter the United States. Additionally under the Bioterrorism Act (Public Law 107-188), the FDA requires foreign companies producing, packing, or storing food products, including alcoholic beverages, to register with the agency prior to shipping the product to the United States.

By law, foreign producers must use importers. These firms play a crucial role because they serve as a liaison between the producer, U.S. government agencies, and distributors. Importers buy beverages from producers in other countries, negotiate prices, and ship to the United States. They know the applicable federal and state laws and make sure that imported beverages meet the necessary requirements. Once the product arrives in the United States and clears customs, they sell it to a distributor who then sells it to retail establishments.

Importing laws do not apply to small amounts of alcoholic beverages brought into the country for personal use. However, customs laws limit the quantity of alcohol individuals may carry into the country or ship duty free, and state laws may vary. Some states may prohibit you from shipping wine from a foreign country to your home.

Table 2.1 Examples of Federal Alcoholic Beverage Taxes

TYPE OF BEVERAGE	TAX	TAX PER PACKAGE
Beer	$18 per barrel (31 gallons)	$0.05 per 12 oz. can
Wine	Per wine gallon	Per 750 ml bottle
14% alcohol or less	$1.07	$0.21
Over 14%–21%	$1.57	$0.31
Over 21%–24%	$3.15	$0.62
Naturally sparkling	$3.40	$0.67
Distilled Spirits	$13.50 proof gallon*	$2.14 per 750 ml bottle, 80 proof

* 100 proof or 50 percent alcohol.

Note: Dollar amounts are examples only and should not be used to determine actual taxes.

Source: TTB

State Laws

The 21st Amendment granted states broad powers to regulate the alcoholic beverage industry, including production, distribution, and sales. Each state determines who receives a license to sell alcohol, how to assess alcoholic beverage taxes, where and when alcoholic beverages can be sold, and how they are distributed. All 50 states have established some form of alcoholic beverage control board, a state agency that enforces state liquor laws. These agencies issue liquor licenses and monitor retailers to ensure that they follow the law.

Each state sets up its own system for determining tax rates and collecting taxes. Often taxes are levied based on a percentage of retail sales, but like the federal government, states can establish tax rates based on a product's alcohol content, with higher alcohol levels taxed at a higher rate, on a percentage of sales, or on the type of beverage.

State laws are often protectionist. They are implemented to keep liquor sales in the state, which, in turn, keeps tax revenues in the state. This is why they may prohibit alcoholic beverages from being brought into the state from other locations, or may require residents to purchase their beverages from a state-operated store, creating a direct source of revenue.

Although some states have similar laws, no two are exactly alike. Each state has developed its own regulations based on the values of its citizens. While selling alcohol on Sunday might be acceptable in California, it might be unacceptable in Utah. Some states allow counties or municipalities within their borders to establish their own laws for licensing, distributing, and selling alcoholic beverages, again to reflect the values of their communities.

State laws are often designed to limit consumption by making it more difficult for a consumer to make a purchase. Here are three examples:

Utah You must buy wine, beer with an alcohol content exceeding 3.2 percent, and liquor from a state-operated store, but bottled or canned 3.2 percent beer can be sold in retail stores.

New York State You cannot buy wine or spirits in a grocery store; you must make your purchases at a state-licensed liquor store. However, beer cannot be sold in liquor stores; you must buy it at a grocery or convenience store.

TIP

Information on state laws is available at the Alcohol and Tobacco Tax and Trade Bureau Web site, www.ttb.gov/alcohol/index/htm. In addition, each state alcohol control board maintains a Web site.

Sunday Sales of Distilled Spirits

Source: Distilled Spirits Council, 2010

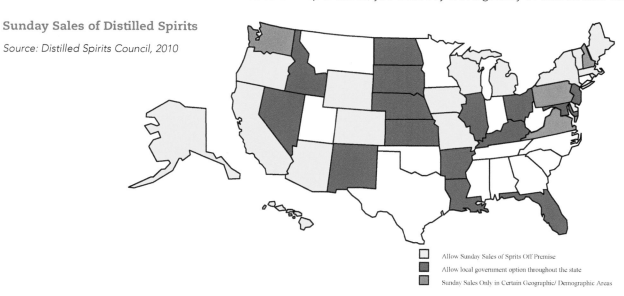

- Allow Sunday Sales of Sprits Off Premise
- Allow local government option throughout the state
- Sunday Sales Only in Certain Geographic/ Demographic Areas

Alabama Beer at 6 percent or less in a 16-ounce or smaller container and wine with less than 14.9 percent alcohol content is sold in retail stores. All other alcohol must be purchased at state stores. However, 26 counties prohibit the sale of any type of alcoholic beverage.

Local Laws

Adding to the difficulty of understanding all the state laws, each local jurisdiction may also be authorized by the state to implement laws controlling alcoholic beverages. Local laws are often designed to control alcohol consumption in the interests of protecting the public health, although local governments might also find the tax revenues helpful. As with state laws, local laws are based on local mores. One community might be comfortable with a bar on the same block as a church or school, while another might think they should be miles apart.

Local governments have many options at their disposal when it comes to curtailing alcohol sales. One way to accomplish this is to limit the number and density of liquor licenses granted, whether **ON-PREMISE** or **OFF-PREMISE**. For example, a small city may only allow 10 licenses within its boundaries. Another technique is to limit the location or density of establishments selling alcohol. For example, a local law may allow only one bar within a five-mile area. Zoning restrictions, special types of cabaret or entertainment permits, and nuisance abatement statutes are other methods that can be employed to restrict the sale of alcoholic beverages.

THE 3-TIER SYSTEM

The 3-tier system for regulating the production, distribution, and sale of alcoholic beverages was established following Prohibition to ensure checks and bal-

Blue Laws

Blue laws are state or local laws that forbid specific behaviors on Sunday, such as selling cars or alcohol. Like sin taxes, blue laws date to colonial times. They were enacted to restrict all but religious activities on Sundays, including drinking, and they are still used in some places to discourage drinking on Sunday. Each state can decide whether or not to restrict Sunday alcohol sales, and some states that have not imposed a statewide ban grant cities and counties the authority to write their own blue laws.

Most states permit alcohol sales in bars and restaurants on Sundays but limit the hours that it can be sold. As indicated in the map at left, by 2010, 36 states had authorized Sunday retail sales of spirits. Some of these states allow Sunday sales at all licensed retail outlets, and others license sales only in selected locations.

The result is a hodgepodge of laws that seem arbitrary and, in some cases, downright silly. For example, you might not be able to buy alcohol in the county where you live, but

you can drive five minutes over the county line, make your purchase in a liquor store, and drive back home. In Michigan, you cannot buy alcohol before noon on Sunday, presumably when you should be in church or at home with your family.

In the past few years, more states have moved to repeal blue laws, citing customer convenience as the primary reason for changing the law. However, desire for increased tax revenue is often the driving force behind approving Sunday alcohol sales.

TIED-HOUSE LAW

A tied-house exists when a producer also operates retail businesses, enabling the producer to control the brands that are sold and effectively remove all competition. Under tied-house law, which was passed following Prohibition, producers may not own retail shops, give retailers something of value, or pay retailers for promotional activities related to their product. The law is designed to free retail establishments from the undue influence of producers. It also protects the 3-tier system because retailers must buy from a distributor rather than directly from a producer. However, tied-house law creates a potential problem because it enables distributors to influence retailers to buy the distributor's preferred brands.

ances were in place to prevent organized crime from controlling all segments of the industry. In addition, the system was designed to prevent producers from unduly influencing retailers (see Tied-House Law, at left).

Wine, beer, and spirits move from one tier to the next in the following stages as defined by law:

Tier 1: Production This tier includes vintners, brewers, and distillers. Importers are also included at this level. Virtually all sales from these producers are handled by distributors.

Tier 2: Distribution The wholesaler or distributor buys the product from Tier 1, producers, and sells it to Tier 3, retailers. Distributors know the various state regulations for selling alcohol and serve as the tax collection point. Some distributors conduct business in only one state, others in multiple states.

Tier 3: Sales Retail shops or restaurants sell wine, beer, or spirits to customers. Retail businesses are required to buy from a distributor and must be licensed to sell liquor by the state. Restaurants and bars receive on-premise licenses for consumption at the location. Retails stores, such as supermarkets, wine shops, drugstores, and liquor stores, receive off-premise licenses because the beverage is consumed at another location.

Most states use the 3-tier system, but there are exceptions. In some states wineries and brewpubs can both produce and sell onsite, and some states allow small producers to act as their own distributors.

Control and Competitive States

Generally states adopt one of two methods for managing the 3-tier system: control or competitive.

CONTROL STATES are directly involved in distribution, sales, or both. There are 19 control states, as shown on map at right, each with its own method of controlling the distribution and sales tiers. For example, the state of Pennsylvania controls the retail level by operating state-run liquor stores and barring sales in any other type of store. State-owned liquor stores have been criticized because they reduce competition and increase consumer costs. However, they are a lucrative source of income for states.

The competitive (or license) method is used in 32 states. These states do not participate in distribution and sales. Instead, they license the private sector to distribute and sell wine, beer, and spirits. California, New York, and Florida are **COMPETITIVE STATES**. Competitive markets result in lower prices and better selection due to competition among private businesses.

Role of Distributors

Distributors are the middlepersons between producers and retailers—in many cases, the sole intermediary. They buy large quantities of wine, beer, and/or spirits, and then store, transport, and sell the beverages to retailers. They help retailers maintain product and inventory control and help states monitor sales and collect taxes. Distributors are in a powerful position because they control sales up and down the chain.

The wholesale industry is changing. Thirty years ago distributors served a useful purpose because they could get product to the consumer, who had limited options for obtaining alcoholic beverages. With the advent of computer and catalog sales, winery and brewery visits, and changing laws, producers and consumers are challenging the role of distributors as a critical link in the supply chain.

Pros and Cons of the 3-Tier System

Many state governments along with distributors staunchly defend the 3-tier system, while many producers and retailers are opposed to it.

Those in favor of the 3-tier system argue that it helps ensure a safe, crime-free environment by controlling the way that alcoholic beverages move from producers to consumers. They say that the system makes it easier for states to collect taxes and ensures that no one tier has complete power, thereby providing retailers and consumers with better prices and selection. In addition, distributors can get products to markets that would otherwise not have access. For example, retailers in Nebraska might not be aware of a new offering by a California winery without the distributor's input. Finally, one of the most frequently cited arguments in support of the 3-tier system is the ability to keep alcohol out of the hands of minors because distributors limit the number of people shipping and delivering alcoholic beverages.

The majority of the arguments against the system are targeted at distributors, who critics say have a monopoly on the industry. They claim that any movement to change the system is stymied by distributors' interests in protecting their tier. Because distributors manage movement from the first to the third tiers, they can influence the brands that reach the consumer and the prices they pay for those brands. Additionally, many small distributors are consolidating, putting the control of the market in the hands of a few large companies, which tends to further reduce competition and limit consumer selection.

Alcoholic Beverage Control States

Source: National Alcoholic Beverage Control Association, 2010

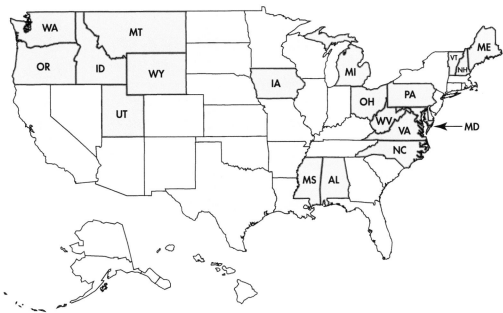

FRANCHISE LAWS

Franchise laws grant distributors sole control of a brand in a state. The laws may also prohibit a producer from terminating a contract with a distributor and force producers to use distributors who do not necessarily represent their interests.

Opponents of the 3-tier system also say that it makes it more difficult for small producers to compete. Small winery, brewery, and distillery owners are at a particular disadvantage because if distributors will not buy and sell their products, they have no way to get them to consumers. Many wineries in the United States and in other countries have expressed concerns about distribution because they cannot get distributors to handle the small quantities of product that they have to sell.

Retailers are at a similar disadvantage in that if their distributor does not carry a brand that they would like to sell, they are not allowed to buy directly from the producer or to use another distributor (see Franchise Laws, at left). Retailers and producers argue that under the 3-tier system, distributors can become complacent and provide poor service because they have such strong controls over the system. And in the end, consumers also pay for the 3-tier system because the cost of a beverage is marked up additionally as it passes through each of the three tiers.

DIRECT SALES

Direct sales occur when a producer ships beverages directly to a consumer, either within a state or between states, bypassing the distribution link in the 3-tier system. States have placed restrictions on interstate shipments that range from open, where goods can flow freely into the state, to closed, where the goods are blocked from entering the state.

Rules concerning interstate shipments fall into one of three categories:

Limited Direct shipping is allowed under certain conditions.

Reciprocal Both shipping and receiving states agree to allow direct sales.

Prohibited Out-of-state producers and retailers may not ship directly to consumers.

The map at right shows the direct shipping laws for states for wine sales only.

Intrastate shipments are also regulated by state laws. Many states allow producers and retailers to ship directly to consumers within the state where they do business, while prohibiting interstate shipping.

States cite two reasons for limiting direct shipping: the state may have problems collecting taxes, and minors may obtain alcohol shipped to their homes. The alcohol wholesale industry has vehemently opposed direct sales. Distributors support laws limiting shipping because they want to control the lucrative channel for moving alcoholic beverages.

Producers who support direct shipping argue that there are ways to collect the taxes from either the producers or the customers. In addition, methods are in place to ensure that alcoholic beverages do not get into the hands of minors. Age verification is required when beverages are purchased. Carriers are trained in procedures for delivering these packages to households, and they will not deliver alcoholic beverages without an adult's signature. Some states and some shipping companies require packages containing alcoholic beverages to be labeled as follows: CONTAINS ALCOHOL: SIGNATURE OF PERSON 21 OR OLDER REQUIRED FOR DELIVERY.

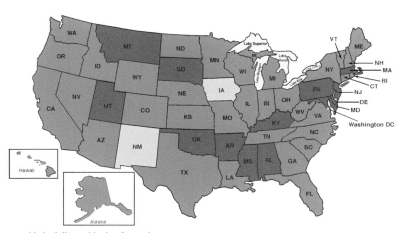

Direct Shipment Laws by State for Wineries

SOURCE: WINE INSTITUTE OF CALIFORNIA, 2010

Limited direct shipping & permit states: allowance of limited shipments:

Alaska (a reasonable amount)
Arizona (permit required for off-site sales only)
California (permit required - taxes paid)
Colorado (permit required - taxes paid)
Connecticut (permit required - taxes paid)
Florida (excise taxes/ consumer taxes paid)
Georgia (permit required - taxes paid)
Hawaii (permit required - taxes paid)
Idaho (permit required - taxes paid)
Illinois (permit required - taxes paid)
Indiana (permit required - taxes paid)
Kansas (permit required off-site sales only)
Louisiana (permit required - taxes paid)
Maine (permit required - taxes paid)
Michigan (permit required - taxes paid)
Minnesota
Missouri (permit required-taxes paid)
Nebraska (permit required - taxes paid)
Nevada (permit required -taxes paid)
New Hampshire (permit required - taxes paid)
New York (permit required - taxes paid)
North Carolina (permit required - taxes paid)
North Dakota (permit required - taxes paid)
Ohio (permit required - taxes paid)
Oregon (permit required - taxes paid)
Rhode Island (on-site sales only)
South Carolina (permit and report required - taxes paid)
Tennessee (permit required - taxes paid)
Texas (permit required - taxes paid)
Vermont (permit required - taxes paid)
Virginia (permit and report required - taxes paid)
Washington (permit required -taxes paid)
Washington D.C. (one case per person)
West Virginia (permit required/ taxes paid)
Wisconsin (permit required/ taxes paid)
Wyoming (permit required - taxes paid)

Reciprocity states:

Iowa (transitions to permit state on July 1, 2010)
New Mexico

Direct Shipment Not Permissible:

Alabama
Arkansas
Delaware
Kentucky (felony)
Maryland (special interstate by 3-tier only)
Massachusetts
Mississippi
Montana (consumer permit - no carrier)
New Jersey
Oklahoma
Pennsylvania (special interstate by 3-tier only)
South Dakota
Utah (felony for winery to direct ship)

Federal Law Allows Shipment to These States Only for Purchases Made Onsite at the Winery:

Delaware - no limit
New Jersey - up to 1 gallon
Oklahoma - up to 1 liter
South Dakota - up to 1 gallon

Buying for Personal Use

As a consumer, there are some things you should know about buying alcoholic beverages directly. The U.S. Postal Service bans mailing alcoholic beverages anywhere in the country, regardless of state laws. Some states allow alcohol to be shipped using a common carrier, such as UPS or FedEx, and some allow consumers to transport alcohol in a car or on a plane if it is packed in luggage. Other states ban bringing alcohol across state lines even in an individual's own car, and wine sent as a gift or for charitable purposes may also be prohibited. Although these laws are on the books, consumers are rarely penalized for buying or transporting alcoholic beverages, except perhaps to have them confiscated.

Instead, producers and shipping companies receive the penalties, which can mean losing licenses to operate. In a few states, it is a felony to direct ship to consumers across state lines. Some large shippers and producers refuse to ship to some states because of the fines that may be imposed if they violate a state law.

Free Trade Issues

Free trade is currently the most contentious topic in the industry because it bypasses the second tier of the 3-tier system. Since 2000, numerous lawsuits, largely spearheaded by the wine industry, have challenged several state laws restricting shipping on the basis of the federal Commerce Clause, which prohibits exclusionary tariffs or laws on products from other states.

In May 2005, the U.S. Supreme Court, hearing *Granholm v. Heald, Michigan Beer and Wine Wholesalers Association v. Heald,* and *Swedenburg v. Kelly,* overturned laws barring out-of-state wine shipments in some situations. The Court found that the laws in Michigan and New York were exclusionary and violated the Commerce Clause. The laws in both states prohibited out-of-state wineries from sending wine directly to customers but allowed in-state wineries to sell directly to customers. The Court said states cannot permit in-state wineries to ship wine while restricting shipments from out-of-state wineries. However, those states that ban direct shipping both in- and out-of-state can continue to prohibit direct sales. Under the ruling, states can ban all direct shipping, but they cannot permit in-state shipping while prohibiting out-of-state shipping. The court ruling only applied to beverages sold by wineries. Retail outlets are not covered.

DRINKING AND THE LAW

Driving while under the influence of alcohol and underage drinking are major public policy issues that have helped shape alcohol laws. Federal law has defined explicit standards to deal with both of these problems, setting a maximum level of alcohol consumption for driving safely and a minimum age for drinking alcohol.

Just as states establish laws governing the production, distribution, and sale of alcoholic beverages, they also implement their own laws concerning drunk driving and underage drinking. Some laws are directed at businesses, while others are directed at consumers. Each state's statutes are too complex to discuss individually, but we will take a look at general laws pertaining to these issues. For more detailed information on state-specific laws, contact your state's Department of Motor Vehicles.

Drinking and Driving

Almost everyone in this country has been impacted in some way by drunk, or impaired, driving, whether you know someone who has been stopped for a suspected DUI or someone who was killed by a drunk driver, or whether you yourself have been involved in an alcohol-related car accident. More than 41 percent of traffic fatalities in 2006 were alcohol related, and more than 15,000 deaths were attributed to a driver who was at or over the legal alcohol limit.[1] These are only the statistics for drinking and driving. Alcohol impairment also contributes to injuries and deaths from other causes, such as boating accidents, drowning, and falls.

Because of the serious health and economic consequences of drunk driving, laws have become increasingly stringent, and many groups advocate even stricter laws and penalties for violating them. As the laws have become

Drunk driving is responsible for a large percentage of fatal auto accidents.
PHOTODISC/GETTY IMAGES

more restrictive, the terms used to reflect the physical and mental condition of a person who has been drinking have changed as well. The term *drunk driving* implies that a person is noticeably intoxicated, maybe stumbling and slurring his or her words. *Impaired driving*, which is now the preferred term, implies that whether or not a person appears drunk, his or her ability to drive or perform other activities safely is affected by the alcohol that he or she has consumed.

MADD (Mothers Against Drunk Driving) was founded in 1980 by two mothers, one whose daughter was killed by a drunk driver and another whose daughter was seriously injured by a drunk driver. MADD, along with SADD (Students Against Destructive Decisions) and many other public and private organizations, has helped to shape policies and laws that have tightened restrictions on drinking and driving over the past three decades.

The Legal Limit

Prior to 2000, each state determined the maximum blood alcohol concentration (BAC), the amount of alcohol in the bloodstream, at which it was considered legal to drive, often 0.10. For several years MADD, in partnership with other groups, successfully lobbied for national drunk driving standards. In 2000, President Clinton signed a bill requiring states to set the BAC limit at 0.08 or lose federal highway funds (Department of Transportation, 2001 Appropriations Act, HR4475). This was a powerful motivator, and by July 2004, all states and the District of Columbia had passed state laws establishing 0.08 as the legal limit.

How Much Is Too Much?

Generally it only takes two drinks, even less for a small woman, to reach 0.08 BAC. However, the number of drinks it takes for a person to reach the legal limit depends on a number of factors, including weight, gender, level of fatigue, and overall health, as well as how fast the drinks are consumed and whether or not they are consumed with food.

On average the body can metabolize about one drink per hour. But the question of what constitutes "a drink" is not a simple one. The standard has long been 1.5 ounces of 80-**PROOF** spirits, a five-ounce glass of wine, or a 12-ounce bottle of beer. Today, however, that is not always an accurate definition. The standard five-ounce glass of wine is based on a wine with 12 percent alcohol content; however, a five-ounce glass of Zinfandel at 16 percent alcohol contains one-third more alcohol than the standard drink. The standard 12-ounce bottle of beer has about 5 percent alcohol, but the typical 16-ounce pint might contain 8 percent alcohol. Clearly, the higher the alcohol percentage of any given beverage, the less you can drink before reaching the legal limit.

Effects on Behavior

Even one drink affects a person's ability to drive. At 0.02 BAC, reaction slows, judgment declines, and you lose the ability to divide your attention among several tasks. Studies report that drivers are impaired at 0.05 BAC, with decreased physical and mental skills. This means a much slower response in emergency situations, reduced inhibitions, and impaired judgment. Although you may not appear intoxicated at 0.08 BAC, your concentration, judgment, self-control, coordination, vision, hearing, and speech are significantly impacted, and you are at great risk of being involved in a traffic accident. At 0.10 BAC, the ability to use judgment and think clearly is severely diminished, and signs of this impairment can include slurring, difficulty focusing, and problems with balance and coordination.[2,3]

Driving Under the Influence

There are almost 1.5 million Driving Under the Influence (DUI), also called Driving While Impaired (DWI), arrests in the United States annually.[4] Each state or lo-

Table 2.2 Sample State Impaired-Driving Penalties

STATE	ADMINISTRATIVE LICENSE SUSPENSION, 1ST OFFENSE	RESTORE DRIVING PRIVILEGES DURING SUSPENSION	IGNITION INTERLOCK	VEHICLE FORFEITURE
Arizona	90 days	After 30 days*	Yes	Yes
California	4 months	After 30 days*	Yes	Yes
Florida	6 months	After 30 days*	Yes	Yes
Georgia	1 year	Yes*	Yes	Yes
Iowa	180 days	After 90 days*	Yes	No
Nevada	90 days	After 45 days*	Yes	No
New Jersey	No	Not applicable	Yes	No
New York	Variable**	Yes	Yes	Yes
Pennsylvania	No	Not applicable	Yes	Yes
South Dakota	No	Not applicable	No	No
Virginia	7 days	No	Yes	Yes

* Must demonstrate special hardship to have privileges restored.

** Suspension lasts until prosecution is complete.

Source: Insurance Institute for Highway Safety, DUI/DWI Laws, January 2008

cal jurisdiction imposes its own sanctions, which increase in severity as a driver's BAC level increases or if a driver has a record of prior offenses. If convicted of a DUI, a driver may pay up to several thousand dollars in fines and legal costs, and may end up in the local jail. In some states continued DUI offenses may result in severe restrictions on a person's right to receive a driver's license.

If a person is convicted of a DUI, his or her license is usually suspended or revoked. Administrative license suspension allows licenses to be taken before conviction if a person refuses to take a chemical test. By January 2006, 41 states and the District of Columbia passed laws permitting administrative license revocation. In 46 states and the District of Columbia, some offenders are permitted to drive only if they have an ignition-locking device on their car. The device disables the ignition if it detects alcohol on the driver's breath. In 32 states, drivers may have to forfeit their vehicles for multiple offenses.[4,5] Table 2.2 provides examples of impaired-driving penalties in selected states.

Underage Drinking

Teen drinking is a serious public health problem that often goes hand-in-hand with impaired driving. Each year, more than 5,000 young people die in alcohol-related motor vehicle accidents, injuries, homicides, and suicides.[6]

Rising death tolls involving underage drivers and lobbying by MADD led to passage of the National Minimum Drinking Age Act of 1984 (U.S. Code, Title 23, Section 158), which established 21 as the legal minimum drinking age. Like the BAC limit, states had to comply with the law or risk losing federal highway funds. Consequently by 1988, every state had set the minimum drinking age at 21. The act prohibits underage drinkers from purchasing, possessing, or consuming alcoholic beverages, and prohibits adults from providing minors with alcohol.[7]

Within the broad parameters of the federal law, state regulations and sanctions for underage drinking vary considerably, and counties and municipalities can also establish their own laws. Adults who allow underage drinking on their property or who buy alcohol for young people are usually in violation of the law. However, states allow exceptions. In most states, parents or guardians may serve small amounts of alcohol to their own children, and some states allow alcohol to be served for educational, religious, or medicinal purposes. These exceptions extend to private individuals only, and under no circumstances are restaurants, bars, or retail stores allowed to sell it to or serve it to anyone who is under 21.

Imposed Sanctions

Regulations against underage drinking target either the underage drinker, adults who supply the alcohol, or retailers who sell alcohol. Sanctions against an underage drinker who drives, uses a false ID to obtain liquor, or consumes alcohol include loss of driver's license (use and lose laws), community service, fines, and required participation in alcohol treatment programs.

Sanctions targeting adults include fines and jail terms. Retailers and restaurant owners are subject to compliance checks to ensure that they are not selling

INTERNATIONAL REGULATIONS

Although some Americans may perceive U.S. laws as harsh, worldwide impaired-driving laws are generally stricter than those in the United States, with a lower BAC limit established for driving. Most countries have a limit of 0.05, and a few have zero-tolerance laws, which mean no drinking and driving under any circumstances. Here are representative samples of BAC limits in other countries:

BAC 0.08
Britain, Canada, Mexico, New Zealand, Singapore, Zimbabwe

BAC 0.05
Argentina, Australia, Belgium, Bulgaria, Cambodia, Denmark, Finland, France, Germany, Greece, Israel, Italy, Peru, Portugal, South Korea, Spain, Thailand

BAC 0.03
India, Japan, Russia

BAC 0.02
Estonia, Mongolia, Norway, Poland, Sweden

BAC 0
Armenia, Colombia, Croatia, Czech Republic, Ethiopia, Hungary, Nepal, Panama, Slovak Republic

Source: Blood Alcohol Concentration Limits Worldwide, International Center for Alcohol Policies, February 2007

to minors, are trained to recognize false IDs, and may be required to establish minimum ages for sellers and servers. They face loss of their business license for violations.

Third-Party Liability

In some states, every person responsible for selling alcohol can be held liable for underage drinking, from the bar server to the restaurant where the patron was drinking to the restaurant's parent company.

Dram Shop Laws

Dram shop laws hold commercial sellers of alcohol liable for injuries caused by their patrons under various circumstances. The majority of states have some form of dram shop law, although provisions vary greatly.

Dram shop laws say that if the seller serves a patron to the point of intoxication, the seller can be held responsible for any injuries caused by the patron to others. This may include injuries resulting from drunk driving as well as those incurred in fights or other actions of the patron.

These laws obviously pose great liability potential for bars, restaurants, and retailers. In just one example, in 1999 a New Jersey court awarded $135 million in damages to the family of a girl paralyzed in a traffic accident caused by a drunk driver. The jury found the driver and the concessionaire at Giants Stadium where the driver had been drinking equally responsible for her injuries.[8] There are many similar cases in other states.

Is the seller responsible for injuries to the intoxicated patron himself or herself? No, according to a majority of states: The law considers drinking to be a voluntary act, so the drinker will not be rewarded for his or her own carelessness by being allowed to sue the seller. But there are a few states where the opposite is true, and drinkers are allowed to sue for their own injuries.

Some states also impose special liability when the intoxicated patron is underage. In California, for example, an alcohol seller who served an intoxicated minor would be responsible for injuries suffered by the minor, but would not be liable for injuries suffered by an intoxicated adult patron (Cal. Business and Professions Code Sec. 25602.1). Carding policies are obviously important to commercial sellers of alcohol. Asking anyone who looks under age 30, or asking all customers, to produce identification is a policy that any seller should institute.

Social Host Laws

By 2004, 35 states had implemented some form of social host law, a form of third-party liability law that holds any adult who serves alcohol to a minor liable if the underage drinker harms himself or herself, kills or injures someone else, or damages property. In 15 of those states, adults can be penalized for serving alcohol even if no one is harmed.[9] Social host laws apply not just to drinking in a private home but to any place that the adult host has reserved—a beach, a boat, a hotel room—whether or not the adult is present. Penalties for hosting underage-drinking parties range from probation to fines to jail.

Industry Responsibility

Restaurants, bars, and retail stores have a responsibility to help keep drunk drivers off the road and prevent underage drinking. In some cases, states mandate licensees to provide responsible-service training programs. In other states, many business establishments have taken voluntary steps to implement programs as a way to protect themselves and their employees from third-party liability lawsuits. Responsible service programs may include recognizing signs of intoxication, effective ways to check IDs, and appropriate interventions if a customer is drunk or underage.

Alcoholic beverage industry associations, including the Beer Institute, the Educational Foundation of the National Restaurant Association and its state affiliates, the Distilled Spirits Council of the United States, and the Wine and Spirits Wholesalers of America, are aware of the seriousness of impaired driving and underage drinking. The industry has formed partnerships with MADD, medical associations, and government agencies to address these problems, and is actively involved in discouraging underage drinking. Industry associations conduct media campaigns, run public service announcements, and host educational Web sites. Some companies promote designated-driver programs, free taxi rides, and hotel sleepovers. All of these programs are designed to encourage responsible adult drinking.

SUMMARY

In the past 25 years, alcohol use has been a significant public policy issue and laws regarding alcohol have become more stringent. The trend continues toward even more restrictive laws, particularly those related to drinking and driving and underage drinking.

Consumers have the responsibility to know how the state or municipal laws where they live impact their alcohol use and to comply with those laws. This usually involves knowing the rules about and the penalties for drinking and driving, allowing young people to drink, and shipping alcoholic beverages.

Producers, distributors, and retailers must know the regulations not only of the state in which they operate, but also in every state in which they do business. Understanding direct shipping laws can be particularly confusing, especially as they are likely to continue to change as producers challenge the laws. Retailers must also know the laws related to selling and serving customers because violating those laws can be costly.

Healthy Drinking

When considered from a health standpoint, alcoholic beverages present a paradox. On the one hand, they are viewed as one component of a healthy lifestyle with recognized health benefits. On the other, alcohol abuse is a major problem and is a leading contributor to illness and accidents. Some have gone as far as saying that alcoholic beverages are a food and should be included in the food pyramid. Others believe that alcohol is a drug that should be prescribed just as antibiotics are prescribed. Which of these perspectives is accurate? Or are both aspects of drinking alcohol valid?

PUBLIC POLICY ISSUES

For millennia, alcoholic beverages were used for medicinal purposes in addition to their use in serving social and religious functions. Hippocrates, the father of medicine, used wine as a disinfectant and as a medicine, and believed it should be included in a healthy diet. Alcoholic beverages have been used to aid digestion, disinfect wounds and injuries, combat disease, and sterilize water. People have consumed alcoholic beverages instead of water in places where the water was too contaminated and dangerous to drink. Numerous home remedies, such as cough syrup made from lemon juice, honey, and hot whiskey, have been staples of the household medicine chest, and medical professionals have prescribed alcoholic beverages as a cure for a variety of diseases. During the early years of this country, the *United States Pharmacopoeia* listed some alcoholic beverages for their medicinal benefits.

In the early 1900s, however, public policy shifted. Health professionals began to discourage the use of alcohol for medicinal purposes, and alcoholic beverages were removed from the *Pharmacopoeia*. This change in attitude developed from a growing awareness of alcohol-related social problems, disease, and death. The social implications of alcoholism led to the passage of laws prohibiting the production, transportation, and sale of alcoholic beverages.

Although many of the laws were later repealed, their impact on public policy remains. Health professionals, the government, and public policy agencies have been reluctant to consider the positive health effects of alcoholic beverages. Particularly in the 1970s and 1980s, growing awareness of lifestyle issues related to health, such as diet and exercise, focused policy makers on the dangers of drinking, spurred by fears of overindulgence leading to alcoholism. Most physicians advised patients against drinking, and medical organizations cautioned against drinking as well.

During this time, government agencies assumed responsibility for alerting the public about potential harms from chemical carcinogens, especially in food products. This led to the addition of warning labels on alcoholic beverages regarding the apparent negative health effects of alcohol because it was considered a potential carcinogen.

Public Awareness of Health Benefits

By the late 1980s, public policy began to shift in favor of moderate drinking. The shift was attributable, in part, to the French Paradox (see page 31), which led to increasing interest in the health effects of alcoholic beverages and research about these effects. Numerous studies in the following years indicated that alcoholic beverages might provide health benefits and could be included in the diet as part of a healthy lifestyle. Yet, the battle still rages between those who believe policy should emphasize abstinence and those who believe in advocating responsible drinking.

POSITIVE HEALTH
BENEFITS

Over the past several decades, medical researchers have conducted hundreds of studies, which have shown that alcoholic beverages—whether wine, beer, or spirits—taken in moderation might provide health benefits. Overall, the evidence indicates a positive relationship between consuming alcoholic beverages and lowered risk of heart attacks and strokes as well as lower rates of death from all causes.

One might assume that nondrinkers have a lower risk of heart disease and a longer life span than those who drink. However, this is not the case. The relationship between drinking alcoholic beverages, heart disease, and overall mortality is defined by a J-shaped curve. Moderate drinkers, in fact, have lower mortality rates than either nondrinkers or heavy drinkers, but heavier drinkers are at much greater risk than abstainers. Heavy drinking, in particular, greatly increases the chances of chronic poor health resulting in death from a number of causes. The decrease in mortality for moderate drinkers is largely attributed to a decrease in heart attacks.

Alcohol and Mortality:
A J-shaped Curve

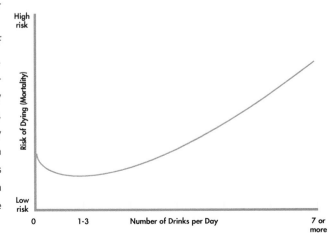

Alcoholic beverages can contribute to a healthy lifestyle, especially when shared with friends around the dinner table.

The evidence is not as clear about alcohol's effects on cancer, diabetes, dementia, and other common diseases, but it appears that moderate drinking may lower the risk for some of these conditions.

Other Factors Affecting Health

Drinking in itself may not be responsible for improved health. Other lifestyle factors, including not smoking, maintaining a healthy weight, eating a balanced diet, and exercising, can change the effects of alcohol on a person's health either positively or negatively.

The age of the individual is another factor that influences the potential health effects of drinking. Research supports the benefits of drinking for middle-aged drinkers who are at risk for heart disease and other illnesses. However, younger adults may not receive the same health benefits because they are not yet at risk for heart disease. At the same time, they are at greater risk of death or injury from heavy drinking due to increased accidents and violent behavior. Research has also shown that the health gains that can result from drinking are less pronounced for both men and women over 65 years of age.

Red Wine and Antioxidants

Many wine industry advocates and researchers report a slight advantage to drinking red wine over other types of alcohol due to the **ANTIOXIDANTS** present in red grapes. In particular resveratrol, an antioxidant found in the skin, leaves, and twigs of the grapes, has received a great deal of media attention because some research has indicated that it may help fight cancer, delay aging, and prevent heart disease. However, the evidence for the role of antioxidants such as resveratrol is mixed, and the relationship between antioxidants and health has not been proven.

Why not eat red grapes or drink grape juice to obtain the potential benefits of antioxidants? It is the properties of the alcohol that provide health benefits, so red grapes do not contain the same health properties as grapes fermented to produce alcohol.

Furthermore, it may not be the alcohol or the antioxidants in wine at all that give it a slight edge in some studies. It appears that wine drinkers exercise more, smoke less, eat healthier diets, and maintain a healthier overall lifestyle than nonwine-drinkers. They also tend to drink with food, so the alcohol is metabolized before it reaches the blood stream, and they tend to sip rather than gulp wine.

ALCOHOL AND HEALTH CONDITIONS

Life Span

There is little doubt that heavy drinkers have shorter life spans than nondrinkers. They die younger, both from disease and from accidents. However, there is a definite positive correlation between moderate drinking and a longer life. When researchers study all causes of death, they find that moderate drinking increases life expectancy regardless of heart disease risk.

Heart Health

One in three Americans has some form of cardiovascular disease. Heart disease, including heart attack, strokes, and high blood pressure, is the leading cause of death in the United States. In 2005, almost 500,000 deaths were attributed to heart attacks and angina, a form of heart disease.[1]

The primary benefit of drinking alcoholic beverages is heart health. Study after study has demonstrated that coronary heart disease is less likely to develop in moderate drinkers, especially in those over 35 years old. Moderate drinking lowers the risk of heart disease by 20 to 50 percent, according to researchers. A World Health Organization project monitored heart disease worldwide and found that mortality rates for heart disease were lower in France, where drinking wine is the norm, than in other industrialized countries, a fact that has become known as the French Paradox.

THE FRENCH PARADOX

The term the French Paradox is based on the observation that the French, who eat more saturated fat, smoke more, and exercise less than Americans, have a considerably lower incidence of heart disease. The concept took the world by storm when Dr. Serge Renaud, the director of the French National Health Institute and a foremost researcher into the relationship between health and wine consumption, appeared on the CBS broadcast 60 Minutes. Dr. Renaud attributed the differences to the French habit of drinking wine, usually red wine, on a daily basis.

Professionals began to take a closer look at the relationship between health and wine. The heart benefits of wine drinking were not only apparent in France but also in Spain, Italy, and Greece, all grape-growing regions where wine is a part of a healthy diet. But people in these countries also consume more fruits, vegetables, and grains, and fat comes from the use of olive oil rather than animal products. Furthermore, people typically drink wine with a meal in the company of family and friends, a healthier way to drink.

Stroke

Ischemic stroke is the third-leading cause of death in the United States. In addition, about 30 percent of stroke victims are permanently disabled and many require permanent institutional care.[2] Ischemic stroke, the most common form of stroke, results from a clot forming in an artery and cutting off the blood supply to the brain. Hemorrhagic strokes are caused by blood vessels bleeding into the brain. Alcohol affects these types of strokes differently.

Just as drinking lowers the risk of heart attacks, light to moderate drinking reduces the risk of ischemic stroke. However, high levels of alcohol increase the risk of stroke significantly, and abstaining results in a slightly higher risk. A little alcohol provides a protective effect against ischemic stroke. Studies on alcohol and stroke replicate the J-shaped curve (see page 29) established for heart attack and mortality.[3,4]

It is a different story for hemorrhagic stroke risk. This form of stroke seems to have a direct relationship with alcohol consumption. As intake increases, the risk for stroke increases. Even one drink a day might increase the risk.[5,6]

High Blood Pressure

Blood pressure is the measure of how hard the heart pumps blood through the body. High blood pressure, another form of cardiovascular disease, is prevalent in the United States. Elevated blood pressure is of concern because it can lead to stroke and heart attack and can shorten a person's life span. It appears that alcohol increases blood pressure in direct proportion to the amount of alcohol consumed.

Alcohol's Role in Disease Prevention

The ETHANOL in an alcoholic beverage works to provide heart protection in two ways: by preventing the formation of plaque in the arteries and by preventing the formation of blood clots.

Low-density lipoprotein (LDL), or "bad" cholesterol, carries excess cholesterol into the blood stream and deposits it in the arteries, where it forms plaque. As plaque accumulates in the arteries, it blocks them and cuts off the flow of blood to the heart, which can cause a heart attack. High-density lipoprotein (HDL), or "good" cholesterol, gathers up cholesterol from plaque and carries it to the liver where it is metabolized and excreted from the body. Alcohol raises HDL levels, which facilitates the removal of plaque from the arteries and should lower the risk of heart attack.

Platelets, the clotting elements in blood, sometimes stick together to form clots, which can lead to heart attack and stroke. Alcohol has anticoagulant properties, which keeps the platelets from sticking to each other and reduces the formation of blood clots, a leading cause of ischemic strokes.

SOBERING STATISTICS

Almost one-quarter of people admitted to general hospitals have alcohol-related problems.

Each year approximately 183,000 rapes and sexual assaults, 197,000 robberies, and well over a million assaults are associated with alcohol use.

One-third of all traffic fatalities and half of all boating fatalities are alcohol-related.

Alcoholics are 16 times more likely to die in falls and 10 times more likely to become fire or burn victims.

Alcohol is associated with between 47 and 65 percent of adult drowning incidents.

Up to 40 percent of industrial fatalities and 47 percent of industrial injuries are linked to alcohol.

Source: National Council on Alcoholism and Drug Dependence, "Alcoholism and Alcohol-Related Problems," June 2002

Cancer

When it comes to the relationship between drinking and cancer, the data are inconclusive. Some studies show that drinking alcohol increases some types of cancer, while other studies indicate that there is no relationship between alcohol and cancer. No one has yet identified the mechanisms at work that relate cancer to alcohol. It is possible that resveratrol and other antioxidants in alcoholic beverages may help fight cancer, but at the same time, the alcohol itself may cause cancer.

In general, it appears that alcohol consumption may slightly increase the risk of cancers of the liver, mouth, pharynx, larynx, and esophagus. The relationship between breast cancer and alcohol has not been definitively established. Research seems to indicate that moderate drinkers do not have increased risk for other forms of cancer. To some extent, overall cancer risk appears to depend on the amount of alcohol consumed, with heavy drinkers at greater risk.

Alzheimer's and Dementia

Alzheimer's and dementia have become major health concerns as the number of adults who are living longer increases. Alcoholic beverages, especially red wine, have been considered as a possible aid to improve cognitive function. But as with the link between alcohol and cancer, the relationship between drinking and these diseases is not clear.

Some studies report that drinking alcohol in small amounts lowers the risk of dementia, but not Alzheimer's. Some indicate that any type of alcoholic beverage provides benefits, while others find that wine is the only beverage that helps to ward off dementia, perhaps due to the presence of resveratrol and other antioxidants.

Diabetes

Type 2 diabetes, the form of the disease that develops in adults, is a major health problem in the United States. Therefore, it has been the subject of much research, including studies investigating the relationship between diabetes and alcohol consumption. The preponderance of evidence indicates that the risk of developing type 2 diabetes is lower in moderate drinkers than abstainers, by as much as 33 to 55 percent.[7–10] However, as with heart disease and mortality, diabetes risk increases with heavy drinking.

Other Diseases

The health benefits of drinking alcoholic beverages, particularly wine, have been attributed to preventing the development of a number of other diseases, including:
- the common cold
- macular degeneration

NEGATIVE HEALTH EFFECTS OF HEAVY DRINKING
alcoholism
brain damage
some cancers
heart failure
pancreatitis
high blood pressure
stroke
skin infections
infertility
bone density loss
liver damage
alcohol poisoning
depression
memory loss
poor mental health and suicide
violence
auto accidents
injuries

THE 24-HOUR EFFECT

Alcohol's anticoagulant properties last for about 24 hours. During this time, the alcohol in the body helps to keep the blood from clotting, which reduces the chance of a heart attack occurring. This is known as the 24-hour effect.

It appears that one of the advantages of drinking wine instead of other types of alcoholic beverages is the manner in which it is consumed. Wine drinkers tend to drink a limited amount with a meal on a regular basis (for example, two glasses of wine daily with dinner). This drinking style keeps a steady amount of alcohol in the system, maintaining its anticoagulant effect. By reinforcing the 24-hour effect, wine drinkers lower their risk of the blood clots that can cause a heart attack.

However, beer and spirits drinkers are more likely to consume their drinks binge-style, with heavy drinking in one session followed by days of abstinence. This form of drinking maintains an unbalanced amount of alcohol in the body. In the period immediately following a binge session, a person is overly anticoagulated due to excess alcohol in the system. This can cause bleeding and lead to a stroke. In the days following the binge, the person is at increased risk of heart attack until the next binge because there is no alcohol in the system to moderate clotting.

- loss of bone density and osteoporosis
- gallstones
- peptic ulcers
- tooth decay

However, the limited research for these diseases presents both positive and negative findings and is far from conclusive. Before health decisions can be made based on linking alcohol with the prevention of these conditions, much more investigation is needed.

ALCOHOL-RELATED PROBLEMS

When it comes to drinking alcohol, it is not all a rosy picture. Alcohol consumption results in about 100,000 deaths each year in the United States.[11] The ill effects of heavy drinking are at best neutral, and at worst can result in severe consequences to the drinker, to family and friends, and to society at large. Drinking alcoholic beverages has been implicated in many health and social problems (see Sobering Statistics, page 32). One of the reasons for the severity of these problems is the fact that heavy drinking lowers inhibitions and impairs judgment, leading to risky behaviors such as unprotected sex and drug use. Alcohol can markedly affect worker productivity and absenteeism, family interactions, and school performance. Alcohol abuse is often involved in violent behavior, murder, robbery, assault, domestic violence, and date rape. Furthermore, alcoholism and alcohol-related problems take a major toll on the economy. The economic costs attributed to drinking exceed $184 billion annually. This includes costs due to lost productivity caused by illness or death, health care expenditures, motor vehicle accidents, property damage, and crime.[12]

Alcohol-Related Health Concerns

Although research shows that light to moderate drinking provides health benefits, heavy drinking can cause a long list of diseases (see Negative Health Effects of Heavy Drinking, page 33). In fact, the incidence of disease and the number of people dying from disease are highest among heavy drinkers. In many instances a little wine, beer, or whiskey might help reduce the risk of a health condition, but too much might exacerbate that same condition. For example, light drinkers experience fewer strokes than nondrinkers, but the incidence of strokes increases in heavy drinkers.

While moderate drinking provides heart benefits, overconsuming places a heavy toll on the cardiovascular system, increasing cholesterol, raising blood pressure, and weakening the heart. Increased calorie intake from drinking may also offset alcohol's benefits by causing weight-related health problems. Furthermore, heavy drinkers often substitute calories from alcohol for calories from healthy food, resulting in malnutrition.

Alcoholism

Alcoholism is a serious health issue in the United States, affecting more than 8.1 million people.[13] Alcoholics develop a physical dependence on alcohol, and suffer withdrawal symptoms if they stop drinking. They cannot limit their drinking to a safe amount, and continue to drink even if their behavior breaks up their family, costs their job, or destroys their health.

Some people are heavy drinkers, but they are not considered alcoholics because they continue to function at work and at home. However, heavy drinking frequently leads to alcoholism, and even a small proportion of light users eventually become alcoholics.[14] For those at risk of becoming alcohol dependent, abstinence far outweighs the benefits of drinking.

Binge Drinking

What comes to mind when you hear the words *binge drinking*? Probably an image of a wild, out-of-control college party, where young people drink until they pass out. This is one example of binge drinking, and an all too common one. But what about having a few martinis after work on Friday night? Or guzzling several beers while watching the Super Bowl? Or drinking an entire bottle of wine at a bachelorette party? Each of these situations is an example of binge drinking.

Although binge drinking is often associated with the young, a significant number of people over age 26 binge drink, and their numbers are rising each year.

Although binge drinking is often associated with young people, it is increasingly a problem for older adults.

What Is Binge Drinking?

Binge drinking is commonly characterized by occasional heavy drinking between periods of total abstinence. Unlike heavy drinkers, who are more likely to drink regularly, binge drinkers imbibe irregularly, be it once a week or once a month.

Health professionals have different opinions about the amount of alcohol that constitutes a binge, but it is lower than you might think. The Centers for Disease Control and Prevention defines excessive alcohol use as more than three drinks in about two hours for women and more than four for men.

Risks Associated with Binge Drinking

Binge drinking carries serious risks, including death from alcohol poisoning. This style of drinking leads to a shorter life span and significant health problems. Binge drinking does not contribute to positive heart-health effects and, in fact, it increases the chances of heart attack and stroke (see The 24-Hour Effect, at left). In addition, binge drinkers are more likely to engage in risky behaviors, such as having unprotected sex, leading to unwanted pregnancies and sexually transmitted diseases, and are more likely to incur injuries, both intentional and unintentional.

To drink safely, it is important to limit the amount consumed in any one session. The effects of two drinks daily (14 over a week) are not the same as the effects of 14 drinks in one evening. Rather than saving up drinks during the week and then having them all on the weekend, if you are going to drink, it is wiser to drink a little bit each day.

WOMEN'S HEALTH ISSUES

It is a fact: Women get drunk faster on fewer drinks than men. Research has established that women metabolize alcohol differently than men and, therefore, are affected more severely by the negative effects of overdrinking. There are several physiological factors responsible for this. Women tend to be smaller and, therefore, have less water in their bodies to dilute the alcohol. They have less muscle mass and more fat, which metabolizes alcohol less efficiently. Women also have less alcohol dehydrogenase, a stomach enzyme that breaks down alcohol before it reaches the blood stream. Estrogen also appears to play a role in the way that alcohol is processed.

These characteristics put women at greater risk of alcohol-related problems even if they drink less than men. Furthermore, while women generally receive some health benefits from moderate drinking, they do not gain as much of a beneficial effect from drinking as men. Drinking may decrease the risk of heart disease and stroke in women over 50 years of age. However, younger women do not receive the same heart benefits because they are at less risk of having a heart attack in the first place. Study results are mixed for other diseases. Some find that drinking reduces cancer risk, and others that it increases risk, especially for breast cancer. Although there is little evidence that moderate alcohol intake increases cancer, medical professionals advise women who are at high risk for breast cancer to be cautious.

Until more information is available regarding alcohol's impact on women's health, health professionals advise women to drink less than men. So far the research indicates that unhealthy effects begin and risks increase at two or more drinks a day.

Women are advised to have fewer drinks than men because their bodies process alcohol differently.

PURESTOCK

Alcohol and Pregnancy

Drinking during pregnancy is a controversial topic because it involves the health of a fetus. The effects of heavy drinking and alcoholism are well documented, but the relationship between light drinking and fetal development has not been well established. Research has not linked light drinking during pregnancy, defined as a couple of glasses a week (less than the one glass a day considered beneficial for nonpregnant women), to an unhealthy fetus. However, evidence is mounting that binge drinking is very harmful to the developing fetus.

There is no doubt that heavy drinking during pregnancy impacts the developing fetus, often causing fetal alcohol syndrome. About 4,000 U.S. babies a year are born with fetal alcohol syndrome.[15] Children who are exposed to excessive alcohol while in the womb may develop severe health problems, including birth defects, brain damage, and low birth weight, and may lag in mental and emotional development as well as motor skills.

In response to concerns about the seriousness of fetal alcohol syndrome, the U.S. Surgeon General required that a warning label appear on all alcoholic beverage containers sold in the United States beginning in the 1980s. The label states: "According to the Surgeon General, women should not drink alcoholic beverages during pregnancy because of the risk of birth defects." Continuing concerns led the Surgeon General to release an advisory in 2005 stating that pregnant women should abstain from drinking any alcoholic beverages.

Most health professionals have followed the Surgeon General's lead and take a cautious approach to drinking during pregnancy. On the other hand, some physicians believe that a limited number of drinks during pregnancy will not harm the fetus because the body can rid itself of alcohol before it reaches the womb. Still, the prevailing advice is to abstain from alcohol while pregnant.

MODERATE DRINKING AS PART OF A HEALTHY LIFESTYLE

Cultural norms greatly influence people's attitudes toward drinking. In Spain, Italy, and France, people drink a glass of wine with a meal on a daily basis. Even children are sometimes given a glass of watered-down wine with lunch or dinner. Public drunken behavior is frowned upon. In other places, such as the United States and Great Britain, public drunkenness is more common, and children are generally discouraged from drinking under any circumstance. As the Mediterranean cultures demonstrate, alcoholic beverages can be an important and pleasant part of a healthy lifestyle.

What Is a Healthy Lifestyle?

Research over the last two decades has shown that for most people moderate consumption of alcoholic beverages is beneficial to health, especially heart health and life span, and can be included in the prescription for healthy living. However, it is important to remember that other lifestyle factors must also be taken into account. Eating a healthy diet, exercising, maintaining a healthy weight, avoiding stress, and refraining from smoking are important components of a healthy lifestyle.

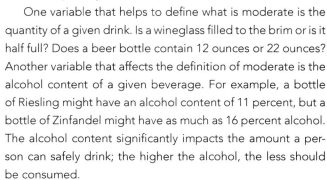

1.5 oz.
80-proof
Spirits

5 oz.
Wine

12 oz.
Beer

What Is Moderate Drinking?

Moderate is a difficult term to define because it is subjective and involves several variables. The number of drinks considered moderate varies from country to country. Even within the United States the recommendations vary from one to five drinks per day, with one to three as the norm.

One variable that helps to define what is moderate is the quantity of a given drink. Is a wineglass filled to the brim or is it half full? Does a beer bottle contain 12 ounces or 22 ounces? Another variable that affects the definition of moderate is the alcohol content of a given beverage. For example, a bottle of Riesling might have an alcohol content of 11 percent, but a bottle of Zinfandel might have as much as 16 percent alcohol. The alcohol content significantly impacts the amount a person can safely drink; the higher the alcohol, the less should be consumed.

U.S. government agency recommendations are two drinks per day for men and one drink per day for women. Most health organizations consider a moderate drink to be:

– 12 ounces beer
– 5 ounces wine
– 1.5 ounces 80-proof spirits

Alcohol content is not taken into account for beer and wine in these recommendations (see Standard Drink Sizes, at left).

Additional factors, including body size, gender, current and past medical conditions, genetic makeup, drug intake and medications, and family history, can also influence the impact that drinking has on physical and social well-being. For example, one glass of wine might be too much for a small woman, while a large woman might safely tolerate three glasses of wine. Research is also finding that tolerance for alcohol may decrease in older people, so the number of drinks considered moderate is lower for those over age 65.

How You Drink

How you drink is just as important as how much you drink. Drinking can be a part of social bonding and is included in many social situations. Quaffing several drinks alone at a bar is not part of a healthy lifestyle. How quickly you drink—sip or gulp—has an impact on how rapidly your body can metabolize the alcohol out of its system. Slow drinking is definitely healthier.

Blood alcohol levels do not rise as high when alcohol is consumed with food; therefore, drinking and eating together is less harmful than having a drink by itself. Finally, drinking every day is a key to beneficial health effects (see The 24-Hour Effect, page 34).

Making an Informed Decision

Many medical professionals and health organizations, such as the American Heart Association, cautiously support moderate drinking. *The Dietary Guidelines for Americans: 2005*[16] recommends no more than one drink a day for women and two drinks a day for men for those who already drink, which is the generally accepted drinking level. But the guidelines also identify situations in which people should never drink.

Many people have moral and religious reasons for not drinking and should never be pressured into trying a drink. Others have health conditions that preclude alcohol or are taking medications that can interact with alcohol to create a health-threatening situation. Anyone who is unable to limit the amount that he or she drinks should avoid alcohol. Those who have a history of drug or alcohol abuse, or who are genetically disposed to alcoholism, or have a family history of alcoholism are generally advised not to drink.

Each adult has to make an individual choice about whether or not to drink based on his or her beliefs and values, personal taste, and medical history. The information that is available about the benefits and dangers of drinking and discussions with your personal physician can help you to make an informed decision about whether drinking is the right choice for you.

SUMMARY

As part of a healthy lifestyle, alcoholic beverages can provide health benefits and add to the pleasure of social occasions. If you choose to drink, however, you must drink responsibly and avoid putting others or yourself at risk. Alcohol impairs judgment, motor coordination, balance, and perception. Therefore, anyone who is drinking, even the recommended amount, should avoid activities that require skill and attention, such as driving, boating, swimming, skiing, and operating machinery.

For the best health and the greatest enjoyment, limit alcohol to moderate amounts, avoid binge drinking, follow the guidelines for healthy living—control weight, do not smoke, exercise, manage stress—and enjoy a glass of wine with dinner.

Viticulture: Growing Grapes

We treat the making of wine as two components: **VITICULTURE**, the science and process of growing grapes, and **VINICULTURE**, the art and science of making wine. Viticulture is growing grapes separate and apart from making wine. Viniculture focuses on winemaking, **VINIFICATION**, but also includes the process of growing grapes as a key component in making wine.

Although winemaking techniques can influence the taste of wine, the underlying characteristics of most wines come from what occurred in the vineyard. Learning about viticulture helps students and consumers of wine understand the most important factors that make a wine taste the way it does. From grape-variety selection, to characteristics of the place where the grapes are grown, to how the grapes are managed in the vineyard, each aspect of viticulture influences the final taste of wine.

WINE GRAPE VARIETIES

Most wine grapes come from the genus of grapes *Vitis* and the species *vinifera*. There are a number of species of grapes within the genus *Vitis*, but only the European wine-grape species *Vitis vinifera* and a few other species produce grapes that are commercially made into wine (see Wine Grape Hierarchy, at right).

Less common species of wine grapes and **HYBRID** grapes crossbred from two different species are sometimes used to produce wines in regions where *vinifera* grape varieties are unable to grow because of climate. As an example, *vinifera* varieties do not typically do well in the cold winters of the Eastern and Central United States and in the humid heat of the South. American grape varieties such as Concord and Norton and hybrid grape varieties such as Seyval Blanc and Baco Noir can be used to make wines in regions like these. Perhaps the greatest achievement of American grape species is their rootstock, which is used as the base for **GRAFTING** most *vinifera* vines throughout the world (see Phylloxera, page 53, and Rootstock, page 52).

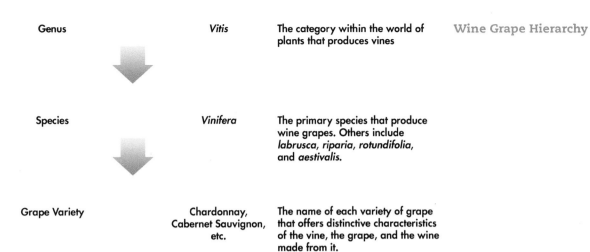

Genus	*Vitis*	The category within the world of plants that produces vines	Wine Grape Hierarchy
Species	*Vinifera*	The primary species that produce wine grapes. Others include *labrusca, riparia, rotundifolia,* and *aestivalis.*	
Grape Variety	Chardonnay, Cabernet Sauvignon, etc.	The name of each variety of grape that offers distinctive characteristics of the vine, the grape, and the wine made from it.	

Vinifera produces virtually all of the world's commercial wines. Hundreds, if not thousands, of *vinifera* varieties exist, but most wine is made from just a few dozen grape varieties. *Vinifera* are commonly divided into two broad categories, labeled by color: white (green-skinned grapes) and red (red- or black-skinned grapes). The categories of red and white grapes are further broken down into varieties, each with its own distinct characteristics. Table 4.1 lists some common grape varieties.

Table 4.1 Common Grape Varieties

VINIFERA		NATIVE AMERICAN	HYBRID
WHITE	RED		
*Chardonnay	*Cabernet Sauvignon	Catawba (W)	Baco Noir (R)
*Sauvignon Blanc	*Merlot	Concord (R)	Cayuga (W)
*Riesling	*Pinot Noir	Delaware (W)	Chambourcin (R)
Chenin Blanc	*Syrah/Shiraz	Isabella (R)	Chardonel (W)
Pinot Gris/Grigio	Cabernet Franc	Muscadine (R or W)	Edelweiss (W)
Viognier	Malbec	Norton/Cynthiana (R)	Marechal Foch (R)
Gewürztraminer	Grenache/Garnacha		Seyval Blanc (W)
	Sangiovese		Traminette (W)
	Tempranillo		Vidal Blanc (W)
	Zinfandel		Vignoles (W)

* International grape varieties, which are covered extensively in this book.

Key: R = red grape, W = white grape.

ANNUAL GROWING CYCLE OF WINE GRAPES

Many factors impact a grape grower's ability to produce quality grapes. In order to analyze these factors, it is essential to first understand the lifecycle of a grapevine. The length of the annual growing cycle can vary at each stage because of weather conditions or desired grape ripeness. As an example some harvests during some vintages in some appellations might start as early as late July or continue into November.

Dormancy (December to February in the Northern Hemisphere) This is the period after harvest when the vines have finished the prior year's lifecycle and before the start of the next lifecycle. Dormancy can last for two to three months. The primary task for the grower during dormancy is pruning, which establishes the amount of vine and grape growth for the coming year. Branches are trimmed back, leaving only the number of shoots that will become active and be able to produce a crop in the coming year. Because freezing temperatures can kill the vines during dormancy, in very cold climates, like the Midwest and Eastern United States, the vines must be protected from a deep freeze. Mulching and mounding the roots help the vines resist freezing at low temperatures.

Bud Break (March in the Northern Hemisphere) Temperatures rise, soils warm, roots become active, and small buds break out. The buds sprout for eight to 10 weeks, developing long shoots. Cool to warm temperatures and adequate soil moisture from rain or irrigation are most desirable at the start of a vintage. During early parts of the growing cycle, the vines are susceptible to frost and at high risk when temperatures approach freezing. Spraying a fine mist of water, using large fans to create air circulation, or even using a helicopter to force down warmer air may, but does not always, keep the buds from freezing during a frost. Sometimes nothing can be done—the buds freeze and the crop is lost for that vintage.

Flowering and Fruit Set (April to May in the Northern Hemisphere) Following shoot growth, clusters of flowers develop at the shoot base, self-pollinate, and begin to develop grapes. This period is called fruit set. Coulure, or poor fruit set, can occur following flowering if all of the fruit does not develop properly. On the positive side, poor fruit set helps control yield if it is not

Dormancy

Bud Break

Flowering or Fruit Set

Grape Growth

too extensive. When excessive, it causes poor yields for that vintage. Certain grape varieties, like Merlot, Grenache, and Malbec, are more susceptible to poor fruit set than other varieties. Bunches that do not form properly are pinched off at this time. Heavy rain, wind, or hail can also have a damaging impact on the grapes during this early period of growth.

Grape Growth (May to August in the Northern Hemisphere) Following fruit set, the **CANOPY** of leaves continues to grow. The grapes develop and increase in size, producing **ACIDS** and **TANNINS**, but they remain hard and dark green. The combination of water, sun, and carbon dioxide from the atmosphere begins a biochemical reaction in the leaves called photosynthesis, which creates sugar in plants. During the summer months, when grapevines are in full leaf, warm and even hot temperatures increase photosynthesis.

Veraison (Late July to August in the Northern Hemisphere) Veraison is the point in grape development when shoot and leaf growth slows and grapes begin to change color from hard green to light yellow-green for white grapes and red or black for red grapes. The grapes also begin to soften as they increase in sugar content and size. Although the number of shoots was determined during pruning, the number of bunches are counted at this point to deter mine the potential yield.

Green Drop or Green Harvest To control yield and improve grape quality, one of the most common techniques used in the vineyard is what is called a green drop. It is a method of fruit thinning that gives each vine the right amount of fruit to achieve the desired yield at harvest. Eliminating unripe grapes reduces grape variability, increasing quality. Each vine is checked, and bunches that are slow to develop are dropped to the ground. Fruit may also be dropped at other times during the ripening phase to control yield or to remove grapes damaged by mold or rot.

Final Ripening (August to September in the Northern Hemisphere) The grapes grow in size and become softer, and the sugar content increases until the desired levels of acid, sugar, ripeness, and flavor development are achieved.

Harvest (August to October in the Northern Hemisphere) Harvest occurs approximately 100 days after flowering. The fruit is picked when the vineyard manager or winemaker decides that it has reached its optimum point of ripeness. Weather and the availability of harvest labor and equipment play a role in this decision. Once harvested, the fruit is taken to the winery and the vines begin a new period of dormancy.

Veraison
PHOTOGRAPHY BY CAROL ANN
THOMAS/CAROLANNTHOMAS.COM

Green Drop or Green Harvest

Final Ripening

Harvest

WHERE GRAPES GROW

Vinifera vines survive and thrive in two fairly wide zones in the world, a band of latitude from approximately 30° to 50° in both the Northern and Southern hemispheres. Attempts to cultivate *vinifera* varieties outside of that band have generally been unsuccessful. Wine grapes cannot generally be grown in polar, tropical, or severe desert conditions, or at 2,000 feet/650 meters elevation or higher. Global warming, however, will likely cause shifts in the range where wine grapes may be grown in the future (see Climate Change, page 49).

Why are some places suited to growing grapes and others not? Within the grape-growing zone, general geographic characteristics of topography, water access, soil, and climatic conditions, like Gulf Stream warming, the pattern of rainfall, and temperature, may make a region particularly good or particularly bad for growing grapes. The French coined a term, **TERROIR**, to describe this combination of factors that makes each wine region or vineyard location unique (see Terroir, page 50). Human intervention in the form of irrigation and soil enhancement can also change grape-growing sites and affect the quality of the wines produced from those sites.

For grapes to grow and ripen, they need moderately warm to very warm temperatures, sunlight, and access to water. Soil composition is important to wine quality, but grapevines do not require fertile soil or level ground. Some world-class wines come from steeply sloped vineyards with miserable soil, rocky sites that appear to have no soil, and terrain so rugged that it is suitable only for grapes and goats.

HIGH-ELEVATION VINEYARDS

There are limited parts of the world where vineyards have been planted successfully as high as 8,000 feet/2,500 meters above sea level. South America has the greatest number of high-elevation vineyards in the Andes of Argentina and Bolivia. In Europe there are plantings at over 5,000 feet/1,700 meters in Spain's Canary Islands and over 4,000 feet/1,350 meters in Switzerland and Northern Italy. In spite of these successes, long-term results to determine whether high-elevation vineyards can consistently produce quality grapes are not yet available.

Topography

Vineyard topography refers to the physical characteristics of the land where grapes are planted. The key characteristics of topography are elevation, slope, aspect to the sun, and proximity to natural water, like oceans, lakes, and rivers. These factors have historically played an important role in vineyard-site selection.

Elevation

Elevation is important because it contributes to the temperature in the vineyard. For every increase of 330 feet/100 meters, the ambient temperature decreases approximately 1.1°F/0.6°C. Grapevines do best within a band of elevation extending from sea level to 1,650 feet/500 meters. When elevation is too low, heat may be too intense to produce quality grapes. If vines are planted at too high an elevation, temperatures may not be high enough to ripen the grapes.

Some grape varieties, like Merlot, Syrah, and Grenache, do well in warmer temperatures at lower elevations, while others, like Chardonnay and Pinot Noir, perform better in cooler temperatures at higher elevations. The same grape variety planted at different points on a slope produces wines significantly different from one another, in part because of the elevation. For example, Cabernet Sauvignon grapes planted in Oakville or Rutherford in Napa Valley, California, are typically ripe and produce fruity wines because the vineyards are on the warm valley floor.

On the other hand, Cabernet Sauvignon grapes from Mt. Veeder in Napa Valley are generally less ripe, lower in sugar, and higher in acid, and produce wines that are lower in alcohol than other Cabernet Sauvignon grapes because Mt. Veeder is at a higher and cooler elevation.

Slope

The slope of a particular vineyard also has an impact on the grapes. Some of the best vineyard sites are sloped, even steeply sloped. In some locations vineyards can have as much as a 30-degree slope. Slope is important because it inhibits water collection and enhances drainage. Sloped vineyards are commonly close to a good natural water source, and grapes in such vineyards rarely freeze because the cold air settles in the valley below the vineyards and is replaced by warm air. The Mosel region in Germany and the Douro Valley in Portugal are some of the most famous and steepest wine regions in the world that benefit from this topographical characteristic.

Aspect

Vines are planted on steep slopes and flatter valleys along Portugal's Douro River.

Aspect is the direction that the vineyard faces: north, south, east, or west. A vineyard's aspect, along with **CANOPY MANAGEMENT** (see page 60), determines the length of time that grapes are exposed to the sun and the intensity of the sunlight. Growers look for sites that provide the right amount of direct sun on each vineyard. Southeast- to southwest-facing slopes north of the equator (northwest- to northeast-facing south of the equator) get the longest sun exposure and are generally more desirable. Heavily sloped vineyards in cool climates that face the afternoon sun benefit from longer periods of direct sunlight. Heat created by the sun is more intense in the afternoon, which is why in cooler regions south-facing slopes are more desirable. Too much sun exposure can be harmful to grapes, causing sunburn and overripe flavors.

Natural Water

Many of the world's grape-growing regions are located close to oceans, seas, or rivers. Much of the continent of Europe and parts of California, Chile, and Argentina, and South Africa, Australia, and New Zealand are directly influenced by natural bodies of water. This is critical in areas with little rain because grapevines have the ability to seek out water as much as 30 feet/10 meters below the soil surface, where they tap the water table.

Vineyards in the Old World have historically been planted close to a water source, or are located where there is ample rain. When natural water sources are available roots tend to travel in a vertical direction in search of water. In the Old World growers believe that the search for water takes roots well below the topsoil, down through the strata of subsoil, which may have a greater mineral content that adds to the complexity of the wines.

Conversely, many vineyards in California and other parts of the New World are found in arid regions. Growers cannot rely on rain or natural groundwater found by deep root growth and must irrigate. In many New World vineyards,

In the Southern Rhône, France, soil is composed of a deep layer of large round stones.

Nick Mills at Rippon Winery in Wanaka, New Zealand, uses organic soil that he makes in his vineyard.

Blue slate retains heat in a Riesling vineyard along the Mosel River in Germany.

Red clay soil is typically found in vineyards in France's Côte d'Or.

particularly with newer drip irrigation systems, the roots tend to spread out in a horizontal fashion, seeking the water that is provided to them and not achieving as much depth as those that rely on a natural water source. The roots remain closer to the surface and do not penetrate the layers of subsoil, and, therefore, may not absorb as much mineral content.

Soil

Soil is a combination of disintegrating rock, organic matter composed of the remains of plants and animals, and mineral nutrients that are key to vine health, including nitrogen, phosphorus, potassium, sulfur, and calcium. Chalk, granite, gravel, limestone, and shale are some of the different types of soil that are found in vineyards around the world. There is no one perfect soil for grapevines because each soil has its own unique combination of components that contribute to the qualities of the resulting grapes. (See the tables in Appendices 2 and 3, beginning on pages 404 and 438, respectively, for information on the soil types found in each grape-growing region.)

The organic matter found in any given soil determines soil fertility. Although most agricultural crops require rich, fertile soil, wine grapes generally need low soil fertility to produce good wines. The soil should contain just enough organic matter to let the vines grow. If the soil has too much organic matter, the vines will overproduce the vegetative canopy, decreasing the fruity character of the grapes. Mineral content must be present in very small amounts. If it is too low the vines will not grow, but too much can be toxic.

The physical makeup of a particular vineyard's soil can range from rocks, stones, pebbles, and gravel to sand, silt, and clay. The makeup of the soil is important because it directly affects the soil's ability to hold water and supply it to the vines and to act as a reflective heat source. Well-drained soils that maintain just enough water to support the vines are most desirable. Clay is an important component in many soils because it holds moisture, but a soil with too high of a clay content may not allow water to drain properly. Large rocks and stones hold heat, benefiting vine growth in cool climates. Growers generally prefer deep soil layers because they allow roots to extend farther to access water.

Some researchers believe that soil components do not add specific flavors to wine but are important only for maintaining the proper level of vine growth. Others believe that different soils introduce distinctive characteristics to the grapes, and therefore to the wine. For example, many believe, particularly in Germany, that slate contributes mineral characteristics to the wine.

Climate

Climate refers to the prevailing weather patterns that occur in a particular location. It includes temperature, precipitation, wind, and cloud cover. Weather, perhaps the most dynamic factor affecting vineyards, is also the only factor that is completely outside the grower's control. The topography and soil in a vineyard do not change on a yearly basis, but weather changes yearly, monthly, weekly, or even daily.

Some grape varieties can only survive under limited climatic conditions. Albariño, for example, grows best in the cool, damp environment of Northwest Spain. Other varieties, like Merlot, perform well in a range of climates but create dramatically different wines depending on the climate. Grapes grown in cooler climates typically have higher acidity levels, create less sugar, and have a lower level of potential alcohol; whereas when grown in warmer climates, they generate higher sugar levels, lose acid, and have a riper fruit character and more alcohol, leading them to produce a heavier, richer, more concentrated wine.

Temperature

Grapes perform best in locations where there is a significant differential between daytime and nighttime temperatures. They need warm, even hot, temperatures during the day to maintain growth and ripening, while cool temperatures at night allow the grapes to rest. Either extremely hot or extremely cold weather can damage crops. As the ambient temperature approaches 100°F/38°C, vines are likely to shut down, protecting themselves from stress and conserving their energy. For example, while the average temperature during the months of July and August in Napa Valley is 82°F/28°C, the summer of 2003 was anything but average with temperatures well over 100°F/38°C lasting for days. No growth or ripening occurred during that time period, contributing to lower yields and an early harvest in Napa Valley that year.

In the winter when vines are dormant, freezing is generally not a problem for short amounts of time. However, if the mean temperature for the coldest months stays below 30°F/–1°C for extended periods, the vines will die. This happens in marginal regions like Northern Germany, the Central Otago in New Zealand, and recently in the United States from Texas through Missouri to Virginia in the East.

When spring arrives, typically March or April in the Northern Hemisphere and September or October in the Southern Hemisphere, the increasing temperatures cause the vine to begin its growth cycle. Soils warm, roots become active, and bud break occurs. Bud break is an anxious time for growers. Temperature can make the difference between the start of a successful crop or a season of failure. If freezing temperatures cause the young shoots to die during this time, the entire crop may be lost. A warm early spring, bringing about premature bud break, can also lead to crop damage if temperatures drop and result in a late spring frost. In 2007 in Missouri and elsewhere in the Central United States, for example, warm weather in March caused an early bud break. Vines had six-inch shoots when a blast of freezing temperatures in May decimated thousands of acres of vineyards. In some locations as much as 95 percent of the crop was destroyed.

Heat Summation Scales

The Winkler-Amerine **HEAT SUMMATION SCALE**, developed over 60 years ago by Professors Albert Winkler and Maynard Amerine of the Department of Viticulture and Enology at the University of California, Davis, has long been used as the benchmark for classifying climate by region in California based on the sum of heat units that is typical from April 1 to October 31. A heat unit is each one degree of temperature over 50°F, and the heat units recorded for a given day yield a unit of measure called the degree day.

Table 4.2 Winkler-Amerine Heat Summation Scale

REGION	EXAMPLES	SUGGESTED VARIETALS
Region I: Below 2,500 degree days	Côte d'Or and Champagne, France Rhine, Germany Willamette Valley, Oregon Carneros, Napa Valley, California Anderson Valley, Mendocino County, California	Chardonnay, Pinot Noir, Gewürztraminer, Riesling
Region II: 2,500–3,000 degree days	Bordeaux and Loire, France Piedmont, Veneto, and Tuscany, Italy Russian River Valley, Sonoma County, California	Cabernet Sauvignon, Merlot, Cabernet Franc, Sangiovese, Sauvignon Blanc
Region III: 3,000–3,500 degree days	Rhône Valley, France Napa Valley, California	Syrah, Zinfandel, Barbera, Gamay, Grenache
Region IV: 3,500–4,000 degree days	Central and Southern Spain Parts of San Joaquin Valley (Central Valley), California Apulia, Italy	Malvasia, Zinfandel, Syrah
Region V: Over 4,000 degree days	North Africa San Joaquin Valley (Central Valley), California	Table grapes, Muscat, Verdelho

The scale breaks climate into five distinct regions, with Region I being the coolest and Region V the hottest. Region I is identified as the best for cool-climate grapes, like Chardonnay and Pinot Noir, and Region V works best for hot-climate grapes, like Thompson Seedless table grapes, but can also produce several other varieties if there is enough of a water supply. The scale demonstrated a connection between a vineyard's temperature and the type of grape varieties that would be successful and has been a dominant factor in determining which grape varieties to plant in a particular vineyard. Many grape varieties can be successful in multiple regions. Although first used in California, other wine regions of the world have also used its rationale in selecting which grapes to plant in which location. Table 4.2 includes examples of regions with their suggested grape varieties.

Precipitation

Rain is of critical importance to vineyards because it drives growth. However, it must come at the right time. Rain is desirable during the early and middle parts of the grape's development period leading up to the ripening phase. Heavy rain at flowering can severely reduce the amount of new berries that become grapes. Rain is not desired later in the ripening phase and at harvest. It adds moisture to the grapes and causes dilution of their acids, sugar, and flavors. And too much moisture at any time creates an environment for mold and rot to attack the vines.

Hail is a form of precipitation that is never desirable. A five-minute hailstorm can wipe out an entire crop, leaving the fruit smashed on the ground. Nothing can be done to protect a vineyard from a hailstorm. For example: In August 2007 only a few weeks before harvest, a hailstorm hit the Côte-Rôtie in the Rhône Valley, France, an appellation that produces some of the finest Syrah in the world. A great vintage had been expected, but the entire crop was lost.

Harvest, which starts as early as August, is another anxious time for growers. They must balance hopes of attaining peak ripeness against the chance of inclement weather. Harvest should ideally occur before fall rains begin. If it does rain significantly at harvest, the grower must decide whether to pick a mediocre, diluted crop or wait and hope for late-summer weather to give the grapes a chance to dry out and ripen fully. There are growers who make the right decision and those who make the wrong decision. The key is how much risk the grower is willing to take. The earlier the harvest, the less likelihood of experiencing inclement weather, but grapes picked early might not achieve the maximum desired ripeness. The later the harvest, the riper the grapes, but rain becomes a much greater threat to the entire crop.

Wind

Wind is another element that can be beneficial or damaging to a vineyard. Wind can moderate temperatures, prevent a frost or freeze, and help to keep grapes dry, which prevents mold and mildew. However, wind can be damaging if it is too strong during flowering and early fruit growth, as it can cause flowers or tiny fruit to blow off the vine, destroying the crop.

Cloud Cover

The time of day or the time of year when cloud cover occurs determines whether or not it is beneficial to the vineyard. Because they limit the amount of sun that reaches the vines, clouds early in the day help to control the temperature, cooling the vines and moderating growth and ripening. But if the cloud cover extends into the afternoon or does not lift for several days during the important ripening phase, the grapes may not ripen adequately. And clouds are also a threat to growers and winemakers because they are a precursor to potential rain.

Fog also plays a major role in many wine regions, particularly in coastal areas or near lakes and rivers. Cool-climate grape varieties, like Riesling, Chardonnay, and Pinot Noir, require the cooling influence of fog to develop the best grapes. Furthermore, it provides the vines with a significant amount of moisture. As with clouds and rain, there is a risk of too much fog providing too much moisture and causing rot or keeping the vineyard too cool.

Climate Change

There is no doubt that temperatures are rising on our planet. A wealth of scientific data indicates that climate change is a huge worldwide issue. While latitude 50° north (or south) has long been considered the limit for cultivating grapes, vineyards are now being planted successfully beyond that in both hemispheres.

Marginal regions, like Northern Germany in the north and Central Otago in the south, are achieving good to excellent vintages more frequently. While in the past some German regions were lucky to get three good vintages out of 10, in the last 10 years there have been eight great vintages throughout Germany. Growers are already testing the limits of this range, planting *vinifera* in countries like Brazil, Thailand, and India. Over the next 100 years it may not be far-fetched to find *vinifera* grapes being produced in significant amounts in the Nordic countries of Europe or in U.S. states like Montana.

As the climate continues to change, growers may need to change grape varieties, rootstock, and vineyard-management techniques in many regions, and climate change may begin to assert itself as one of the most important components of grape growing as old regions adapt and new regions are found.

Terroir

The French term *terroir* describes the distinctive combination of topography, soil, and climate characteristics of a particular region, appellation, or vineyard, which cannot be duplicated elsewhere. In English we might call it its distinctive place.

In much of Europe terroir is a key factor in determining where grapes are grown and what varieties are planted. Old World growers believe that the distinctiveness of their terroir is a major factor in producing their particular wines. We have yet to meet a grape grower in Europe who does not think his or her terroir is perfect for the grape varieties he or she is growing and the wines being produced. European winery owners use the concept of terroir as a selling point for their wines, describing their vineyards' unique soil, elevation, aspect to the sun, temperature variation, and other factors with a great deal of local pride.

The Napa Valley in California is one of the prime grape-growing regions in the United States.

Terroir is a concept that has historically had far more meaning in Old World wine regions, particularly France, than in New World regions, but that is changing. Growers and winemakers now seem to be focusing on terroir in many New World countries including the United States, Australia, New Zealand, Chile, and Argentina. Only a few decades ago it was not unusual to see several grape varieties planted widely in many different regions of the United States with no regard for terroir.

Today, growers and wine producers realize the importance of the connection between geographic characteristics and grape-variety performance. Examples of varieties that are being grown in the United States guided by consideration of terroir include Pinot Noir and Chardonnay in the regions of the Carneros in Napa Valley and Sonoma County, the Russian River Valley and Sonoma Coast in Sonoma County, and the Santa Rita Hills of Santa Barbara, all in California; Cabernet Sauvignon in Rutherford in Napa Valley, the Alexander Valley in Sonoma County, and Red Mountain in Yakima Valley, Washington State; and Riesling in such far-afield locations as the Finger Lakes in New York; the Old Mission Peninsula north of Traverse City, Michigan; the Columbia Valley in Washington State; and the Anderson Valley in Mendocino County, California.

Appellations

APPELLATION, which developed following the concept of terroir, is now an accepted international term that describes a geographic area with specific boundaries (see Appendices 1, 2, and 3, pages 376, 380, and 422, respectively). Controlled by governmental organizations, the parameters for determining what constitutes an appellation differ from country to country. Most European countries have stringent controls and regulations. In the New World, the United States has emulated this practice only to the extent of establishing defined areas of production. Other New World countries, like Chile, Argentina, and New Zealand, do not have formal geographic appellations, but do identify particular regions where grapes are grown. Still others have recently established appellation systems, like Australia which has regions known as Geographic Indications (GIs).

In theory, a controlled appellation is identified because of the particular characteristics of the defined area that makes the wines special—in other words, the terroir. In reality, it also takes into consideration political motives that may not be related to the unique characteristics of the growing area. This can have a significant impact on wine prices. Although the establishment of an appellation is intended to qualify certain geographic characteristics, the inclusion or exclusion of an adjacent area can mean a significant difference in the price of a piece of property and a wine sold that comes from within that designation.

The best example of a change to property value occurred in France recently when the Champagne region added formerly nonappellation land to the Champagne appellation. The property values of the additional land skyrocketed from a few thousand dollars per acre to about $750,000 per acre. Many believe that the newly attached land does not produce the same distinctive character in its wines as that of the traditional Champagne appellation. Even though wines made from those grapes may be inferior to those of the original appellation, producers in the area will be able to label their wines as Champagne and charge Champagne prices.

Appellations in the United States

Appellations in the United States range in size from the 150-acre Cole Ranch in Mendocino County to the 26,000-square mile Ohio River Valley, which takes in property in four states. They are analogous to a bull's-eye (see Appellation Bull's-Eye Chart, at right) in that there are larger appellations that can be subdivided into ever narrower **AMERICAN VITICULTURAL AREAS** (AVAs). The largest appellation is American, encompassing the entire country. Within the American appellation, each state (e.g., New York) and county within a state (e.g., Sonoma County) is considered an appellation. Appellations can be further divided into AVAs that have more homogeneous characteristics. Finally, in the center of the bull's-eye are individual vineyards that have specific identities, but are not AVAs. Theoretically, the closer to the center of the bull's-eye and the more defined the appellation, the better the grapes—and so the better the wine.

The only federal regulation governing appellations and AVAs is that 85 percent of the grapes in a designated wine must come from that area. Individual states can have more stringent regulations, but those are rare.

Appellation Bull's-Eye Chart

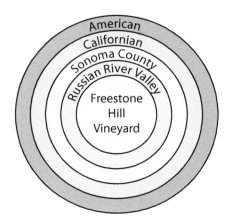

MATCHING GRAPE
VARIETY TO VINEYARD

Grape-variety selection for a particular vineyard is one of the most critical decisions a grower must make. When planting grapes, a producer must consider the rootstock, variety, and clones that best suit the vineyard's terroir. It takes up to three years for a vineyard to start producing usable grapes. It takes five to seven years before a vineyard reaches full production, and it may take 10 to 12 years or more before grapes achieve enough maturity to make the best wines. Thus, there is a great deal of pressure on growers to choose the best possible varieties for each vineyard, in order to avoid years of lost income from a vineyard that does not produce usable grapes. Making the correct grape-variety choice can ensure many years of great wines and consistent income for a vineyard. Making a bad grape-variety choice can be devastating if the variety is unsuited to the location or there is no market for the selected variety. If a vineyard is already planted and the variety there does not meet the desired quality or match market demand, another variety can be grafted onto the old vines to save years of lost production.

Rootstock

Rootstock is the part of the plant that grows the root system that brings water and nutrients to a new vine. Until the 1860s new vines were created by taking plant cuttings from existing vines. The cuttings developed their own roots and a new vine was established. The European phylloxera epidemic of the 1860s changed how grapevines were grown (see Phylloxera, at right). Growers could no longer use cuttings because doing so would spread the aphids that cause the infestation to the new vines. Native American rootstock was able to resist the pest, and by grafting new *vinifera* vines onto the native American rootstock, they could avoid infestation.

In a few locations, like Chile, parts of Australia, and parts of the United States where vines are planted in sandy soil, *vinifera* vines are still **OWN-ROOTED** because the aphid does not inhabit sandy soil. But elsewhere in today's world, genetically improved non-*vinifera* rootstock is planted and a new *vinifera*-variety vine is grafted onto it.

Growers must select the right rootstock to accomplish their goals. This is an important decision because it impacts disease resistance, vine vigor, yield, ripening, and ultimately grape quality. Some types of rootstock work best in particular soil types and climatic conditions, while others are more resistant to disease and pests. When growers choose rootstock, it is not as important to consider the grape variety and clone that will be grafted to it because most grape varieties will adapt to whichever rootstock is used.

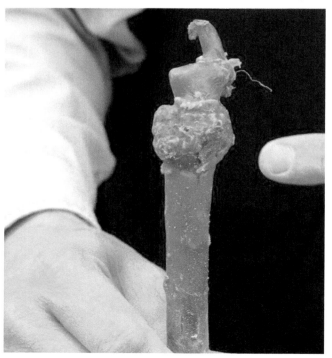

Rootstock and graft of Tempranillo at Can Feixes in Catalonia, Spain

Grape-Variety Selection

Choosing the right grape variety for a vineyard requires much of the same evaluation process as choosing a rootstock. Each grape variety has unique traits that allow it, when grown under certain conditions, to give the resulting wines desirable colors, aromas, and flavors. Certain grape varieties are also better suited to certain climates, topography, soils, and levels of water accessibility. Grape varieties also vary in terms of disease resistance, grape quality, and grape volume. The market and the vineyard owner's individual preferences must also be considered. Experts are often brought in to evaluate each of these factors and to aid in selecting the right grape varieties for a vineyard.

When replanting a vineyard, a grower who has had success with a particular grape variety is likely to continue using that same variety because its performance in that particular vineyard has been established. However, if a grower is planting a new vineyard or has decided to change varieties, more analysis may be required. In some countries, particularly France, there may be limited choices as to what the local appellation law allows because only certain varieties are permitted in each appellation.

Financial and Personal Influences

Vineyard owners also are influenced by financial and business concerns. The cost of planting and growing grape varieties can influence the choices made. As an example, Pinot Noir is more difficult and expensive to grow than Sauvignon Blanc. The decision is made more difficult because the vineyard owner must try to anticipate market conditions three to five years before the first usable harvest. This can be devastating if market tastes change before the wine is even on the store shelf. Merlot is an example of changing tastes. When the American public's desire for red wine escalated dramatically in the early 1990s, consumer demand for Merlot exceeded supply. Many growers grafted Merlot onto the rootstock of white varieties. Today, it is less popular, and with all of the added vineyard acreage, the supply of Merlot far exceeds demand. In some vineyards more popular grape varieties, like Pinot Noir, are now being grafted to rootstock previously used to grow Merlot.

Sometimes growers plant new grape varieties with great anticipation that customers will rush to buy the resulting wines, only to have sales fall short of expectations. Viognier and Sangiovese, for example, were thought to be sure bets for success in California, but they never broadly caught on. On the other hand, lesser-known Italian and Spanish grape varieties are now crowding out space on retail shelves and restaurant wine lists previously designated for varieties like Merlot.

PHYLLOXERA

Phylloxera is a small aphid (insect) called a root louse. It eats the roots of grapevines, stopping the flow of water to the plants, which kills them. The European phylloxera infestation began in the 1860s when native American vines were taken from the United States to France in hopes that they might create new, innovative wine styles. The importers of the native American vines unknowingly brought in vines that carried phylloxera. The American vines were resistant to phylloxera; the *vinifera* vines were not. Once in France phylloxera was spread by man and machine throughout the European continent, and eventually it made its way to Australia. Within a decade, the epidemic wiped out almost all of the vineyards in Europe, nearly destroying the wine industry.

A wide array of cures were attempted and all were dismal failures until it was noticed that the American vines were thriving, whereas *vinifera* vines were dying. Early tests showed that *vinifera* vines grafted onto native American rootstock developed vines that produced wines with pure *vinifera* characteristics but whose roots were resistant to phylloxera. A solution had been found. Although there were skeptics, the grafting technique took hold, and eventually European vineyards recovered.

A vineyard owner or grower's personal preferences, which may not follow sound scientific or business practices, can also influence grape-variety selection. Several years ago a new vineyard and winery owner in the Penedès region of Spain was enthralled with Pinot Noir and planted 60,000 vines only to have them all pulled out within five years. The wines were total failures because Pinot Noir, a cool-climate grape, was planted in a moderately hot climate. The vineyard was replanted primarily with Cabernet Sauvignon and now produces some of the best wines in that region.

Clonal Selection

Almost any agricultural crop can be bred to produce particular characteristics, and grape varieties are no different. Every grape variety started from a single vine, but over time, natural genetic changes (mutations) created different characteristics in different vines of that variety. These characteristics can include varying sugar levels, acid levels, aromas, flavors, ripening, yields, and disease resistance. New vines can be propagated by growers, who use cuttings or buds from a parent vine with desirable characteristics to recreate those characteristics in the propagated vines. A population of vines propagated from a mother vine to retain the characteristics of the mother vine is called a clone.

The French are meticulous in developing their clones, and after a lengthy, controlled process give each clone a number, like Pinot Noir 115 (abbreviated PN115), that can be traced back to the mother vine. This long-standing exacting process, university- or appellation-driven, matches a characteristic to a particular vine rather than to the traits created by a vineyard's location or terroir. The characteristics of these specific clones are well identified. American vintners are less meticulous, and only in the last couple of decades has university research in the United States started to mimic standards found in France. Clonal names, like Wente (for Chardonnay) or Calara (for Pinot Noir), come from known producers or vineyards in the United States. Although considered clones, they might better be identified as "selections" because the vines can only be traced to a particular vineyard, not to a mother vine.

Producers typically grow vines from multiple clones of the same grape variety to add complexity to a wine. Each clone produces grapes that showcase a different characteristic, like acid, tannins, or sweetness, that will complement the final blend of the finished wine. For example, a winemaker may make a single wine from Burgundian Dijon Pinot Noir clones PN115, PN114, PN667, and PN777, each of which adds its own distinctive characteristics to the wine, such as a wine with beautiful fruit flavors and good acid balance that has long-aging potential. Grape varieties like Pinot Noir and Chardonnay, which may be thousands of years old, have developed far more identifiable characteristics than younger varieties, like Cabernet Sauvignon, that may only be 300 years old. These older varieties have produced a greater number of clones than younger varieties.

VINEYARD MANAGEMENT

Vineyard management is what historically most would call farming—taking care of the vineyard and its grapes to maximize productivity and quality. It has come a long way from simple farming, evolving into a sophisticated process with many components. Vineyard-management decisions, including planting, **VINE TRAINING**, irrigation, and canopy management, represent the art and science of bringing in a crop of grapes of the right quality and size to produce a particular wine.

Planting and Density

Growers must decide how many grapevines to plant per **ACRE** or **HECTARE**. Historically, Old World vines have been more densely planted than New World vines, because Old World growers believe that dense planting controls vine vigor by forcing vines to compete for nutrients. Some areas, like the Burgundy and Champagne regions of France, have been planted with more than 4,000 vines per acre/10,000 vines per hectare.

Older vineyards in New World regions, like California, have historically had much lower density plantings, sometimes as low as 450 vines per acre/1,120 vines per hectare. Vines there were planted in widely spaced rows, with more space between each vine, in order to accommodate the type of tractors available in the United States. Additionally, because most early vineyards in the New World were not irrigated and rainfall was scarce, growers believed that each vine needed maximum spacing to produce grapes from the available precipitation.

Over time, research and improvements to vineyard equipment changed the rules in the New World. Today, following the lead of Old World growers, New World growers generally plant new or replanted vineyards more densely. Equipment has been developed that requires less space, allowing for much narrower rows and therefore yielding more vines per acre, and the increased density helps to promote vine competition, hopefully resulting in higher fruit quality.

Slope also impacts the density that is possible in a vineyard. On flat, rolling terrain or moderately steep slopes, vineyards can be planted in narrow, evenly spaced rows with high density. On steeper slopes, found predominantly in the Old World, the spacing of vine plantings is driven by the contour of the slope, so vineyards on steep slopes tend to be less densely planted.

Crop Size

The size of a wine-grape crop, or its yield, is affected by a wide range of factors. Some of these are controlled by nature and some are controlled by the grower or producer. Both have an impact on the quality of grapes and hence the quality of the resulting wine. Crop size is measured by tons of grapes per acre in the United States and some other New World countries, and by **HECTOLITERS** per hectare of wine in Europe. The rule of thumb is the smaller the crop (fewer tons per acre), the better the quality of the grapes and the wine. The larger the crop (more tons per acre), the lower the grape quality and the less distinctive

KEY PARTS OF A GRAPEVINE

Vine the entire plant, including all parts from the rootstock to the grapes

Rootstock the part of the vine that extends into the ground

Trunk the part of the vine that extends vertically from the ground to the head

Bud the point on the vine where growth begins and shoots develop

Shoot the new growth sprouting from the bud that develops stems, leaves, and, eventually grapes

Cordon a woody part of the vine that comes from the top of the trunk and can be trained horizontally along a wire to support the arms

Arm a woody part of the vine where canes and spurs develop

Cane a mature shoot containing the buds that is used to form the following year's shoots

Spur the stub of a pruned cane that is used to produce the following year's shoots

the wine because the grapes lose concentration. But there are many exceptions to this general rule.

Contributing Factors

What are the factors that contribute to crop size? First, and most important, is the vineyard site itself. The best grapes come from lean sites with soil that is lacking in fertility. The lack of fertility prevents a large crop from developing, but produces higher-quality grapes. A vineyard with soil that is rich in organic matter and nutrients and has access to plentiful water has the ability to produce a larger crop, but the quality of the grapes is generally lower. Growers on all types of vineyard sites can manage crop size and reduce the yield using an array of vineyard-management techniques; however, human intervention alone rarely produces a crop equal to the quality of grapes from a lean site.

The color of the grapes also has an effect on the size of the yield. White grape varieties left on their own normally generate larger yields and red grapes generate smaller yields, with the exception of Syrah. Red grape varieties are typically managed by growers to achieve an even smaller yield to develop better color, concentrated flavors, and tannins. In white grape varieties color and tannin development is less important.

Yield is naturally controlled by soil nutrients, water, and the age of the vines. The more nutrients and water are used, the larger the yield. Older vines eventually start producing less. As an example, if well fed with nutrients and water, Zinfandel from the San Joaquin Valley (Central Valley) of California can yield 15 or more tons per acre, which results in thin wines. By contrast, 100-year-old Zinfandel vines in various locations throughout California make wonderful concentrated wines, but may only produce one ton per acre. Beyond what naturally can occur in the vineyard, yield is controlled by vineyard-management techniques, like pruning and fruit drop, based on the wine quality desired by the producer.

Business Effects

Each grower or vineyard owner generally works with the wine producer to determine the desirable yield from a vineyard. The expected yield is often established in a contract between the grower and producer, and price per ton is agreed upon based in part on the yield. Higher-yield grapes are generally priced lower per ton and subsequently used for lower-priced wines, while lower-yield grapes are generally priced higher per ton and used in higher-priced wines. If the yield is higher than the contract stipulates, penalties may be assessed against the grower.

Vine-Training Systems

Many centuries ago trees served as frames to support grapevines, which grew up their trunks. Over time, this proved impractical, as the trees took up a lot of space and harvesting was difficult. So growers let the vines develop naturally close to the ground in the shape of a bush. For the most part newly planted vines today are trained on some type of artificial **TRELLIS**, which combines wood, metal, or stone stakes and wire to train the vines to grow in a certain way. They enhance the vines' growth and impact the quality and quantity of the grapes. They also facilitate tractor access and make harvesting easier.

Vine-training systems are selected to coincide with the objectives of the wines' producers. A vineyard that produces popular priced wine uses a training system that focuses on high-yield capability, mechanization including machine harvesting, and pruning to maximize productivity and minimize cost. An estate vineyard that produces high-quality grapes, by contrast, employs a system that promotes grape quality over quantity.

Training Systems

A number of different training systems are available. These help control vine productivity by determining the level of sun exposure, the height of a vine, and the number of grape bunches capable of being produced. Each system enhances certain growing characteristics that are believed to be appropriate to a particular grape variety or appellation. Low training systems, like those used in Burgundy, are preferred in cool climates because part of the vine's heat source comes from the ground as well as the sun. In the rainy Rías Baixas region of Galicia, Spain, keeping the grapes high in a pergola allows the windy climate of the region to dry the grapes quickly, preventing mildew and rot. Other systems allow the vines to spread out on wires, making the vines and grapes more accessible to the sun.

Gobelet, Head Pruned, Bush Vine, or Albarelo This is one of the oldest and most rudimentary training systems. The vines are simply left to grow in a bush shape close to the ground. However, when vines are planted in the United States in this style today, they are staked and allowed to grow taller. At the top of the trunk, spurs are left unpruned, creating the start of the next vintage's buds. This system is used in low-vigor areas. Excessive foliage (or canopy) shading the fruit can be a problem. The only way to harvest is by handpicking, which can be backbreaking work because the vines are low. It is often seen with old vines in both the Old World and New World, yielding wonderful grapes that make great wines.

Cordon Trained A cordon is the woody permanent branch extending horizontally from the head of the vertical trunk that has been trained along a wire. It is possible to have a single cordon running in one direction or two bilateral cordons running in opposite directions each trained along a wire. Each cordon is spur pruned to produce two to eight buds. The distance from the ground to the cordon is fairly low in Europe but generally higher in New World countries. Height varies depending on climate and historical grape-producing influences. This type of training system is used for low- to moderate-vigor vines. The single cordon is commonly used in the Old World, while the bilateral (double) cordon is more prevalent in the United States. Both are found throughout the rest of the world.

Gobelet

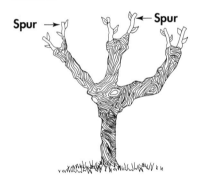

Cordon Trained (Double Cordon)

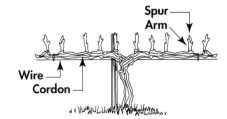

Guyot (Single Cane)

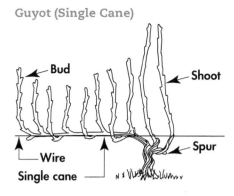

Bud — Shoot — Wire — Single cane — Spur

Vertical Shoot Positioning

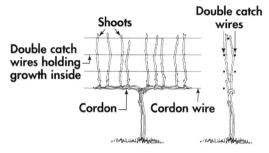

Shoots — Double catch wires — Double catch wires holding growth inside — Cordon — Cordon wire

Scott Henry

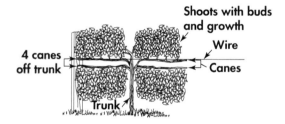

Shoots with buds and growth — Wire — Canes — 4 canes off trunk — Trunk

Lyre

Shoots with growth in double catch wires — 2 cordons

Guyot This is a single or double cane-pruned system where one or two canes extend horizontally along a wire from the head of the trunk. One or two bud spurs are left that will form shoots for the next year's canes. In Europe the number of buds for each cane is typically controlled by appellation law.

Vertical Shoot Positioning (VSP) or Vertical Training This is a heavily used system in both Old World and New World vineyards. Taut double wires are run along a vineyard row from post to post. The shoots from each vine's canes or arms are trained to grow upward between the wires. The shoots can grow quite tall but are typically severely pruned. Cane and spur pruning are both possible. This type of system can be effective for both handpicking and mechanical harvesting.

Scott Henry This is a more complex system that uses four canes, two on each side of the trunk. The canes are held in place by two foliage wires as in the VSP system. The two upper canes are trained so the shoots grow upward. The lower canes produce shoots that are trained downward. It is used in moderate- to high-vigor vineyards with potentially high yields. Originally developed for cane pruning, a variation called the Smart-Dyson has been developed for spur pruning.

Tendone or Pergola System In this system, vines grow on stone or wooden frames. The tops of the vines come together, creating a covered garden effect, with the grapes hanging down under the canopy of leaves. The height of a pergola can be as much as six to seven feet/two meters. This old style of vine training is found in Italy, Spain, Portugal, and South America and is used for high-vigor vines. In humid areas, pergolas allow air to circulate, drying the vines and preventing mildew and bunch rot. Underripeness due to the lack of sun exposure can have a negative impact on the grape quality.

Geneva Double Curtain At the head of the trunk two cordons are trained horizontally but perpendicular to the row of vines, producing two separate, parallel canopies. Shoots from spurs are trained downward, creating the appearance of a curtain minimizing the canopy and maximizing exposure to the sun. This system, developed at Cornell University for non-*vinifera* grape varieties in the cool Northeast United States, is now used in high-vigor, cool-climate situations in other regions of the world.

Lyre As in the Geneva double curtain, the vine is cordon trained to create two separate parallel canopies. Spurs are facing up and the shoots are trained upward rather than downward. This system optimizes leaf and grape exposure to sunlight, promoting heavier yields and riper grapes. It is used for moderate-vigor vines.

Pruning

Vines bear fruit on one-year-old growth. Pruning is a continual renewal of this growth. Pruning takes place during dormancy at the end of one vintage and before the start of the next one. Vineyard workers remove the dormant vegetation from the prior year's growth and prepare the vines for the coming year. Many growers prefer to wait for the first freeze to prune, but that is not necessary if vine dormancy occurs before then. Vines are usually pruned between December and February in the Northern Hemisphere and June through August in the Southern Hemisphere.

Pruning balances vine growth and reduces the future potential quantity of grape bunches with the intent of increasing the quality of the remaining grapes. The more buds left on a vine, the more bunches are likely to grow. The more bunches, the smaller the grapes and the longer it takes for them to ripen. Cool climate regions, like Burgundy or the Anderson Valley, do better with fewer bunches per vine because of the short growing season. Grapevines also tend to sprawl, and pruning helps keep a degree of order in the vineyard and helps in the management of the vines.

Cane Pruning

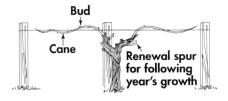

There are two types of pruning: cane pruning and spur pruning. The selection of pruning method is based on grape variety, vigor of the vineyard, training system, method of harvesting, and the grower's personal preferences. With cane pruning canes are removed each year by pruning and new canes are trained on wires (see Cane Pruning, at right). The pruner selects the best one or two shoots from the harvest that was just finished. Those shoots become the canes for the current year's growth. These new canes sport six to 12 buds. At the top of the trunk two spurs are selected that will form the shoots for the canes that will become the next year's growth. All the other branches are removed. Selecting and pruning the best canes is a very skilled job that requires extensive experience to ensure the desired growth occurs the following year. This type of pruning is used for high-vigor vines.

Spur Pruning

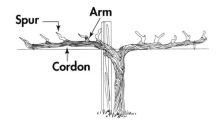

Spur pruning removes all shoots from the just completed harvest back to the top of the trunk in head-pruned vines or to the arm in cordon-training systems (see Spur Pruning, at right). Spur pruning is used for low-vigor vines. It takes less skill than cane pruning because all the branches are cut back. Although it is not as clean and pretty as cane pruning, it is significantly less expensive because of the reduced labor cost. Traditionally, canes are removed by hand, but mechanized spur pruning is becoming more prevalent.

Irrigation

Irrigation has not typically been used in the Old World, where grapes are planted close to water sources or in areas with ample rainfall. Even in hot, extremely arid climates, like that of Central Spain, vineyards are usually dry-farmed; that is, farmed without irrigation. In these regions, only the hardiest vines have survived. They have developed resistance to drought, but yields are quite low.

Irrigation is more common in the warm, dry climates of the New World, like California, Australia, and South America. Many large New World growers prefer

Drip irrigation systems give the vineyard manager control over the amount of water each vine receives.

to set up irrigation systems to guarantee that the grapes receive a constant amount of water at the appropriate time.

Quite often, the decision to irrigate is determined by a country's wine regulations. Historically in Europe, irrigation has been illegal even under the driest conditions, except when planting new vines. However, governing agencies have been changing the rules to allow watering in some appellations. Producers in the Old World are now more likely to employ irrigation where permitted.

Both sprinkler and drip systems are used for irrigating vineyards, but drip systems are preferred. Drip systems use emitters that release measured amounts of water to each vine. This gives the vineyard manager maximum control over moisture levels. Sprinkler systems use excessive water and do not water each vine with the same amount of water. With the cost and availability of water for irrigation becoming a major concern in many New World appellations, sprinklers are becoming less desirable. As an example, as a result of a 10-year drought in Australia, many vineyard owners have not been able to irrigate their vineyards.

Canopy Management

Canopy management is a New World term coined in the United States in the 1960s, but techniques of canopy management have been used worldwide for hundreds of years. The term canopy refers to all components of a vine above the rootstock or trunk. The canopy includes the vines, shoots, leaves, and fruit. Canopy management is the process of maximizing and controlling the amount of sun exposure the leaves and fruit receive. It can improve yield and quality, and reduce a range of vine diseases. The techniques of canopy management include shoot thinning, which removes the entire shoot; trimming back excessive shoot growth; and removing leaves to provide more sun exposure to the grapes. Shoot positioning, which places shoots in the desired location, makes the other techniques easier and less costly to perform by properly training the shoots. Pruning is not considered a canopy-management technique, but can have a tremendous impact on canopy management.

Canopy management is more appropriate for high-vigor vines, like those found in New World vineyards where more sun, fertile soil, and irrigation are common. Because of cool weather and lack of irrigation, most Old World vineyards have moderate- to low-vigor vines, and growers do not have to rely on canopy management as much.

VINE HEALTH

Maintaining vine health requires the right mixture of soil, nutritional enhancements, and pest and disease control. The range of methods used by grape growers is quite diverse, and there are widely differing views on the best approaches to producing the right quality and quantity of grapes.

Fertilizers

Grapevines are generally less demanding than other agricultural crops. In fact, some of the best vineyards are the least fertile. It looks like nothing could grow on the rocky soils of Châteauneuf-du-Pape in France or the slate slopes of the Mosel, but they produce incredibly wonderful wine grapes.

In the typical monoculture vineyard, where only grapes grow, nutrients must be added to the soil either with the use of cover crops, or artificially by using fertilizers. Cover crops are plants other than grapes that grow between the rows of vines. They add nutrients to the soil, prevent erosion, and compete with the vines for moisture, resulting in improved grape quality.

Fertilizers are organic and nonorganic mixtures that are used to add or replace lost or nonexistent nutrients in the soil. Most desirable are natural fertilizers, such as manure, that are a requisite in organic and biodynamic viticulture (see Organic Viticulture, page 63, and Biodynamic Viticulture, page 64). In most vineyards, however, nonorganic fertilizers are used. These are a combination of minerals, such as nitrogen, potassium, or phosphorus.

Pests and Diseases

Grape growers face threats from insects, animals, viruses, and bacterial diseases. Pests that affect vine growth range from the tiny aphid phylloxera to large animals, such as deer, kangaroo, and wild boar. The list includes leaf hoppers, mites, nematodes, bees, birds, and rodents. Not only can pests do damage by eating the roots, vines, and fruit, they also carry an array of diseases. Fences, netting, and low-voltage electric wires are used to keep animals out of vineyards and away from the grapes, but insects and diseases can be more difficult to control.

Phylloxera

The best-known and most-feared vineyard pest, is a tiny aphid called phylloxera, a root louse that feeds on the roots of grapevines (see Phylloxera, page 53). A permanent cure for phylloxera has yet to be found. Outbreaks continue to occur, with one of the most recent taking place in Northern California in the 1990s.

Downy Mildew

This fungal disease was first identified in Bordeaux, France, but is now found worldwide. Downy mildew attacks the green parts of the vine, reducing photosynthesis and thereby slowing the ripening process. Appellations with warm and humid summers are most susceptible to this disease, as are those with heavy rain. To protect against downy mildew, a copper-based mixture with lime and water is acceptable, even in organic vineyards. Another preventative measure is to plant grape varieties that are resistant to the fungus. Those most resistant are hybrids or non-*vinifera* vines, neither of which have achieved broad popularity.

Powdery Mildew

Another fungal disease that created havoc throughout Europe and other parts of the world is powdery mildew, also known as oidium. It is spread through the air and attacks the green parts of the vine. Because it favors heavily canopied vines, many producers use extensive leaf removal to minimize damage. Dusting with copper or sulfur is a standard practice to prevent the mildew, and is considered acceptable in organic vineyards. Certain *vinifera* grape varieties, such as Pinot Noir, Malbec, Merlot, and Riesling, are more resistant to powdery mildew than other varieties, such as Chardonnay and Carignan.

Pierce's Disease

Pierce's disease, a bacterial disease carried by the insect known as the sharpshooter, is totally debilitating to the grapevine and there is no known cure. Initially known as Anaheim disease, it was first identified in the late 19th century in Southern California near Anaheim, which at that time was widely planted with grapes. The disease was so devastating that all grapevines were destroyed and never replanted.

The bacteria were originally carried by the blue-green sharpshooter, which had a relatively narrow geographic range. Recently another major infestation started in Southern California and created grave concern because the bacteria was being carried by a much larger insect, called the glassy-winged sharpshooter, which has a much wider range. This new insect was brought to Southern California from the southeastern part of the United States on ornamentals used in landscaping. It found a home in citrus groves, which are resistant to the disease. Because citrus groves are often planted adjacent to vineyards, it was easy for the insect to move to vines and infect them. The Southern California infestation gradually started spreading to the more extensive grape-growing regions of Central and Northern California, but it seems to have stopped its migration.

Leafroll Virus

This disease appears everywhere grapes are grown. Spread by mealy bugs, it does not kill the vine but it damages productivity and quality, and can totally devastate a vineyard. The spread of leafroll is caused by humans, probably during the grafting process.

Pesticides and Fungicides

Pesticides and fungicides remain the most common way of combating vineyard pests and diseases. Growers find the use of chemicals the most cost-effective way of managing the health of a vineyard in a very competitive wine market. Although the use of alternative vineyard-management methods, such as organic and biodynamic farming, is gaining in awareness and popularity, the tried-and-true method of using chemicals continues to improve grape quality, yield, and resistance to pests and diseases.

Pesticides are mixtures of natural or chemical ingredients that, when applied to grapevines, allow the vine to resist or kill pests. Pesticides are usually toxic chemicals that can be harmful to humans or the soil. They work very well, but over time the pests may develop resistance to the pesticide, requiring heavier treatments or the use of more toxic chemicals. This, along with environmental responsibility, has moved many growers to take a closer look at natural, nontoxic alternatives.

The earliest fungicide, known as Bordeaux mixture, was developed in Bordeaux in the 1880s. It is composed of lime, copper sulfate, and water. Today it is still being used along with other forms of copper and sulfur to control mildew. These preparations are allowed in organic vineyards.

In the cases of some pests, there is no cure and the only way to fight the pest or disease is to pull out a vineyard and replant.

New Approaches to Vine Health

While chemical agriculture, based on grape monoculture and the use of chemical pesticides, fungicides, and fertilizers, has long been the dominant practice worldwide, in recent years more holistic types of viticulture have been gaining ground. Growers and producers that use these holistic approaches perceive that over time these practices are better for the vineyard and grapes. They also gain a more favorable response from consumers who see the resulting wines as natural and wholesome.

Organic, sustainable, and biodynamic viticulture are alternative methods of growing grapes. Less than 10 percent of vineyards worldwide are classified as organic or biodynamic currently, with Italian producers (who receive government support) generally considered to be the leaders in alternative viticulture. However, with the dramatic increase in consumer demand for organic products, overriding concerns about the quality of products that humans consume, and concern about climate change, the use of organic vineyard practices is likely to increase.

Organic Viticulture

Organic viticulture is a term that has various meanings in different grape-growing countries around the world. However, the standards are quite similar. According to the 1990 U.S. Farm Bill, organic viticulture "is a system of grape growing which does not employ industrially synthesized compounds as additions to the soil or vines to maintain or increase fertility, or to combat pest problems." The basic goal of organic farming is to eliminate nonnatural (chemical) fungicides, pesticides, and fertilizers. In the United States, wines from grapes grown organically can be labeled as "made from organically grown grapes." Organic wine is something quite different and quite a bit less popular because sulfites cannot be added to the wine, resulting in wine with a very short shelf life. Only a small number of producers in the United States make and sell organic wine.

European countries have practiced organic agriculture longer than New World countries, with France being the first country to adopt these techniques.

Italy, Germany, and Spain also strongly support this type of viticulture. The less restrictive sustainable agriculture and more restrictive biodynamic viticulture both developed from organic viticulture. In Europe wines made from organically grown grapes can be called organic wine.

Sustainable Viticulture

Sustainable viticulture, as it is known in the United States, is very similar to integrated pest management (IPM) or *lutte raisonée*, the French term. This type of viticulture encourages the move from a monoculture vineyard where only grapevines grow and all other vegetation is eradicated to one where other crops or even weeds are encouraged to provide a more balanced environment. This diversity creates conditions that sustain both pests and their predators, providing natural controls. Cover crops improve the soil quality, reducing the need for chemical fertilizers. They also create an eco-friendly environment, encouraging healthy vines and discouraging diseases.

In sustainable vineyards the use of chemical herbicides and pesticides is greatly reduced, although not altogether banned. When severe infestation occurs, chemical pesticides may be the most rational choice. The overall guiding principle is to maintain the integrity of the vineyard and its crop for the long term by taking a more holistic view.

Biodynamic Viticulture

Biodynamic viticulture is a more extreme form of organic viticulture that is also spiritual in nature. Jamie Goode in his book *The Science of Wine* considers it "an altered philosophy or world view that impacts on the practice of agriculture." Biodynamism not only encompasses all of the requirements of organic viticulture, but goes beyond with what some consider mystical preparations that have little scientific meaning. Ingredients such as cow horns, stags' bladders, or yarrow flower heads are used according to cosmic factors, such as the moon's phase, seasons, and even the time of day. The entire vineyard and its environs are treated as one living system.

Scientists keep at arm's length from its philosophies, but some biodynamic vineyards show positive results. In France several of the most highly respected producers who practice biodynamism have extremely healthy vineyards and are making stellar wines. Biodynamism was strongly driven by Nicholas Joly (initially a naysayer) at Clos de la Coulée de Serrant in the Loire. Today, no one doubts the quality of wines from other French producers such as Domaine des Comtes Lafon, Domaine Leroy, and Domaine de la Romanée-Conti in Burgundy; Zind-Humbrecht in Alsace; Huet in the Loire; and M. Chapoutier in the Rhône Valley.

RIPENING

The objective of the grape-growing process is to ripen grapes to achieve the desired state of maturity before harvesting. Ripening begins with veraison and concludes with harvest. The ripening of grapes is affected by many factors: the grape variety, terroir, vine training, and certainly the vintage's weather. Decisions about grape ripeness and harvest timing are made by measuring and evaluating several key components in the grapes, including sugar content, total acidity, pH, seed color, and the more subjective issue of determining the ripeness based on flavor. Assessing ripeness is not an exact science. One winemaker or grower may believe one vineyard of grapes is fully ripe and ready to be picked, whereas another winemaker or grower may believe the same grapes are underripe or overripe.

Sugar

As grapes ripen their sugar level increases. In the United States, **BRIX** is the measure of sugar concentration in the grapes. It is a percentage by weight of sugar compared to the entire volume of liquid in the grapes. Brix is measured using a refractometer in the vineyard or a hydrometer in the lab. Samples from the vineyard or the scales at weigh-in are usually taken to the lab where the hydrometer provides greater accuracy.

Sugar level at harvest can range from a low of 18° to 20° Brix (18 to 20 percent sugar) for grapes being harvested to make sparkling wine to a high of 26° to 28° Brix (26 to 28 percent sugar) for a ripe Zinfandel table wine. Grapes for sweet dessert wines might not be harvested until the grapes have achieved a sugar level as high as 40° Brix (40 percent sugar) or more. The amount of sugar in the grapes determines the amount of potential alcohol in a wine. Lower-alcohol wines are harvested at lower, less ripe sugar levels. Higher-alcohol wines are harvested from grapes that have been left on the vine to ripen longer, achieving a higher sugar level. The conversion rate of sugar to alcohol is approximately 58 percent. Therefore, if grapes are harvested at a Brix of 24° (24 percent sugar) and the wine is fermented totally dry (no sugar), the alcohol level will be 13.9 percent (24 x .58 = 13.9). If 1 percent of sugar is not fermented and left in the wine, the alcohol level will be 13.3 percent (23 x .58 = 13.3) and the **RESIDUAL SUGAR** 1 percent.

Acidity

Acid is the other major component measured along with sugar at harvest. Acidity gives wine its brightness and crispness and is a key contributor to balanced wine. It also plays a major role in matching wine with food. There is an inverse relationship between acid and sugar. As grapes ripen sugar increases and acidity decreases (see Acidity-Sweetness Relationship, at right).

Acidity is measured in two ways, total acidity and pH. Total acidity is the amount of both fixed and volatile acids in the **MUST** (see Chapter 5, Must, page 72) or in a finished wine. It is a measure of the total grams of acid per liter (g/l), and ranges between 4.5 g/l and 8.0 g/l. Tartaric, malic, and citric

Acidity-Sweetness Relationship

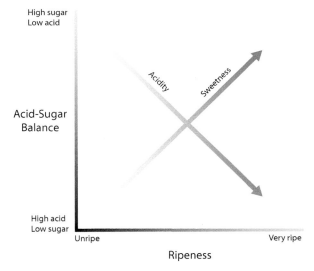

are the dominant fixed types of acid in wine. Acetic acid, typically found in very small amounts, is an example of a volatile acid. At the start of ripening, the ratio of tartaric and malic acids is fairly balanced. As ripeness develops, tartaric acid levels remain constant, but malic acid decreases and total acidity declines. If grapes become overripe, they lose acidity and do not have enough to provide proper balance to a wine. To compensate, New World wineries are allowed to add acid to the wine during fermentation.

The pH is the measure of the concentration of acidity. As total acidity increases, pH decreases. Low pH indicates a high concentration of acidity, and high pH indicates a low concentration of acidity. The pH scale runs from 0 to 14 with 7 being neutral. Zero represents pure acid and 14 represents pure alkalinity. Water is neutral at 7; neither alkaline nor acid.

The pH scale is logarithmic, much like the Richter scale for measuring earthquakes. An earthquake of 6.0 is 10 times more intense than an earthquake of 5.0. Similarly with wine, a pH of 3.0 is 10 times more acidic than a pH of 4.0. The pH of a finished wine ranges from 2.8 to 4.2. The standard for white wines is generally a pH level of 3.1 to 3.4; for red wines it is 3.3 to 3.6. Today some wine critics and consumers desire intense, fruity, ripe, and powerful wines. This drives pH to a high level when wine is fermented dry. With this style, white wines can have a pH as high as 3.6 to 3.7 and red wines can hit 3.8 to 3.9. The result is a much softer, less acidic wine.

The pH level provides insightful information about a finished wine. Wines low in pH have a crisp, tart, sour, even sharp character, while those with high pH have a round, flat, or flabby personality. The pH level also impacts color (particularly in red wine). As pH climbs, the color of red wine begins to turn from red to a bluish purple. Lower pH retards bacterial growth. Wines with a pH of 3.7 or higher are more unstable, with a tendency to allow bacterial growth. Wines also require a moderately low pH in order to age well. Red wines tend to age best in a pH range of 3.4 to 3.6. The pH level can be reduced during the winemaking process by adding acid, typically tartaric. The pH level is not usually found on a wine label, but wineries will provide it if asked. It is often worth the trouble to get the information because pH is one of the most helpful indications of a wine's character.

Other Ripeness Factors

Physiological ripeness, or grape maturity, is also determined by grape-variety aroma and flavor. The appropriate level of maturity is subjective and opinions differ widely on what constitutes maturity for each variety. This is usually decided by the winemaker, the grower, or by consensus between them, and skill at making this determination comes with experience.

The color of the grape seeds can also be a distinguishing marker of ripeness. Though opinions vary on the subject, some believe that underripe seeds remain green, whereas fully ripe seeds turn brown. Physiological ripeness in concert with the appropriate sugar and acid levels are the primary factors that determine when to harvest.

Different grape varieties have different maturation dates, which permit the harvest to occur over a period of time. For example, Sauvignon Blanc is ready for harvest before Chardonnay, and Merlot is ready for harvest before Cabernet Sauvignon. But weather, harvest-crew availability, and other factors sometimes throw a wrench into the harvest plan, forcing some grapes to be harvested earlier or later than scheduled. This can change the style of wine that is produced.

HARVESTING

Harvesting is the process of picking grapes and moving them to the winery. The process can be completed by hand or machine, depending upon the training system and vineyard-management techniques that are used in the vineyard. At harvest the most critical decision is determining the optimal time to pick. The producer must be confident that the grapes have achieved the desired ripeness in flavor and sugar content and that those components are in balance with the acid levels. Weather plays a major role in the harvest date and can be the determining factor between a great vintage, an average vintage, or even a poor vintage.

Timing Harvest

Grapes are ready to harvest when the decision has been made that they have achieved the desired ripeness level with sugar, acid, and **PHENOLIC** compounds at optimum levels. Timing is critical at this stage because full ripeness lasts only a few short days and as grapes remain on the vine in the vineyard they continue to ripen. The producer must be ready to pick as soon as the decision to begin harvesting is made.

Weather can alter the timing decision. Early-fall rainstorms can play havoc with a harvest (see Climate, page 46). Rain adds moisture to the grapes, reducing their intensity and flavor and making watery, thin wines. It also promotes the growth of mold and rot, which can quickly destroy a crop.

Weather patterns and weather forecasts are constantly monitored as each grower and winery closes in on the harvest season. A grower must decide whether to harvest a bit early to miss a storm or to delay harvest until after the storm passes and hope for a late summer warm spell to dry the grapes. Waiting can be risky if weather does not improve. It may rain again, or keep on raining, heavily damaging the crop. Once picking has started, it becomes a difficult decision whether to continue harvesting or stop if the weather changes. The cost of starting and stopping the harvest must also be factored in. For large growers, once picking a particular grape-variety crop has started, it is usually finished no matter what the weather.

The temperature of the grapes is also critical during harvest. If the fruit is too warm when it arrives at the winery, the juice ferments prematurely, oxidation occurs, and there is a greater chance for bacterial infection. Nighttime and early morning harvesting prevents this damage because of cool temperatures, so grapes arrive at the winery fresh and ready for fermentation.

WHAT MAKES A GREAT GRAPE CROP?

1 Poor soil fertility
2 Access to just enough water (naturally or through irrigation)
3 Well-drained soils
4 Sloped vineyards facing south in the Northern Hemisphere or north in the Southern Hemisphere, except in hotter climates
5 Grape-variety selection that is complementary to the climate and soil of the vineyard
6 Large temperature variation between day and night
7 Vineyard-management techniques that are adapted to the variety, climate, and soil
8 Minimal use of chemicals
9 Low yields based on the grape variety
10 Luck

Hand-Harvesting

Handpicking has been used for centuries to harvest grapes. The worker cuts the stem of each individual bunch with a sharp curved knife or clippers. For late-harvest wines or grapes that ripen unevenly, such as Zinfandel, individual berries can be picked. The worker places the fruit into a small container that is dragged along the row. When the small baskets are full, they are stacked and transported to the winery, or the baskets are emptied into a larger bin that is carried to the winery by tractor or truck.

An advantage to hand-harvesting is that a skilled picker selects only good bunches, leaving those that are unripe or rotting on the vine or on the ground. Hand-harvesting can be done in any vineyard, under any condition. Because of the slope- or vine-training system used, some vineyards can only be hand-harvested. Productivity can be enhanced by the selection of training systems, the type of canopy management, and the height of the fruit zone, where the grape bunches hang. Hand-harvesting is far more expensive and time consuming than mechanical harvesting, but there is less damage to the fruit in the process and it arrives at the winery in better condition. It is the preferred method for grapes that will be used in high-quality wines.

Increasingly there is a shortage of workers who are willing or able to do this backbreaking work. Worker cost is increasing and immigration issues in both Old World and New World countries are beginning to prevent workers willing to do the harvesting from entering the countries where the grapes are grown. Because of these problems, mechanical harvesting is becoming more common throughout the world.

Hand-harvesting is a hot, backbreaking job, but it ensures only the best grapes are used.

www.taggphotography.com

Mechanical Harvesting

Mechanical harvesting replaces man with machine. Mechanical harvesting first started in California in the 1960s but did not take hold in California and other parts of Old World and New World wine regions until the 1990s. Early efforts were aimed at increasing productivity by assisting manual pickers. Today the most common type of mechanical harvesting uses rods or bows to slap the foliage, shaking the clusters of grapes, or more often the individual grapes, from the vine. The grapes are caught on a conveyor belt and deposited in a bin that follows the harvester.

The advantages of mechanical harvesting start with cost. After the capital investment in machinery, the cost of labor can be reduced dramatically. It takes only one or two workers driving the harvester and collection bin, as opposed to a dozen or more pickers, to harvest the same amount of grapes. In addition, the mechanical harvester is significantly faster. This is particularly acute when the crop must be harvested quickly because weather reports are predicting a storm that will impact grape quality. Mechanical harvesters work at night when the grapes are coolest and grape quality is best for harvesting. Manual harvest generally requires daylight, with its much higher temperature potentially minimizing the benefit of full-bunch harvesting.

A major disadvantage to mechanical harvesting is the inability of the machines to be selective. A handpicker leaves behind poor or rotting bunches; machines pick unripe or rotten grapes along with good grapes. Most wineries specify, by contract, that defective fruit must be dropped prior to machine harvest.

Other disadvantages to machine harvesting center on the rough treatment the grapes receive in the harvesting process. Bunches and grapes are shaken very sharply, causing grapes to disengage from bunches and in many cases split open and release their juice. This juice is susceptible to oxidation. Cool-temperature harvesting and the addition of sulfur dioxide to the picking bin slow spoilage, but the winery must be close to the vineyard to avoid spoilage in transit; otherwise hand-harvesting is recommended. However, new models of mechanical harvesters are improving significantly, causing less bruising or damage to the grapes.

Mechanical harvesting is faster and less expensive than hand-harvesting.
www.taggphotography.com

In addition, some producers prefer whole bunches of grapes for fermentation, particularly in the case of Pinot Noir and other thin-skinned grapes, and sparkling wine producers require whole bunches to create a gentle first press for their best sparkling wines. However, the intense shaking that occurs with mechanical harvesters leaves few bunches, and most grapes are separated into individual berries. Finally, mechanical harvesters only work well on flat or rolling terrain. They are not capable of working in steeply sloped vineyards.

With the quality of mechanical harvesters improving and the difficulty of finding workers for hand-harvesting, it is almost assured that there will be a significant increase in mechanical harvesting. There will always be a need for handpicking in locations where mechanical harvesters cannot go, for grape varieties too fragile for mechanical harvesting, and for wine styles that require delicate handling of grapes and multiple harvests from the same vineyard. In addition, there will always be those growers and producers who believe hand-harvesting creates the best grapes to start the process of making a great wine.

SUMMARY

Growing grapes is a complex process that largely determines the quality of the finished wine. Decisions are required prior to planting a vineyard, and more are required during the grape-growing season.

Grapes grow in a broad range of locations that are determined by topography, access to water, soil composition, and climate. Of the hundreds of grape varieties that are grown, only a few dozen are commercially made into wine and only a handful achieve broad international popularity.

Today growers use sophisticated methods of viticulture including site and vine selection and vineyard-management methods. Environmental factors come into play with the impact of climate change and the pressure to make vineyards more eco-friendly. But the underlying factors that produce great grapes remain the same: planting the right varieties in an appropriate site, providing the ideal balance of sun, nutrients, and water, then harvesting at the right time.

Viniculture: Making Wine

Following the first stage of making wine, viticulture or grape growing, we now focus on viniculture, the actual making of wine. The process of making wine today remains fundamentally the same as it has for centuries: Sugar in grapes and yeast that adheres to their skins or is found in the vineyard or winery work together to create alcohol and… Voila! We have wine. But over the last few decades, significant developments in winemaking techniques have improved wine quality dramatically.

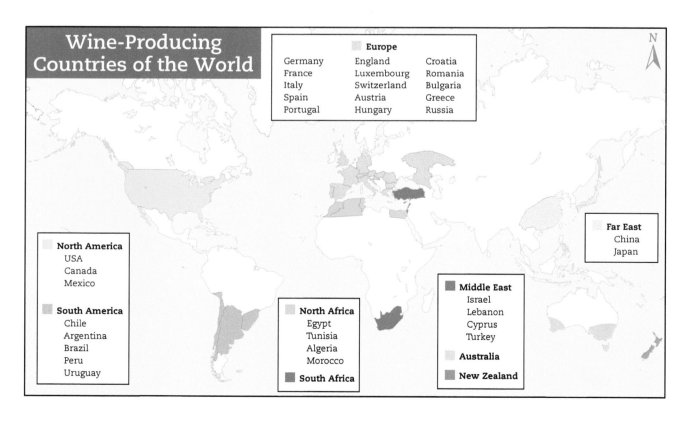

Wine-Producing Countries of the World

Europe

Germany	England	Croatia
France	Luxembourg	Romania
Italy	Switzerland	Bulgaria
Spain	Austria	Greece
Portugal	Hungary	Russia

North America
USA
Canada
Mexico

South America
Chile
Argentina
Brazil
Peru
Uruguay

North Africa
Egypt
Tunisia
Algeria
Morocco

South Africa

Middle East
Israel
Lebanon
Cyprus
Turkey

Australia

New Zealand

Far East
China
Japan

N

The process of making white wines and red wines is similar overall, but there are some key differences. In white wines, skins are separated from the juice before fermentation, whereas in red wines skins are left in contact throughout fermentation. White wines also ferment longer at cooler temperatures than red wines, and red wines typically age longer than white wines. (See Chapters 17 and 18, pages 276–309, for a discussion of making sparkling and dessert wines.)

MAKING WHITE WINE

There are nine steps in the process of making white wine (see White Winemaking Flowchart, below), from receiving the grapes at the winery to bottling the finished product.

Grape Receiving White grapes are delivered to the winery and weighed. Samples are taken to determine sugar content and for lab analysis.

Crushing and Destemming Grapes are dropped in a wide collection chute, called a hopper, where a screw or belt conveyor carries them to a crusher-destemmer. **SULFUR DIOXIDE** (SO_2) may be added to the grapes to prevent oxidation and bacterial growth. The grapes are pushed through the crusher-destemmer where they are crushed and stems are removed.

Pressing The grapes, or what is now called must, are conveyed to a press where the grapes are squeezed, and the juice is separated from the skins, seeds, and any remaining stem parts. The juice is pumped to a tank where remaining grape solids will be allowed to settle for 24 to 48 hours. The juice is then racked to a fermentation vessel.

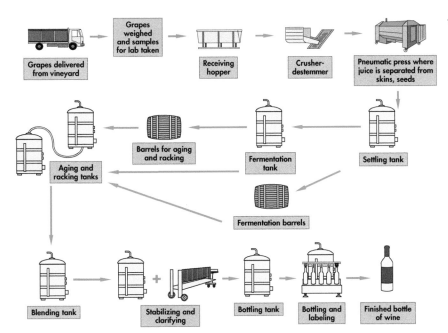

White Winemaking Flowchart

Fermentation When the juice reaches the fermentation tank, it is inoculated with yeast and fermentation begins. Fermentation typically occurs in a refrigerated tank and can take from six days to several weeks.

Aging After fermentation, the wine may be aged on its lees, placed in barrels for oak aging, or racked, blended, stabilized and clarified, and bottled if a fresh more aromatic wine is desired.

Racking The wine is racked two or three times, either in a fermentation tank or an aging barrel. This entails pumping the wine to a clean tank or barrel, leaving the sediment at the bottom of the original container.

Blending At this point, the winemaker decides whether to use each lot of wine as is or to blend it. If blended, it can be mixed with other lots of the same grape variety or it can be mixed with different varieties.

Stabilization and Clarification When the wine is ready to be finished, it is subjected to stabilization, which prevents tartrate crystals, unwanted carbon dioxide (CO_2), and cloudy wine. To further clarify the wine, it is fined and/or filtered using a fining agent and/or a mechanical filter.

Bottling The wine moves to the bottling stage where final adjustments are made. Each bottle receives a final addition of SO_2 to prevent oxidation. After the wine bottle is filled, corked, capsuled, and labeled, it is stored for bottle aging, if appropriate, or shipped if it is ready for market.

MAKING RED WINE

There are 12 steps in the process of making red wine (see Red Winemaking Flowchart, at right), three more than in making white wine. While the procedure in each step is similar whether the wine is red or white, the order of the steps is different.

Grape Receiving When red grapes are delivered to the winery, they are weighed and samples are taken to determine sugar content and for lab analysis. For premium wines, the grapes may also be passed through a sorting table to ensure that only the best grapes are selected.

Crushing and Destemming The grapes are dropped in a hopper where a screw or conveyer belt carries them to a crusher-destemmer. SO_2 may be added to the grapes to prevent oxidation and bacterial growth. The grapes are pushed through the crusher-destemmer where they are crushed and the stems are removed. Some stems may be added back to fermentation in some wines.

Cold Soak Many newly harvested red grapes are allowed to macerate in tanks chilled to below 45°F/7°C for several hours, or even days, prior to fermentation. This is done to enhance fruit flavors and increase color.

Fermentation The must is then pumped to a fermentation vessel, most likely temperature-controlled stainless steel, where it is inoculated with yeast, allowing the fermentation to begin. Red wine fermentation occurs at higher temperatures than white wine and lasts from five to 10 days.

Maceration The skins, seeds, and possibly stems, if they have not been discarded, are soaked in the wine to extract color, tannins, and flavor compounds.

MUST

Must is the slurry created when grapes are crushed and destemmed. It is a combination of the juice and any other remaining solids and includes skins, pulp, seeds, and stem particles. In France it is known as *moût,* and in Italy and Spain it is called *mosto.*

Pressing When fermentation and maceration are complete, the wine is pumped to the press where it is separated from the solids.

Malolactic Fermentation (MLF) Lactic bacteria is added to almost all red wines to convert sharp malic acid to softer, rounder lactic acid.

Aging If no oak is desired, the wine may be aged in an inert vessel (stainless steel, concrete). If the winemaker wants oak characteristics, it is placed in oak barrels for aging, which can last from six to 18 months or more. For less expensive wines, oak alternatives (staves, chips, etc.) can be added to an aging vessel, thereby allowing the wine to absorb the oak flavors.

Racking During aging, the wine is racked two or three times over several days to remove the solids that collect in the bottom of the tank or barrel. The wine is pumped from the current tank or barrel into a clean tank or barrel, leaving the sediment behind.

Blending Most red wines are blended, which may occur before or after racking. Wine lots may be blended with others of the same grape variety or with other varieties.

Stabilization and Clarification When the wine is ready for bottling, it is subjected to stabilization and clarification, removing any remaining particles to ensure wine clarity. Some red wines skip fining and/or filtration.

Bottling The wine moves to the bottling room, where it is put in a tank and final adjustments are made. Each bottle receives a final addition of SO_2 to prevent oxidation. The wine bottles are filled, corked, capsuled, sealed, and labeled. The wine then moves to storage for bottle aging if appropriate or shipped if it is ready for market.

Red Winemaking Flowchart

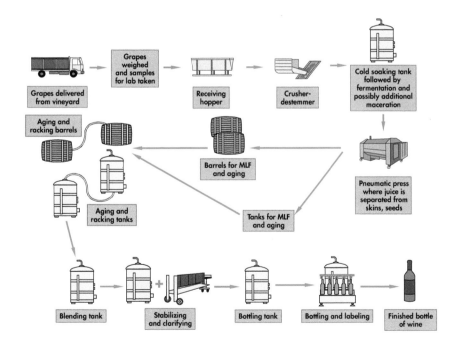

FROM FIELD TO TANK

The process of making wine starts in the vineyard. The old adage, "You can't make a silk purse out of a sow's ear," holds true in winemaking: It is impossible to produce a good-quality wine from poor grapes.

Where does the practice of growing grapes stop and winemaking begin? Technically winemaking begins when grapes arrive at the winery, but there is significant overlap between these two processes. The winemaker is almost always involved in vineyard-management decisions, including when the grapes will be harvested. Sometimes the producer or the winemaker has tremendous influence in deciding how the vines and grapes will be handled during the growing and ripening cycles to ensure that the grapes come to the winery with the desired characteristics. These decisions can have tremendous impact on grape quality and on the style of wine that is eventually produced.

For the purpose of separating these two components, growing grapes and making wine, we start discussing the winemaking process at the point when the grapes leave the vineyard and arrive at the winery.

TOP: *Harvested grapes are delivered to the winery by truckload or in small bins.*
BOTTOM: *At the sorting table, rotten and poor-quality grapes are removed by hand.*
www.taggphotography.com

Receiving Grapes at the Winery

Harvest bins as large as two tons full of machine-picked or handpicked grapes as well as small crates of handpicked grapes are delivered from the vineyard to the scales at the winery. The receiver weighs the grapes, which serves as the basis by which the grower gets paid, and takes sample grapes to inspect for quality, for chemical analysis, and to determine the Brix, or sugar content. Many grape contracts have stipulations about tonnage per acre and sugar level required to receive the full value of the contract. There are times when the grapes that are delivered may be rejected outright for not meeting the standard, in which case the grower must find another buyer, almost assuredly at a lower price. Once they have been accepted by the winery, most of the grapes are dropped into a stainless steel hopper or bin and head to the crusher-destemmer. High-quality grapes may receive a second hand-selection before moving to the crusher-destemmer, or whole cluster grapes may be sent directly to the press.

Grape Selection

When grapes are picked by hand, pickers make the initial selection of grapes in the vineyard. For higher-quality wines, many producers also choose to have a second selection when the handpicked grapes arrive at the winery, a process the French call *triage*. On a sorting table, workers remove any remaining green, moldy, or diseased bunches that were missed during the initial picking, and also take out individual berries that would detract from the quality of the final wine. This is a very expensive process due to the cost of the additional labor, and is typically completed only for high-quality grapes being used to make superior wines that will be sold at premium prices. Mechanically harvested grapes seldom, if ever, go through a second hand-selection.

Crushing and Destemming

Most grapes are sent directly from the scales to a hopper, where a screw or conveyor belt carries them to the next step: extracting the must in the crusher-destemmer. The crusher-destemmer removes the grapes from the stems and crushes the fruit, releasing the juice in preparation for fermentation. The end result is must, the composition of unfermented grape juice, skins, pulp, and seeds that will eventually be converted into wine.

As soon as the grapes are crushed, the juice immediately comes in contact with air and begins to oxidize. To prevent **OXIDATION** and spoilage from occurring before fermentation begins, **SULFUR DIOXIDE** (SO_2) is commonly added to the grapes before they enter the crusher-destemmer. (see The Interaction of Wine and Oxygen, page 79).

From the crusher-destemmer, must for white wine is pumped to a press to remove the skins from the juice because white wine is typically fermented without skins. Must for red wine is allowed to **COLD SOAK** or is directly pumped to a fermentation tank where it is fermented with the skins.

When whole cluster pressing or whole cluster fermentation are used, crushing and destemming is skipped and grapes, including stems, go directly from the sorting table or hopper to the press in the case of white wines or to a cold soak prior to fermentation for red wines. Occasionally, a winemaker will also send white grapes directly from the crusher-destemmer to the fermenter for partial fermentation with skin contact.

The crusher-destemmer removes the stems and leaves and gently breaks the grape skins.

Using Stems

Stems add vegetal aromas and flavors as well as tannins and **BITTERNESS** to the must. Thirty years ago these characteristics were desirable in making most red wine, especially in the fermentation of Cabernet Sauvignon. Not so today. The majority of today's winemakers believe that stems give undesirable characteristics to their wines. Still, some winemakers choose to leave stems on a small percentage of red grapes during part of the fermentation because the stems stabilize color and add the herbal character, tannins, and complexity that they are seeking. Pinot Noir, for example, is sometimes fermented with some stems.

Cold Soak

Cold soak, a fairly new but frequently used technique, allows red wines to macerate before fermentation to increase color and flavor characteristics. It also enables the grapes and juice to homogenize before fermentation begins. Thicker skinned, more tannic varieties benefit from two to four days of **MACERATION**.

Thin-skinned red grapes may macerate as much as five to six days. Some wine-makers cold soak their grapes as long as 10 days.

IN THE TANK

Before fermentation begins, the winemaker must decide which type of yeast to use, which fermentation techniques will be most effective, and which type of fermentation vessel is most likely to yield the results he or she desires. During fermentation, the winemaker is extremely busy monitoring each tank and evaluating the fermentation's progress. Each day, he or she tastes the fermenting liquid from each tank and completes a lab analysis.

Fermentation

Fermentation is a natural biochemical process whereby yeast converts sugar into alcohol. The simple formula for fermentation is:

Sugar + yeast = alcohol + carbon dioxide (CO_2)

In today's wine world, many additional components can enhance this natural process. The type of fermentation vessel and yeast, the amount of skin contact, the timing of pressing, and the addition of SO_2, all affect fermentation. In addition, the fermentation temperature and alternate techniques used, like **MALOLATIC FERMENTATION** (MLF) and carbonic maceration, contribute to the style of the finished wine.

To begin fermentation for white wine, juice is transferred from a settling tank, where it has been allowed to **SETTLE** for 24 to 48 hours after pressing, removing solids from the juice. Red wine fermentation starts when the must has been pumped from the crusher-destemmer or after a cold soak has occurred.

Sugar released from the grapes begins to ferment with natural yeast or with inoculated commercial yeast. White wines are fermented under refrigeration at cooler temperatures (50° to 65°F/10° to 18°C) than reds. Because it takes longer to convert sugar to alcohol at lower temperatures, refrigeration extends the length of the fermentation, which maintains the wine's fresh fruit flavors and allows for a controlled, even fermentation. White wine fermentation can last from several days to several weeks. Red wines ferment at warmer temperatures (77° to 86°F/25° to 30°C) and the process lasts from five to 10 days. As yeast converts sugar into alcohol, **CARBON DIOXIDE** (CO_2) and heat are generated as by-products. For table wines, the CO_2 is allowed to dissipate into the atmosphere. The heat that is generated must be controlled to protect flavor development in the wine and prevent a **STUCK FERMENTATION**, which occurs when heat kills the yeast before it has completed its task of converting all of the sugar into alcohol. Refrigerated fermentation tanks are designed to control the excess heat.

When grapes do not ripen enough, which can occur in many cool climates, it is legal to add sugar to the must to achieve minimum alcohol levels. The sugar is used to increase alcohol, not to add sweetness to a wine. This process is called **CHAPTALIZATION**. In some appellations, like those in California, there is plenty of sun and ripening is never a problem, and in fact in California chaptalization is illegal.

When grapes ripen, they lose acidity, so in warmer climates like those in California, it is legal to add acid to wine, normally before or during fermentation, to create balance. This process is called **ACIDIFICATION**. In many European appellations producers struggle to achieve ripeness and rarely have a problem with too little acid, and in many of these places, it is illegal to add acid.

Yeast

Either wild or cultured yeast can be used to make wine. Wild yeast is naturally found on grapes or is airborne in the winery or the vineyard. The most common wild yeast is *Saccharomyces cerevisiae,* although it has also been domesticated. During fermentation some wild yeast cells may die at quite low alcohol levels (4 percent), while others may continue to thrive in a high-alcohol environment (18 percent). This makes wild yeast less predictable and more difficult to control than cultured yeast. However, it has unique characteristics that are closely tied to the terroir, so some winemakers (primarily European) prefer to use wild yeast.

Commercially cultured yeast strains are more commonly used than wild yeast because they eliminate the unpredictability that is inherent in wild yeast. These yeast strains (there are several dozen) exploit different characteristics of the grapes to achieve a desired result. The winemaker determines the desired outcomes and selects the yeast strain that is compatible with those goals. Commercial yeast also allows the winemaker to obtain consistency in each lot of wine.

Carbonic Maceration

Carbonic maceration is the process of allowing fermentation to occur without crushing or destemming the grapes. Whole grapes are fermented in clusters, allowing alcohol and CO_2 to develop inside each grape. Just as you have popped grapes in your mouth and bitten down to release the fresh, sweet juice, think of popping a carbonic macerated grape in your mouth, biting down, and tasting fermented wine. It is an interesting and fun experience.

Thin-skinned red grape varieties are typically used for carbonic maceration because they enable the yeast to easily penetrate the skins and ferment the sugars. When you drink this type of wine, you find fresh fruit flavors, the aroma of strawberries, and a splash of effervescence from the CO_2. Carbonic maceration is used to create the young, fresh style of Beaujolais and is sometimes used with Pinot Noir grapes.

Jon McPherson, the winemaker at South Coast Winery in Temecula, California, punches down grapes to combine the solids with the juice.

Cap Management

For red wines, the winemaker needs to establish a schedule for cap management. When wine is fermented and the yeast begins to convert sugar into alcohol, heat is created as a by-product. As a result, the solids in a red wine, which are lighter than the juice, float to the top of a vessel, forming a cap—a thick layer of skins, seeds, and other solids. Because these solids add color, tannins, and flavors to the wine, they must be reincorporated into the juice by processes called **PUNCHING**

TOP: *Pumping over is another method of combining solids with the juice.*

BOTTOM: *Rotary fermenters incorporate solids and juice as the wine ferments, eliminating the need for punching down and pumping over.*

DOWN or **PUMPING OVER.** Punching down is done mechanically in large vessels or by hand in smaller vessels. A long pole with a flat plate or open tubing at the end is pushed down on the solids, mixing them in with the juice. During fermentation, this process is completed as frequently as two or three times a day under the winemaker's direction.

Another alternative is pumping over. The juice is drawn from the bottom of a tank and then pumped to the top of the tank and sprayed over the cap. The juice then filters through the cap, picking up desired color, tannins, and flavors from the solids. Another method is maintaining a **SUBMERGED CAP** where the solids are always kept below the top surface of the wine, incorporating color, tannins, and flavors into the wine.

If the wine is made in a **ROTARY FERMENTER**, a cap cannot be permanently formed because the solids are mechanically reincorporated into the wine. A stainless steel rotary fermenter is horizontal rather than vertical and is similar to a cement mixer, with rotating vanes that mix the solids back into the juice for better extraction of color, tannins, and flavors. It eliminates the need for punching down or pumping over and can reduce the fermentation and maceration time required in making red wine. However, rotary fermenters are very expensive and require a tremendous capital investment. In addition, because they are a closed fermentation system, wine is not aerated.

Fermentation Vessel

Fermentation takes place in a vessel or tank that holds the must or juice. It can be as small as a five-gallon plastic bucket or as large as a 250,000-gallon stainless steel, temperature-controlled tank. In ancient times, amphorae (a type of vessel or pot made of clay) were used for this process. Over time clay vessels were replaced by wooden barrels or casks, then by concrete, and finally by stainless steel.

Old World winemakers traditionally used wood and concrete fermenters. When winemakers began producing wines in the New World, they replicated wood barrels and concrete fermenters because they closely followed Old World methods.

When new, more scientific fermentation techniques were developed in the United States and Australia, one of the first innovations was the introduction of stainless steel tanks to replace the old wood and concrete fermentation vats. In time many European producers also changed to stainless steel, and today temperature-controlled stainless steel tanks are by far the most common fermentation vessels throughout the world. But they are not the only type of vessel. Small oak barrels are used for some varietals, such as Chardonnay, and large oak **CASKS** are still used in Europe. In addition, old concrete fermenters are quite common in Europe, and there is even a movement, however small, back to the adoption of newly manufactured temperature-controlled concrete fermenters, large oak casks, and even new amphorae, which look much like the ancient containers.

Winemakers must decide whether to use an open or closed fermenter. Open fermenters allow the wine to be in contact with oxygen, adding a broader range of flavors to the wine but risking oxidation. Closed fermenters help maintain the fresh fruit flavors of the grapes but risk **REDUCTION** (see The Interaction of Wine and Oxygen, below).

Stainless Steel

Stainless steel tanks have become the standard vessel used to ferment wine for several reasons. The first is hygiene. When fermentation is complete, a stainless steel fermenter is emptied of all sediment and washed down. Tartrates, tartaric acid crystals that are naturally formed in wine, do not adhere as much to the walls of stainless steel tanks, making them easier to clean than wooden casks and the walls of concrete fermenters, which must be scraped out by hand. As hard as someone works to clean an oak vessel or concrete tank, he or she never achieves the cleanliness level of stainless steel. In addition, the hours required to clean a stainless steel tank are significantly less than that of an oak or concrete vessel of the same size, so there is a significant cost savings.

Stainless steel fermenters with refrigerated jackets allow the winemaker to control the fermentation temperature.

The Interaction of Wine and Oxygen

Grape juice or wine interacts with oxygen from the time the grape juice is first exposed to air to after the wine is bottled. Oxygen enhances wine's flavors, softens its tannins, and stabilizes its color. Aeration, exposing wine to moderate amounts of air, is desirable in the early stages of fermentation because oxygen helps yeast grow and convert sugar into alcohol. After a wine is fermented, exposure to small amounts of oxygen can enhance aromas and flavors over time. This is one reason for aging wine in barrels. Even after wine is bottled, a cork closure permits the exchange of air. And when you open a bottle of wine, you swirl it in your glass to add oxygen to open the wine's aromas.

While the right amount of oxygen can be beneficial, too much or too little air can cause problems. Oxidation occurs when wine is overexposed to air. It turns white wines brown and creates brick or orange color in red wines as they age. While oxidation tends to be more harmful to white wines than to red wines, it can damage flavor compounds in both. Oxidized wines develop sherrylike aromas and flavors. In addition, whether wine is stored in a barrel or bottle, too much air exposure can convert the wine's acids into acetic acid, namely vinegar.

The winemaker can slow or prevent oxidation by taking steps to minimize or eliminate air contact with the must or wine. Getting grapes from the field to the winery as quickly as possible before crushing retards oxidation and temperature controls during fermentation are extremely important. SO_2, one of the winemaker's most effective tools, can be added before the grapes enter the crusher-destemmer, or during fermentation, aging, and bottling. The winemaker must closely monitor SO_2 because it can bleach red wines and reduce aromas if too much is added. SO_2 is used in almost all wines produced in the United States.

Reduction, the opposite of oxidation, occurs when wine does not get enough oxygen, typically because fermentation tanks are airtight. When wine lacks oxygen, hydrogen sulfide (rotten egg smell) or mercaptans (skunky aroma), or both can develop. Racking reduces or eliminates these odors because it exposes the wine to air.

Furthermore, stainless steel is the best vessel to prevent bacterial infection in wine because it does not have tiny cavities where bacteria can hide and tank refrigeration helps control bacterial growth.

Temperature control is another reason why stainless steel tanks are so widely used. In the last few decades temperature-controlled stainless steel tanks, with a glycol wrap added for even refrigeration, have enabled the winemaker to control fermentation temperature. This is especially important in white wines because lower temperatures slow fermentation, which in turn helps to maintain the aroma and flavor characteristics of the wine. Red wines are fermented at higher temperatures, but these also need to be controlled to maintain desirable flavors and prevent yeast from dying off, which occurs at 113°F/45°C. In recent years, winemakers have developed sophisticated computer programs that can control the temperatures of all of the tanks within a winery.

There is no doubt that stainless steel tanks are invaluable to winemakers. However, they do have a major drawback, reduction. Aeration cannot take place if the vessel is airtight (see The Interaction of Wine and Oxygen, page 79).

Concrete

TOP: *World-renowned winery Cheval Blanc in Bordeaux, France, uses concrete fermenters to make its classic wines.*
BOTTOM: *At Craggy Range in Hawkes Bay, New Zealand, wine is fermented in wood casks.*

A concrete fermenter is nothing more than a large cement container, like a room with four walls; access is gained from the top and through a small opening in the front. To wine producers (mostly European) who use concrete instead of stainless steel, the advantages are aeration and the ease of manually incorporating the solids with the juice through punching down and pumping over (see Cap Management, page 77). Some of the great wineries of the world, such as Château Pétrus and Château Cheval Blanc in Bordeaux, France, still use unlined concrete fermenters to produce their wines.

The potential drawbacks of concrete include the difficulty and cost of cleaning and maintenance, problems controlling bacteria, and the inability to control temperature in most concrete fermenters.

To deal with these drawbacks, new concrete fermenters have been constructed with epoxy or ceramic linings, which enable them to reach almost the same sanitation level as stainless steel. There is even a small group of winery owners who are constructing new temperature-controlled concrete fermenters. A primary reason for reverting to concrete is to eliminate the microvibrations that occur in stainless steel tanks, which are believed to have a negative impact on wine quality.

Wood

Large oak casks or vats are still used widely by traditional winemakers throughout Europe. These typically old wooden vessels provide desired aeration as well as natural **STABILIZATION** and clarification, but they have long ago lost any oak flavors. Some New World wineries, like the Robert Mondavi Winery in California, use large oak casks for some of its wines, particularly Pinot Noir.

Smaller oak barrels are commonly used throughout the world for fermenting some white wines, most notably Chardonnay. The same benefits noted above are obtained but, more importantly, the smaller and usually newer barrels infuse the wine with oak flavors.

Making Oak Barrels

A cooper is a person who makes barrels, and a cooperage is where the barrels are made. Coopers cut the staves for the barrels from oak logs. American oak staves are typically artificially kiln dried for 12 months or less, whereas French oak staves are naturally air-dried for three years. Some American coopers follow this drying method.

After staves are aged, they are made into barrels. In the past, barrel making was strictly an artisan craft, and there are still artisanal barrel producers, but many barrels are now made by a partially automated process.

During construction the barrel is toasted. The inside is heated over a fire or coals to achieve a light, medium, or heavy **TOAST**, adding a new dimension to the aromas and flavors transmitted to a wine. The heavier the toast, the more aggressive the aromas and flavors of the oak. Lightly toasted barrels might produce a vanilla aroma, while a heavy toast might have an aroma of toasted bread or even smoke.

Toasting the inside of newly constructed barrels adds flavors to wine as it ages.
WINE INSTITUTE OF CALIFORNIA

Maceration

Maceration is leaving skins in contact with the must or wine to extract color, tannins, and flavors. The longer the skins are in contact with the juice, the more of these components are infused into the wine. This can occur before fermentation, as with cold soaking, or during fermentation, but commonly occurs following fermentation. Red wine maceration normally lasts a few days, but some wine can macerate for several weeks.

White wines are typically pressed before fermentation and generally (although not always) have no skin contact. Rosé or blush wines made from red grapes obtain their moderate color when the skins are left in contact with the must for a very short period of time to achieve some color, but are pressed long before fermentation is completed. Red wines are fermented with skins and pressed when the desired level of color, tannins, or flavors has been reached, usually after fermentation is complete. Some red wines may be left to macerate for additional time to increase color, tannins, or flavors.

Pressing

Must is pressed to separate the solids from the juice or wine. Pressing can occur before, during, or after fermentation depending on the desired color, tannins, and flavors. White wines are generally pressed before fermentation and red wines are normally pressed after fermentation and maceration.

Two types of presses are used for table wines: pneumatic and basket. A pneumatic press, more commonly used, contains a bladder that inflates with air, gently pressing against the grapes or must. A basket press works by manually or mechanically applying pressure from a plate on the grapes to squeeze out the juice. The basket press has been used for hundreds of years but until recently has had limited application. Today, there is a resurgence in its use for quality winemaking, primarily for grapes to be made into red table wine or sparkling wine.

With either press, about 65 to 70 percent of the juice or wine, called **FREE-RUN**, falls through holes in the press into a vessel sitting underneath. The free-run is separated from the juice or wine that remains in the press. Free-run is bright and fruity and is used for better-quality wines. The remaining juice or wine is extracted through pressing. The remainder, called press wine, contains more color and tannins and is considered of lower quality than free-run juice. Pressed wine is blended into lower-quality wines or blended back into the free-run.

Malolactic Fermentation

MLF is a secondary fermentation that occurs when lactic acid bacteria added to the juice or wine converts malic acid, a sharp and crisp acid found in tart green apples, to lactic acid, a softer and rounder acid, and as a by-product creates diacetyl, which has a creamy, buttery character. If used, the process occurs during or following the primary fermentation. Almost all red wines and some white wines undergo MLF.

In red wine production, MLF starts during or after fermentation. In a barrel that has had prior MLF, it starts naturally from the bacteria left in the barrel. If fermentation started in a stainless steel or concrete fermenter, the wine is inoculated with lactic acid bacteria. Red wines go through MLF as much for stability as for moderating acidity or taste.

MLF is less common with white grape varieties. Most Chardonnay typically goes through partial to full MLF, whereas other varieties, like Riesling and Sauvignon Blanc, usually do not. The buttery characteristic of diacetyl really stands out in white wines. At moderate levels, diacetyl can also add a subtle nutty undertone to the buttery character. Full MLF can produce an overpowering and one-dimensional artificial buttery character.

Because MLF reduces acidity in wine, it is not used if a wine has low acidity and the winemaker is trying to maintain as high acid levels as possible. In warm climates, where acid levels naturally drop as the grapes mature, wine undergoing MLF is monitored carefully to make sure acid levels do not drop precipitously.

TOP: A newly crafted basket press replicates the original style of wine press.
BOTTOM: Winemaker Josep María Huguet i Gusi in front of a pneumatic grape press at Can Feixes in Catalonia, Spain

Aging

Aging is letting a wine sit for a period of time, from a few weeks to several years to several decades, either in a tank, barrel, or bottle. Most wines mature as they age, allowing aromas and flavors to develop and integrate. They may also change substantially over time.

Many white wines, like Riesling and Sauvignon Blanc, do not require aging. These wines are cold fermented in stainless steel and then bottled and immediately released into the marketplace. Young red wines may also be sold immediately after fermentation and bottling. Of the red wines that do not require aging, Nouveau Beaujolais is the most famous. Gamay grapes from Beaujolais, France, are harvested, fermented using carbonic maceration, bottled, and immediately released worldwide on the third Thursday of November. Because of its youth and freshness, the wine should be consumed within one year.

When a decision is made to age a wine, different methods can be used, each for a specific reason. These include sur lies aging, wood aging, and bottle aging.

Sur Lies Aging

LEES are the solids left in wine (dead yeast cells and particles of skin and seeds) at the end of fermentation. Typically, lees settle to the bottom of the tank or barrel, and the clear wine is **RACKED** off from these sediments. *Sur lies aging*, the French term for "on the lees," occurs when the lees are intentionally left in contact with the wine for several days or even weeks. During this time, the lees are continually stirred back into the wine using a large paddle, a procedure the French call *bâttonage*. The rationale for this type of aging is that the lees add flavor and complexity, creating a richer wine. This type of aging is common for white wines and can occur in any type of fermentation vessel. The Muscadet wines of the Loire, France, are the most well-known wines aged on the lees, but the practice is now commonly used in Old World and New World countries. Sur lies is commonly printed on the label if the wine is aged sur lies.

Wood Aging

When the term *aging* is used, it typically refers to oak barrel and cask aging. Many consider barrel use primarily for its addition of flavor components to wine, but other factors are equally important. An oak barrel or cask allows a certain amount of aeration and evaporation to occur. Porous barrels allow a small amount of oxygen to seep into the barrel and also let some of the wine evaporate from the barrel (the "angel's share"). Minute amounts of oxygen integrated into the wine develop aromas, enhance flavors, and soften tannins creating a better wine. Oak also naturally clarifies and stabilizes the wine, thereby eliminating steps in winemaking. Some winemakers use older barrels and casks for aging because they offer the advantages of aeration and maturation in oak without the influence of oak flavors. Some traditional European producers let wine age in casks for several years, believing that it makes a more integrated wine. Smaller barrels are also used for added flavor and aroma, as well as aeration.

Oak Barrels, Casks, and Vats

Although winemakers have experimented with different types of wood, over time oak has become the most desirable choice for wood aging because its subtle aromas and flavors are a complement to many wines. Today, virtually all new barrels and casks are made of oak. However, old casks made of other woods, such as chestnut in Italy and France, pine in Greece, or even redwood in California, are still in use. Galleano, one of the historical wineries in Southern California, still uses 100-year-old redwood vats for aging every vintage.

OAK ORIGINS Oak trees from the various forests around the world are harvested when they are between 100 and 160 years old. There are different types of oak, each giving a barrel distinct character. American oak has a tighter grain, is less porous, and has overt aroma and flavor characteristics. French oak is more

TOP: *James Healy, owner/winemaker of Dog Point Vineyard in Marlborough, New Zealand, barrel samples the development of Pinot Noir.*

BOTTOM: *Galleano Winery in the Cucamonga Valley, California, still uses old redwood tanks to age wine.*

Barrique storage at Bodegas Osborne in the Vinos de Madrid DO of Spain

subtle. Eastern European oak falls between the two. The cost of a barrel is largely determined by where the oak was grown rather than where the barrel was made, just like wine. A French oak barrel can cost more than $1,000, while an American oak barrel costs only about $300. Barrels from Eastern European countries or mixed French/American oak barrels cost somewhere in between.

French oak barrels are highly prized by winemakers worldwide because of their subtler aromas and flavors. American oak is popular in the United States as well as other countries, like Australia, Italy, and Spain, because of its distinctive character and low price. It imparts stronger aromas and flavors, which some winemakers want to include in their wine.

Typically, barrels are made from one type of oak. However, some barrels combine different types of oak, such as barrels that are made of French staves (the sides of the barrel) and American heads (the top and bottom of the barrel).

BARREL SIZE Oak vessels vary in size depending on the country and region where they are used. Barrels (or **BARRIQUES**) are made in a range of sizes. The standard 225-liter barrel, which holds approximately 59 gallons yields about 24 cases of wine (288 750-milliliter bottles), is the most common size, but they can be smaller or larger. The larger the barrel, the subtler the oak character will be because there is a higher ratio of wine to oak. Small barrels provide more surface contact of oak to wine, intensifying oak character. This oak style is used by New World producers and winemakers who are making modern styles of wine.

In the United States, barrel aging usually takes place in the standard 225-liter barrel. In Europe, Italian **BOTTI** and French **CUVES** (400 liters or larger) are frequently used instead of barrels for aging, resulting in a style of wine that is less oak-influenced. Brunello di Montalcino from Tuscany, Italy, and red and white Burgundies are examples of wines that are commonly aged in oversize barrels. Casks are larger than barrels, and vats are even larger. Large oak vats can be used for fermentation or aging in place of stainless steel or concrete.

BARREL LIFE The life of a standard-size barrel is up to 10 years. The aromas and flavors of oak are at their greatest when the barrel is new. Each succeeding year, the aromas and flavors have less oak-flavor impact on the wine, and after four or five years, the barrel no longer gives off any oak essence.

Larger casks are likely to have a life of 20 years or more. In Europe, some large casks that are hundreds of years old are still being used for aging wines.

ALTERNATIVES TO NEW OAK BARRELS Most wines produced and sold fall in the $2 to $12 range. Wine producers cannot afford to use new oak barrels at that price point. As a result, alternatives to new barrels are used because they impart oak characteristics at a dramatically reduced cost. Alternatives to new barrels include purchasing older or recycled barrels or using oak barrel staves, inserts, sticks, chips, and powder.

Recycled barrels are shaved on the inside after two to three years of use to provide some (but not all) of the character of a newer barrel. Adding barrel staves, oak chips, or oak powder to large tanks and inserts and sticks to old barrels infuses oak character into the wine. Alternatives are produced with different levels of toast and can be made of oak from any country of origin.

WHAT IT ALL MEANS The impact of oak on a wine is determined by the origin of the oak, the cooperage, barrel age, barrel size, or the oak alternative used. The amount you pay for a bottle of wine gives you some indication of how oak has been applied to the wine. A $30 bottle of California Chardonnay was most likely fermented and aged in new French oak barrels. A $3 to $9 bottle of Chardonnay was probably produced using one of the alternatives. Ultimately, each wine producer and winemaker can use (or not use) oak to impact the style of their wine.

The size of the barrel, cask, or tank and the length of time the wine is left there help to create the style that the winemaker or producer is attempting to develop. Banfi, a very large producer of Brunello di Montalcino and Super Tuscan wines, uses small barrels and larger casks in different aging programs to obtain different styles of wine. López de Heredia, a very traditional winery in the Rioja region of Spain ages some of its best wines in very large old oak vats untouched for a decade or more.

Each wine made with oak barrel aging has a program that stipulates what age of barrels will be used during its aging period. As an example, an appellation-designate Merlot may call for 20 percent new oak barrels, 40 percent one-year-old barrels, and 40 percent two-year-old barrels. Some producers only use new barrels and then sell them off at the end of each vintage. Others use new barrels for their marquee wines, such as a reserve Chardonnay or Cabernet Sauvignon, and then progressively move the used one-year-old barrels to barrel programs for their other wines.

In Rioja, Spain, López de Heredia's Viña Tondonia is aged in old vats and barrels that have been in use for many decades.

Racking

All wines are racked, a simple but important stage in winemaking. After the natural solids in the wine have settled to the bottom of a barrel, cask, or tank, the wine is siphoned off and pumped (racked) to a clean vessel, leaving the solids in the bottom of the original vessel. Racking usually takes place while wine is aging, but it can also occur before, during, and immediately after fermentation. Racking during fermentation has the added advantage of providing additional aeration, helping to improve the wine's flavors and integration. During the racking process, it may be necessary to add more SO_2 to prevent oxidation. After each racking, the wine is allowed to resettle. Racking may be done two or three times before a wine moves on to the next step of blending.

Blending

Most wines are blended. By combining different wines to create a distinctive final product, the winemaker capitalizes on the unique characteristics of each of the wines used in the blend. Blended wines can be made from a combination of different grape varieties. The most famous blended wines are made with the Bordeaux varieties: Cabernet Sauvignon, Merlot, Cabernet Franc, Malbec, and Petit Verdot. Wines of the same variety taken from different appellations or different vineyards are also blends. For example, a Sauvignon Blanc wine may come from both the Dry Creek Valley and Alexander Valley vineyards in Sonoma County, California. Sometimes blends are even created from different lots of the same varietal wine that have been processed using different methods. A lot of Chardonnay that received MLF might be mixed with a lot that did not in order to achieve a balance of malic and lactic acids.

Blending enables the winemaker to accomplish four goals:
- Improve the quality of the finished wine.
- Develop unique styles.
- Achieve status through the use of highly regarded grape varieties.
- Reduce the cost of a wine by using lower-cost grapes.

Laws in each country govern blending, with some much more stringent than others. In France, Italy, and Spain, regional appellations establish the criteria for blending, including which varieties are permitted and in what quantities. These laws also state how the wine must be identified on the bottle. In Burgundy, France, for example, blending is not permitted; virtually all wine is 100 percent variety-specific. However in Bordeaux, red wines can be blended using six different varieties, and whites can be a blend of three different varieties.

New World countries do not regulate blends as strictly as Europe. Although most wines are identified by variety, there is no guarantee that they are not blended. In California, for example, wine can be identified as Chardonnay even if it contains only 75 percent of that variety.

FINISHING THE WINE

Stabilization and Clarification

Stabilization and clarification combine overlapping techniques that prevent particles, CO_2, or cloudiness from redeveloping in finished wine. These are common practices in all wineries. The level and sophistication of the various techniques used vary depending on the size of the winery, the specific wine produced, and the philosophical approach of the winemaker.

Stabilization

Stabilization is a process primarily used to remove protein, yeast, and bacteria, along with other particles, from wine. Proteins cause cloudiness. Yeast in combination with any sugar remaining in the wine when it is bottled can start another alcoholic fermentation. Bacteria left in the wine can damage its quality. The winemaker can choose from several techniques to avoid this, including naturally occurring stabilization. A completely dry fermentation inhibits the growth of yeast and bacteria, and racking allows solids to settle out of the wine while it is in the barrel or tank. Adding SO_2 just before bottling prevents bacteria from developing when the wine comes in contact with oxygen.

Cold Stabilization

Cold stabilization prevents tartaric crystals, one of the natural forms of acid found in wine, from forming after bottling. Chilling white wine to just above freezing causes the tartrates to crystallize and fall to the bottom of the tank. This process is used in the United States for aesthetic reasons because, despite being harmless, uninformed consumers think tartrates look like tiny slivers of broken glass and refuse to drink the wine. European wines that are not exported to America are typically not cold stabilized and tartaric crystals may be seen on the bottom of the cork or bottle.

Clarification (Fining and Filtration)

Heat stabilization, known as protein **FINING**, is a process that uses a fining agent, such as bentonite (aluminum silicate clay), egg whites, or gelatin to separate solids in the wine that may cause cloudiness and reduce undesirable smells and color. Given enough time, fining occurs naturally through settling, but fining agents speed up the process, and so they are used frequently. Some winemakers are opposed to fining because removing solids from the wine also removes flavor components. Many high-end wines are not fined, which may be stated on the front label.

The winemaker can use **FILTRATION**, a mechanical process that removes particles from wine, to ensure that the wine is fully stabilized and clarified. Very fine pads or membrane filters with tiny microscopic holes are used to eliminate every particle that might remain in the wine. Filtration is a great concern to many winemakers because it also removes flavor components.

Bottling

Bottling is the final step in the winemaking process. Wine can be packaged and sold in a few different ways. In some European wineries, customers arrive with their own container and draw the wine out of a cask. Restaurants, in many cases, sell wine out of a tap, just like a keg of beer. Wine can also be poured into plastic bags and boxed (bag in the box). But the glass bottle is, by far, the most typical wine container.

Ideally, all of the wine that will have the same label should be made and bottled at one time so that every bottle is the same. However, wineries with huge production may actually bottle different lots of wine under the same label. They may use the same recipe for each lot, but there are likely to be small differences in each of them. An expert taster may notice these differences, but the average wine drinker is unlikely to spot them.

Filling the Bottle

The bottling process starts in two locations: one for the wine and one for the bottles. The wine is sent to a bottling tank where it is analyzed to make sure everything is perfect before bottling, a time for final adjustments. The wine may undergo a final filtration before it goes to the filling machine. Once it is placed in the bottle, it is extremely difficult and expensive to make any corrections.

At approximately the same time, a machine removes empty bottles from their cases and sends them to a bottle-washing machine if necessary (this step may be bypassed if the bottles are new). Next, the bottles are filled with nitrogen to remove oxygen, which will extend the shelf life of the wine. The bottles then feed to the filling machine where they are filled. The filled bottles are corked (or closed with an alternate closure) and a capsule that covers the cork is applied. Finally, the bottles are labeled on the front and back, then packed in cases, palletized, and sent to storage for aging or to await shipping.

Robotic bottling, packing, and storing system at a high-tech winery in Rioja, Spain

Bottle Aging

Table wines are subjected to a great deal of handling during the multiple stages of bottling, including the addition of SO_2 and packaging. Immediately following bottling, the wine is likely to be a bit fractured, not really tasting as it should, a condition known as bottle shock. Bottle aging, a period of rest before the wine is released for sale, is required to reintegrate the wine.

As little as a month or two of bottle aging may be all that is needed to achieve the desired aromas and flavors. However, some wines, most commonly reds but also some whites, are aged in the bottle for several months or even several years. This additional time in the bottle accomplishes some of the integration features that also occur during barrel aging. In this case, there is only minute oxygen contact. One of the greatest wine experiences is opening a bottle of wine that has aged for several decades and finding that it is still alive with wonderful characteristics. There are wines in France and Italy dating back to the 1800s that are still drinkable. Top-growth Bordeaux wines are known for having the potential to age for long periods of time.

Wine Bottle Closures

Before wine is even fermented, the producer must decide what type of closure to use. The decision takes into account cost, aesthetics, and practicality, among other factors.

For centuries cork closures were used to seal wine jugs or bottles, regardless of the quality of the wine they contained, and they are still dominant today. In the 1970s, more economical screw caps replaced corks on generic or jug wine bottles, and winemakers can now select from an even greater choice of closures. Why has there been such a change? A number of factors, including the cost, quality, and availability of cork, led the wine industry to explore other forms of closures; however, cork remains the most desirable.

Cork Closures

Traditionally, cork closures have been the hallmark of a fine wine. Why is cork a desirable material for wine stoppers? First, it is a natural product that expands and contracts to fit minute variations in the glass bottle, forming a perfect seal. No other closure has been found that has this capability. Cork also allows a very small amount of oxygen into the bottle, which helps wine, particularly red wine, to age. Additionally, wine corks have a life span as long as 25 years, which is beneficial to long-aged wines. Finally, a wine cork offers a certain pizzazz to the opening of a wine bottle; opening a cork-finished bottle creates its own excitement.

TCA–CORKED WINE Wine corks can often be infected with a compound 2,4,6-trichloroanisole, more commonly called TCA. When TCA comes in contact with wine via a cork, even in tiny amounts, it covers up the wine's fruitiness and gives it a musty smell, creating off-aromas or off-flavors. This condition is known as **CORKED WINE**. The problem is in the cork, not the wine.

TCA contamination is often caused during processing when corks come in contact with chlorine during sanitization. The incidence of cork taint has been reduced by the elimination of chlorine in most sterilizing procedures, but it has not been totally eradicated, suggesting that TCA is found in cork trees and in cork before any treatment at cork-processing facilities.

Given the length of time it takes before a cork tree becomes productive (43 to 45 years for wine corks), growers may look for additional ways to increase cash flow by allowing cattle and sheep to graze where cork trees grow. Although not definitively proven, precursors to TCA may be transferred from the animals to the cork trees.

Additionally, cork that has come in contact with soil may be more susceptible to TCA. Some growers harvest the cork bark all the way to the ground, and others leave harvested bark sitting on the ground until it is delivered to the processing plant, possibly resulting in contamination.

Cork producers purchase virtually all of their cork from independent growers, and often from hundreds of different growers. Though the producers control the cork once it arrives on their property, they have little control over cattle grazing, cork-harvesting practices, or how long the cork sheets sit on the ground until they are shipped to the plant for processing.

WHAT IS CORK?

Corks come from the bark of a specific type of oak tree, *Quercus suber*, commonly called cork oak. Most cork oak forests are found in specific climatic zones of Portugal where there are hundreds of cork growers and producers. In addition, some cork oak is harvested in Spain, and a small amount comes from France, the island of Sardinia in Italy, and a limited number of other locations. Cork is used to make many different products, from shoes to flooring to gaskets for your car, but its most profitable use is in making wine corks.

Cork oak is renewable; a portion of the bark can be removed from the tree and that section will regenerate for a future crop. The first usable (nonwine) cork crop can be harvested from a tree after 25 years of growth. Thereafter, the cut section of bark can only be harvested every nine to 11 years. Wine corks cannot be produced from a tree until the third harvest or until the tree is 43 to 45 years old. After this point, a cork tree has a typical life span of an additional 100 years, producing 10 to 12 usable crops before it dies. The older the tree, the better the cork quality.

Cork is being cleaned at an Amorim plant in Portugal. The company is the world's largest cork producer.

Winemakers became aware of broad TCA problems in the 1980s. Tests indicated that up to 5 percent of all cork-sealed wines had this flaw. In samples from some wineries, entire lots of wine were tainted, but in others it was only an occasional bottle that was flawed. Naturally, wine producers are very concerned about this problem.

Winemakers have responded by investigating and using alternative closures, which has had a negative impact on the cork market. This was a wake-up call to the cork industry, and it has finally moved to control the introduction of TCA in cork. Large cork producers, such as Amorim, have gone to great lengths to improve their ability to minimize TCA. They are supervising sources more closely and transporting bark from the forest to the production facility more quickly. They have created new oak-bark storage and production facilities. When cork arrives at these facilities, it is stored on concrete rather than on the ground. The cork is processed using new methods designed to destroy TCA, and it is tested for TCA at each step during processing. Any cork product infected with TCA is destroyed. But small producers may still use chlorine to sterilize corks. Their corks are entering the marketplace and may well account for some of the TCA found in recent vintages of wine. However, there appears to be a significant reduction in TCA in cork-finished wines.

No doubt the cork industry will continue to monitor TCA and explore ways to control it both in the forest and at the manufacturing plant. In the meantime, wine producers must be vigilant and know their cork source.

TYPES OF CORK CLOSURES

One-piece One-piece cork is punched from a thick slab of cork bark and is generally used for better-quality wine. Corks are cut in a range of lengths and circumferences, which are determined by the size of the bark. The longer the cork, the more perceived quality it has and the more expensive it will be. The bark is rated based on its natural flaws, with better cork having fewer flaws. These corks have two disadvantages. First, they are the most expensive type of cork. Second, they may not be eligible for all of the TCA identification techniques that are available for agglomerate corks.

Agglomerate Agglomerate cork is composed of poorer-quality cork that is ground, mixed with food-grade glue, and shaped in various diameters and lengths to resemble a whole cork. Agglomerates range from those with large chunks of cork glued together to those that are almost a blend of sawdust and glue. The larger the pieces of cork, the better their perceived quality will be. Less expensive than one-piece corks, they are typically used for lower-priced wines. It was thought that TCA was found more frequently in agglomerate corks; however, with new processing methods, they may be freer from TCA than one-piece corks.

Twin-top The twin-top cork has two solid pieces of cork at each end of an agglomerate cork. It is significantly less expensive than a one-piece cork but gives a better top appearance and places a solid piece of cork at the end in contact with the wine. These are generally considered a midrange-quality cork in between one-piece and agglomerate corks.

Champagne A variation of a twin-top cork, a champagne cork has solid pieces of cork at each end and agglomerate in the center. Because of the extensive pressure placed on the cork by the CO_2 in the champagne bottle, which would force out a standard-size cork, they have a greater circumference than table wine corks.

Different types of cork, from left: fortified wine, one-piece, agglomerate, twin-top, Champagne

TOM ZASADZINSKI/CAL POLY POMONA

Synthetic Closures

Synthetic closures have the same shape as cork closures. They are produced in a range of colors, which some producers find is a marketing advantage, and are opened like cork-sealed wine. They are also inexpensive to produce and use the same bottling-line equipment that is used for natural corks. However, synthetic closures do not have the same elasticity as cork, so they do not provide a perfect seal against glass, which lets in more oxygen than is desirable. And once a synthetic closure is removed from a bottle, it is extremely difficult, if not impossible, to re-stopper the bottle. Synthetic corks may also emit undesirable aromas, although not at the same noticeable level as corks infected with TCA. In spite of these problems, they have taken over a significant portion of the wine closure market and are most commonly used for moderately priced wines. The volume of synthetic corks in use has certainly reduced the wine closure market share of the natural cork industry.

Screw Caps

Screw caps are aluminum closures that screw onto a glass bottle. They have been used for decades for low-cost jug wines. Winemakers choose screw caps because they are inexpensive and, most importantly, they do not transmit TCA to the wine.

Although screw caps form an excellent seal for wine, they also present several problems. The biggest is perception; they are associated with inexpensive jug wines, not high-quality expensive wines. They also prevent any oxygen from getting into the wine bottle, which leads to questions about whether the wine can age properly. Unfortunately, at this point in time, no one knows how wines in screw-cap bottles will age over long periods. Switching from corks to screw caps requires a significant capital investment in new bottling equipment. In spite of these problems, screw caps provide a viable alternative to cork because their cost is significantly lower and there is no risk of TCA.

Several producers worldwide have switched to screw-cap closures for high-quality wines. More than 70 percent of all wines from New Zealand are now sealed with screw caps, virtually all of them white wines, including world-class Sauvignon Blancs and an increasing number of world-class Pinot Noirs. Australia, a major wine producer in the world market, is dramatically increasing the use of screw caps for white wines, moderately priced red wines, and even some expensive red wines. Other wine regions, including California, are also beginning to use screw caps for better-quality wines.

Glass Stoppers

A glass closure with a sterile rubber O-ring seal has been developed as an alternative to natural corks and other types of synthetic closures. The O-ring prevents oxidation and bacterial contamination that can occur because glass-to-glass contact does not provide a tight fit. The glass closure looks like the stopper of a glass decanter, so it provides style to the bottle and eliminates the risk of TCA infection.

Zork Cork

Zork is a brand-name cork alternative that was developed in Australia and is used by some producers worldwide. The plunger is made of polyethylene and prevents TCA and oxidation. It also provides the panache of the *pop* of a cork when opening the bottle.

TRENDS IN WINEMAKING

Not too long ago, making wine was primarily an artistic process that required tremendous skill learned over many years. More recently, winemakers with strong scientific backgrounds have implemented many changes into the process that are now becoming standard practice. Some of the most important are micro-oxygenation, removing alcohol from high-alcohol wines using reverse osmosis, and adding coloring agents, such as Mega Purple.

Micro-Oxygenation

Micro-oxygenation, a process developed in Southwestern France by Patrick DuCournau in 1991, is the very slow introduction of oxygen into a wine while it is aging in a barrel, cask, or tank, whether stainless steel, concrete, or wood. Becoming widely used, it softens tannins, improves **MOUTHFEEL**, opens up the wine's aromas, and integrates its components. It enhances barrel maturation, but does not replace it, and can also benefit wines that are not barrel aged. It is used for all quality levels, from fine Bordeaux to mass-produced wines from the New World. The winemaker's only concern with micro-oxidation is the potential to add too much oxygen, creating an oxidized wine.

De-Alcoholization of Wine

In many New World wine regions, it takes extended hang time to achieve desired flavor ripeness. During the period when the grapes are obtaining flavor ripeness, sugar levels are also increasing, which results in a high-alcohol wine. In regions like California, most wines are fermented to an alcohol level well over 15 percent, according to Clark Smith of Vinovation, an internationally recognized company that adjusts wine alcohol levels after it has been made. In a typical vintage, his company removes 1 to 2 percent of the alcohol from a given wine using reverse osmosis, a very fine filtration system that retains the flavor components while reducing the level of alcohol in the finished wine.

Some winemakers elect to de-alcoholize their wine to moderate the sensation of heat that alcohol leaves on the palate. Some winemakers also believe that de-alcoholization creates a more balanced wine that is more acceptable to consumers who do not want to drink high-alcohol wines. Detractors say it is not a natural process and consumers are not informed when de-alcoholization is used.

Color Enhancement

Sometimes wines do not achieve the desired color during fermentation. In those areas where blending is allowed, a dark grape variety may be added to the wine to bump up the color. In Bordeaux, Petit Verdot is often used for this purpose. In California, a bit of Petite Sirah or Alicante Bouschet is typically added to a blend. All of these are considered natural and acceptable.

Recently, it has come to light that in California (and elsewhere as well) a commercial product, called Mega Purple, is being used to enhance wine color. Mega Purple is a grape concentrate made of Rubired grapes, a hybrid grape with red juice that is grown extensively in the San Joaquin Valley (Central Valley) of California. However, this color enhancer and other similar products also add richness and sweetness to the wine. Although they are made from grapes, there is concern that wines with these coloring agents are a less than natural product, something produced by a chemist rather than a winemaker. Very few producers admit to using this additive; however, it apparently has been included not only in inexpensive wines but also in some of the most expensive California wines.

WINE STYLE

WINE STYLE refers to the characteristics that define a wine and make it different from other wines. A wine's style is determined by factors including geographic conditions (terroir), weather, variety, the winemaker's preferences and desires, and marketing (what type of wine will sell). In the not too distant past particularly in the Old World, wine style also took into account the heritage of the wines produced by that winery. Developing the wine's unique style was an art that took generations to perfect.

In more recent times, the study of growing grapes and making wine has become an academic discipline throughout the world. The science of winemaking has become equal to, if not more important than, the art of winemaking. Many more tools and techniques are available to growers and winemakers that allow them to craft finished wines to achieve a particular style.

There are two fundamentally different wine styles: Old World or traditional style, and New World or modern style. You may also hear terms like *international style, Parker style,* or Wine Spectator *style,* but these are permutations of the two major styles.

Old World Style: Traditional

Old World or traditional style emphasizes European tradition and terroir. It includes traditional winemaking practices: the use of classic or indigenous wine grapes; nonstainless steel fermentation vessels, such as concrete and large oak casks or vats; and the restricted use of new barrels, so that oak flavors do not dominate. Grapes are harvested at moderate sugar levels, which allow the resulting wines to retain a higher acidity level and lower alcohol level than wines made from grapes harvested at high sugar levels. This style of wine tends to be less fruit-driven and includes other characteristics that create its aromas and flavors, such as minerals, herbs, and earth. Fruit flavors are part of the complexity but do not dominate. The alcohol level of these wines finishes at 11 to 13.5 percent. Hence, many would say they are made to go with food—a very European perspective.

Over the last 20 years, Old World style has incorporated new technologies and concepts. Managing crop size and canopy and only using quality fruit (discarding rotten or poor-quality fruit) are now standard practices at traditional style wineries. In many cases, stainless steel tanks have replaced concrete and oak, but a significant number of concrete and wood fermenters are still in use throughout Europe. However, in the Old World there are also wineries that have not changed anything in their winemaking process for hundreds of years.

New World Style: Modern

New World style is a relatively recent term, used to reflect the winemaking style of newer wine-growing regions, in particular California and Australia, and their production methods. New World or modern style wines, which are usually from warmer regions and climates, are very different from those produced in colder climates of the Old World. Grapes ripen more easily with more sun; therefore, they contain higher sugar levels. Higher sugar levels create higher alcohol levels and a different balance between acid, sugar, and alcohol. Most importantly, higher alcohol levels result in a wine with riper, fruit-driven characteristics. Therefore, New World wines are more likely to be fruitier, softer, and rounder with more perceived fruit sweetness and alcohol levels from 14 to 16.5 percent, resulting in wines that people enjoy drinking with or without food.

The role of science in New World wine production is perhaps of even greater significance than climate and ripeness. Americans and Australians initiated much of the research that created major changes in the winemaking world. Vineyard management, crop levels, trellising, and innovative canopy- and leaf-management techniques are credited to New World oenology schools and wine producers. A scientific approach also called for control of bacteria and mold in wineries. Use of stainless steel and improved sanitary practices create a much cleaner environment, minimizing potential undesirable bacterial and mold infections. Standardization and industrialization of growing and production techniques led to wines that are more homogeneous. Some would argue that New World style wines are all too similar; that every New World Chardonnay tastes just like the next one and that once you reach a high degree of ripeness and alcohol all red varietal wines taste pretty much the same.

International Style

International style, a relatively new term, echoes the New World style, but it is being implemented by many producers in a broader range of wine regions, including Italy, France, and Spain. International style wines are those that are widely produced around the world, identified by their varietal name (Cabernet Sauvignon, Merlot, Syrah, Pinot Noir, Chardonnay, Sauvignon Blanc), and generally made in modern New World style rather than Old World style.

International varieties that have achieved great popularity in the New World, most notably French varieties, like Chardonnay and Cabernet Sauvignon, have been introduced to regions where they had never before been grown, and they have replaced plantings of many indigenous varieties, particularly in Italy and Spain. Merlot and Syrah, also of French origin, are more recent additions to this category, but historically they were not widely planted in other European countries.

Parker and *Wine Spectator* Style

Parker and *Wine Spectator* style exemplifies how important individual or media opinions have become in the wine world. Robert Parker is an American who has achieved cult status as the best taster in the world, and *Wine Spectator* is the most prominent American wine publication. They, along with a relatively narrow range of other journalists and trade publications, rate wines on a 100-point scale. A score of 90 or more from an "expert source" almost ensures that the wine will sell well. A score of less than 90 makes it more of a struggle to sell, and a wine with a score of 85 or below will be hard to sell without discounting.

Based on the style of wines that most frequently receive the highest scores, and therefore sell the best, some producers and winemakers make wines that are most likely to suit reviewers' palates hoping to receive higher scores. High-scoring wines typically reflect the characteristics of the modern style. Some wineries state that achieving a 90 by a particular author is a goal. Some writers appear to prefer particular styles, appellations, and/or producers, making it difficult for producers from less-established grape-producing regions to gain recognition.

SUMMARY

As you may have figured out by now, making wine is a fairly rigorous process. Historically, producing wine has been considered a craft raised to the level of an art, handed down from one generation to the next. Today's winemakers are part artist and part scientist. They combine the traditions of winemaking with a foundation in biology and chemistry, and study their craft around the world at universities that offer enology courses. In the following chapters, we will focus on how the processes of growing grapes and making wine apply to each of the best-known varieties in the world.

Sensory Evaluation of Wine

When you first begin drinking wine, you will probably evaluate it based on the pleasure it gives you with comments to the effect of: "I like it" or "I don't like it." As you become more interested in learning about wine, however, you should begin to evaluate its sensory components based on established characteristics for a particular varietal and style, as outlined in the varietal chapters that follow.

A number of factors, such as grape variety, geography, growing conditions, grape ripeness, and winemaking techniques, play a role in making wines taste different from one another. Each varietal wine has its own distinct characteristics that set it apart from other varietals, but even two or more wines made from the same grape variety can be dramatically different due to such factors.

THE EVALUATION PROCESS

It is difficult for a new consumer of wine to know how to accurately describe the differences between wines, but with practice, anyone can learn to evaluate wine and describe its characteristics.

The evaluation process is broken down into three steps:
- look at the wine
- smell the wine
- taste the wine

The first step, evaluating a wine's appearance, gives you some initial information about a wine, but the smell and taste are the two principal components of evaluation. Evaluating the appearance, smells, and tastes also allows you to learn to identify faults that make a wine less enjoyable, or even unpalatable. As your wine knowledge and tasting skills develop, a wine's appearance, smells, and tastes can become increasingly informative. They can tell you about its variety or blend, origin, age, and style.

Evaluate wine against a clean white background with good lighting.

TOM ZASADZINSKI/CAL POLY POMONA

Sensory evaluation is complex, and opinions on each wine are widely divergent. You will often find a lack of agreement on most wines among consumers and even among wine experts and wine journalists. When comparing wines, remember that each person has a different palate, perceives a wine's characteristics at different threshold levels, and may arrive at different conclusions or opinions about the same wine. As you gain more experience, you will feel increasingly confident about depending on your own opinions of any wine you drink.

Look at the Wine

Generally, a wine's appearance is subtle, and considerable practice is required to discover the secrets it holds. A wine's appearance can hint at its variety, age, the climate in which it was grown, and the winemaking techniques that were used. For example, pale-colored white wines frequently come from cool climates, while darker, yellower, or even golden white wines can indicate a warmer climate, more extraction, or barrel aging.

In order to assess its appearance, a glass of wine must be evaluated under good lighting against a clean white background. To look at the wine, pick up the glass by the stem and tilt it at a 45-degree angle in front of the white background.

Your observation should focus on:
- clarity and brightness
- color and concentration
- viscosity

Clarity and Brightness

When evaluating appearance, first determine if the wine is clear or cloudy. Clearness is the first indication of a properly made wine. Lack of clarity sometimes indicates a problem with winemaking. It also occurs when the wine contains sediment that has not been allowed to settle to the bottom of the bottle. If this is the case, the bottle needs to rest a day or two before the wine is served.

If the wine is clear, decide whether it is bright or dull. Bright wine is free of solids and has luster or sparkle. Dull wine can contain invisible solids that cut down the brightness. Most wines, particularly in the New World, are stabilized to eliminate any remaining yeast or protein, resulting in bright, clear wines. Some varietals, like Pinot Noir, and many Old World wines may not be mechanically fined or filtered, which can leave the wine with a slightly dull appearance. This may be nearly imperceptible to the human eye, but the wine will lack the brightness of fined and filtered wines.

Color and Concentration

Each white and red grape variety has its own unique color. Two wines made from the same grape variety can also differ in color as well as concentration. It is what happens in the vineyard and the winery that determines the wine's final color and concentration.

FINING AND FILTRATION

Many wines, typically from the New World, go through a sterile filtration and bottling process that removes bacteria or proteins that can cloud wine. However, some producers of high-quality wines believe that only natural fining and filtration should be used, and anything more strips the wine of its flavors. Therefore, you may find wines labeled as "unfined" and/or "unfiltered." Although the wines may look clear to the naked eye, they are not as brilliant as a wine that has been stabilized and mechanically filtered.

The color of white wines ranges from clear to greenish yellow to deep gold.
WILFRED WONG

In general, grapes harvested when less ripe with lower sugar levels will be lighter in color than the same grapes harvested with more ripening and at higher sugar levels. Any moisture loss from the grapes also concentrates color. Wines from cooler climates tend to be lighter in color than those from warm and hot climates.

White wines range from clear to greenish yellow to a deep golden color. As an example, Riesling and most Sauvignon Blancs are clear to greenish yellow to light straw when harvested at low to moderate sugar levels. Barrel aging can concentrate and intensify color to become deeper yellow. A barrel-aged wine, like Chardonnay, can have a lush, golden color. Even a Sauvignon Blanc that is barrel aged shows a deeper yellow color than one that is not.

Red wines range from shades of red to purple or blue. Classic Pinot Noir is typically light red, while Cabernet Sauvignon is dark red, and Syrah is often very deep red. Zinfandel can be deep purple when young and almost inky in some styles. The color of red wines is typically lighter in cooler climates and darker in warmer climates. Cabernet Franc from the Loire in France or a Merlot from the Veneto in Italy has a different color than either variety grown in a warmer climate.

The color of red wines ranges from light red to ruby to deep purplish blue.
TOM ZASADZINSKI/CAL POLY POMONA

With red wines, concentration is assessed by placing your finger behind a glass. How well can you see your finger? With less concentrated varietals, like Pinot Noir and Sangiovese, you should see your finger distinctly. With more concentrated varietals, like Cabernet Sauvignon and Merlot, it fades behind the glass. Red grapes from warm or hot climates that receive extended hang time are very dark and concentrated, and you will not be able to see your finger at all.

A traditional cool-climate light red Burgundian Pinot Noir, for example, is not heavily concentrated. Conversely, some warm-climate modern style Pinot Noirs are deep red, a result of riper grapes and extended maceration making it a darker, more concentrated wine. It may even be blended with a darker grape, like Syrah, to give it more color.

Color can also indicate the wine's age. Tip your glass to the side against a white background. Do you see any color variation from the rim to the center of the glass? Fading rim color and color variation can tell you that it is an older wine. Younger red wines tend to have more consistent purple or blue tones. Older red wines begin to fade and turn brick red then orange around the rim. Orange or brown color in red and white wines is the first indicator of oxidation. However, young wine should not have any visible signs of oxidation.

Viscosity

Viscosity is the level of resistance in a liquid. To observe viscosity, swirl a glass of wine. Is the wine that sticks to the side of the glass thin, like water, or is it thick, even syrupy? Viscosity develops in a wine that is high in alcohol and/or sugar.

Viscous wines have rivulets that run down the side of the glass, commonly called **LEGS** or tears. Although many people consider wine with legs to be of better quality, this is not the case. Viscosity is an indication of style, not quality.

Smell the Wine

Smell is the least developed of our five senses. Many new tasters only detect a few of the thousands of different aromas that exist. By learning the scents associated with different natural and man-made products, we can greatly expand our ability to identify aromas and use this ability to assess wine.

Developing Your Sense of Smell

You can begin developing your sense of smell by increasing your awareness of the aromas in your surroundings. At your local grocery store smell the different fruits and vegetables. Visit a flower shop and learn the scents of different flowers. Smell the grass in your yard, visit a farm, walk into an empty field; all have aromas that you may find in wine. When preparing food, smell each ingredient uncooked and then again when cooked. The aromas will change. For practice, it is possible to purchase an array of less common aroma compounds.

Aroma awareness is not a skill that can be developed in a couple of weeks. It is a lifelong project. However, once you begin to recognize scents in your environment, you can pick many of them out in a glass of wine.

The Wine Aroma Wheel
Using the Wine Aroma Wheel, created by Dr. Ann Noble at the University of California, Davis, helps expand your knowledge and fine-tune your sense of smell. Besides the common aromas of fruits, vegetables, and flowers, the wheel identifies dozens of other scents found in wine, like soy sauce, kerosene, tar, and chocolate.

Noble slices the wheel into wedges, assigning an aroma category to each wedge. Then the wheel is divided into three circles with general categories in the inner circle, like fruity or spicey. The middle circle separates each general category into more specific categories. For example, fruit is divided into berry, citrus, or tropical. The outer circle contains even more specific aromas. Berry fruit is divided into raspberry, blackberry, blueberry, and so on. Although not all scents are identified on the wheel, it provides context to aid in aroma recognition.

In addition, Noble's work has been adapted into other, more specific wheels, like the German Wine Aroma Wheel, Australian Mouth Feel Wheel, and her own Zinfandel Aroma Wheel.

Evaluating Aromas

To evaluate a wine's aroma, first swirl the glass to add air and increase the aroma compounds. Then put your nose deep in the glass and inhale. To help you focus on your sense of smell, try closing your eyes. Look for scents that you are familiar with. You want to identify the different aromas you smell in the glass. They can be simple or complex, faint or intense.

Legs on the side of a glass indicate viscosity.
TOM ZASADZINSKI/CAL POLY POMONA

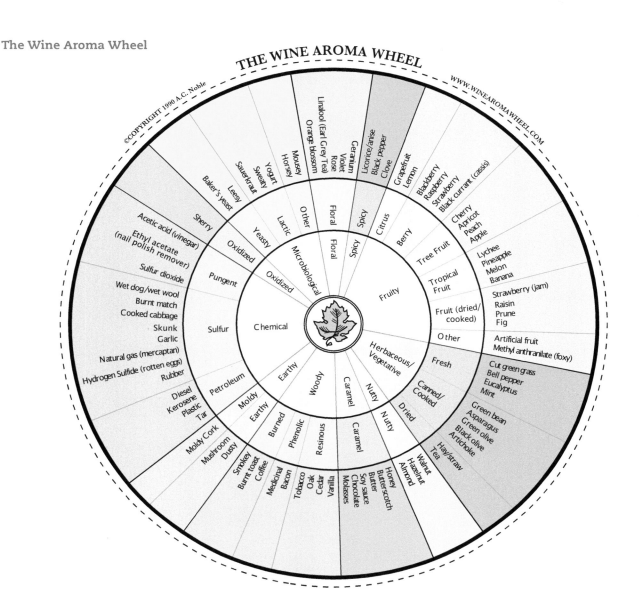

As described in the Wine Aroma Wheel above, aromas are divided into several categories. The most easily identifiable, and highly favored by many consumers, is fruit. This ranges from citrus to berries to tree fruit to tropical fruit. New World wines tend to have more fruit scents than Old World wines because grapes ripen more in warmer climates, developing more concentrated fruit aromas. Another common category, vegetal aromas, like green peppers, green beans, and green olives, tend to develop in cool-climate grapes, which are not as ripe. In addition, floral, spicy, earthy, and nutty scents can be present in the wine. Finally, several other categories, including chemical, microbiological, and pungent, are sometimes detected.

Each grape variety and varietal wine may start with a simple aroma profile that may be generally consistent for that variety, but variations develop quickly based on a number of different factors related to the particular grape variety, the terroir, weather, vineyard techniques, winemaking methods including barrel aging, storage conditions, and age of the wine. White grapes normally have lighter fruit aromas associated with them, like citrus, tree fruits, and tropical fruits. Red grapes normally have darker fruits, like berries, plums, or cherries, used as descriptors. Toasty and smoky aromas come from barrel contact. Chemical compounds normally derive from winemaking. But what sets wine apart from other beverages is the array of aromas that can be found in different wines.

Oak

Oak is a dominant feature of many wines, particularly red wines. It is one of the primary techniques used to manipulate the flavor of wine. Many of the aromas that we think of as being natural to grapes are in fact from oak. Aromas like vanilla, coconut, clove, spice, char, and smoke only come from oak, not from grapes. The ability to identify specific aromas unique to oak is an important step in learning to identify wine styles successfully because it is a distinguishing trait between wines fermented or aged in barrels and those that are not.

Factors that can have a perceptible impact on the amount of oak aromas in a wine are barrel fermentation, barrel aging, the toast level of the barrel, the barrel's age, and the amount of time wine spends in a barrel. Heavier toast, newer barrels, and a longer amount of time spent in a barrel can change oak flavor in wine from a background component to a major factor in its flavor. New oak has a strong influence on the flavors found in oak-fermented or oak-aged wine, whereas old oak provides a much subtler influence and may not impart any oak flavors at all to the wine.

Smelling wine incorrectly
TOM ZASADZINSKI/CAL POLY POMONA

Aromas and Bouquet

As you begin evaluating a wine's perfume, you must be able to make a distinction between aroma and **BOUQUET**. An aroma is a single identifiable smell, like lemon, butter, cherry, or vanilla. Most wines contain several aromas, although one may be dominant, commonly a fruit. Bouquet reflects a collection of these aromas, more than one. A bouquet can change over time and develop through the aging process both in the barrel and in the bottle. Some consider the development of a bouquet to be a distinguishing characteristic of wine complexity.

Simple bouquets are generally found in young wines, most typically white wines. The total bouquet has relatively few aromas, but they can be quite distinct. For example, many Sauvignon Blancs are likely to have a dominant and powerful citrus aroma, like lemon or grapefruit. Other aromas, such as freshly mown grass or fresh herbs, might be less forceful but their presence gives added dimension to the bouquet.

Smelling wine correctly
TOM ZASADZINSKI/CAL POLY POMONA

Red wines usually have more complex bouquets because of increased skin contact, barrel contact, and more aging and blending. A red wine from Bordeaux, France, for example, uses all of these techniques and is generally regarded as one of the most complex wines available.

Taste the Wine

The tasting experience incorporates the five primary tastes recognized by our taste buds, as well as mouthfeel. Tasting a wine normally confirms the first perception obtained from its aroma. A key component of sensory evaluation is developing a taste memory. One method of accomplishing this is to artificially interrupt your sense of smell. By pinching your nose while tasting, you can by-pass aromas and highlight the wine's taste characteristics.

The Five Tastes

There are four basic tastes: sweet, sour (acidity), bitter, and salty. A fifth, umami, or savoriness, is often mentioned in food and wine literature. Sweet, sour, and bitter are commonly found in wine. Umami is thought by some to be found in wine, while salty is rarely present.

Sweet

SWEETNESS comes from residual sugar left in wine after fermentation. The degree of sweetness ranges from nonexistent to barely perceptible residual sugar in DRY wines to a high percentage of residual sugar in dessert wines. Most wines that demonstrate sweetness, other than dessert wines, are white. Riesling is an example of a varietal that is made in a range of styles from dry to off-dry to semisweet to sweet. Off-dry and semisweet wines leave a perceptible sweetness, but are balanced by the acidity in the wine. In California many large producers are leaving a small amount of residual sugar in dry white and red wines to attract consumers who like sweeter, fruitier wine.

Sour

Acidity produces the sour, tart taste found in wine. It comes primarily from tartaric and malic acids. Acidity is highest at the beginning of the ripening phase of grapes and gradually decreases as grapes ripen. Cooler climates producing less-ripe grapes generally result in higher acid levels in wine. In warmer climates grapes fully ripen, and therefore lose acidity. Most very ripe grapes have little acidity, requiring acidification by the winemaker before or during winemaking.

Acidity is tremendously important to the sense of taste. Wines with too much acid can be too sharp, while those that have too little can be flabby. Sauvignon Blanc and Riesling are examples of white wines that are typically harvested at moderate sugar levels and contain high acid levels. Red wines noted for their acidity are cool-climate Cabernet Franc and cool-climate Italian varietals like Nebbiolo and Sangiovese. Grapes more commonly grown in warmer climates, like Syrah and Grenache, generally have lower natural acidity.

Bitter

Bitterness in wines comes primarily from grape seeds but can also come from excessive skin contact during winemaking or from oak. It ranges from mild to harsh and is found in both white and red wines. Some tasters mistake the drying sensation of ASTRINGENCY found in tannic wines for bitterness. However, bitterness is a taste; astringency is a mouthfeel. Most people do not like bitter tastes, so bitterness generally detracts from a wine's perceived quality. However, the bitterness characteristic of some wines contrasts with the richness of some foods, creating an enjoyable tasting experience. Gewürztraminer and Muscat are highly recognizable varietals that can have a bitter taste, as are some Italian varietals, like Negroamaro, which means "black bitter" in Italian.

Salty and Umami

Low levels of salt are present in a few wines that are made from grapes grown in soil or irrigated with water that contains sodium chloride.

A somewhat controversial fifth taste, umami has a pleasant savory quality. It is based on glutamate, which can commonly be found in monosodium glutamate and rich foods, like cream cheese, mushrooms, marbled steak, and Parmesan cheese. Most wine authorities would say that umami is not present in wine, but there has been limited research that indicates that umami may be found in wines that have been macerated and aged on their lees, like sparkling wine.

Mouthfeel

Although most consider that aromas and tastes are the major components of sensory evaluation, mouthfeel, or how wine feels on the palate, also plays an important role. Alcohol, tannins, wine temperature, and carbonation are individually identified in wine but are not usually discussed as separate entities. Rather, they are the components of mouthfeel.

Alcohol

Alcohol, a significant contributor to mouthfeel, is an odorless, colorless, flavorless liquid. But in combination with fruit flavors from grapes, it gives wine **BODY**. Without alcohol wine feels thin and watery. The higher the alcohol content, the more body the wine develops. Because high-alcohol wine is derived from ripe fruit, higher levels of alcohol are normally found in tandem with ripe, fruity flavors. These wines are generally produced in regions with warmer climates. In table wines, high alcohol levels, 14 percent or higher, can add a quality described as **HOT**, literally a hot-on-the-tongue mouthfeel.

Tannins

Tannins come from the grape skins, seeds, and stems. A relatively small increase in tannins occurs when wine is fermented or aged in oak. Tannins give wine astringency, a tactile sensation on the palate that feels like drying or puckering. Often mistaken for the taste of bitterness, astringency does not contribute flavors to wine. Tannins also give wine a rough texture and help stabilize the color of red wines. The astringent characteristic of tannic wine is one of the most difficult wine components to appreciate. By themselves, astringent tannins are aggressive and unfriendly. With food, they fade significantly against richer dishes.

White wines that are normally fermented after pressing contain little or no distinguishable tannins. The few white wines that show tannins have been allowed to macerate extensively with the skins. Thick-skinned red grape varieties, like Cabernet Sauvignon and Nebbiolo, tend to yield tannic wines. Extended maceration during fermentation also results in high tannin levels. Tannins are typically more aggressively astringent in young wines, softening as the wines age. One of the newest and most successful methods of softening tannins is the implementation of micro-oxygenation.

Temperature

The temperature of wine when it is served can have a distinct impact on how it is perceived. Temperatures that are too cold hide aromas, while those that are too hot cause flavors to dissipate and alcohol to overpower the wine. Serving wine at the correct temperature (see Chapter 21, Table 21.2, page 372) can make the difference between an enjoyable or unenjoyable drinking experience.

Carbonation

Carbonation is the feeling of carbon dioxide on the palate. Most commonly found in sparkling wines (see Chapter 17, Bottling and Secondary Fermentation, page 279), CO_2 it is also present in low levels in some table wines. Carbonation adds freshness to young wines that have fresh fruit flavors, like Pinot Grigio and Beaujolais. It can also indicate a flawed wine that is going through secondary fermentation in the bottle.

Other Components of Sensory Evaluation

In addition to appearance, aromas, tastes, and mouthfeel, other factors enter the equation when you are evaluating wine. These include complexity, intensity, and persistence.

Complexity

Complexity describes the array of different aromas, tastes of acid, sugar, and bitterness, and the mouthfeel created by alcohol and tannins. It is identified by the wine's aromas, which range from a simple single scent to an intricate bouquet containing multiple aromas.

Simple wines have one dominant profile, like citrus and green grass, that remains throughout the tasting experience. Complexity develops from the grape variety used, blends of different varieties, the grapes' origins, and winemaking techniques, including barrel fermentation, barrel aging, and bottle aging. As an example, when smelling a wine you might get an immediate scent of cherries, but the longer the wine is in the glass the more aromas develop. The cherries become raspberries, followed by tobacco and licorice. This may be overlaid with a background of vanilla or clove, indicating oak influence. In some complex wines, different aromas can develop over a lengthy period of time.

When tasting wines, it is best to begin with white wines because of their simplicity, and then move to red wines because of their greater complexity. Red wines contain more components that contribute to enjoyment potential. When using the terms *simple* and *complex*, you should not assume that a simple wine is not as good or enjoyable as a complex wine—they are just different.

Intensity

Intensity is the strength, concentration, and depth of wine aromas and tastes. Intensity of aromas varies from barely perceptible to extremely redolent. It can come from the grape variety itself, or from growing techniques, winemaking, and barrel contact. Sometimes wine can be very subtle, containing many flavor components, but none is overpowering. Intense wines have pronounced aromas and tastes that are strong and concentrated. Pinot Noir is a good example of a subtle grape variety. It can be very complex but not show great power in its aromas and tastes. Other varieties, like Syrah, Zinfandel, and Cabernet Sauvignon, can be very powerful with strong, distinguishable aromas and tastes.

The intensity of the aromas and tastes influences a taster's perception of the wine. If the aromas are pleasing, many tasters prefer intense aromas, which add to their overall tasting experience. Others would rather drink wines with subtler aromas.

Persistence

The terms **FINISH** and **AFTERTASTE** are commonly used to describe the persistence of flavor in a wine. How long does the taste remain with you after the wine has been swallowed? Does it leave a pleasant aftertaste? Unpleasant flavors left in the mouth would not be enjoyable, but a pleasing flavor that lingers is a mark of a good wine.

Balance

The final step in sensory evaluation consists of determining whether or not the wine is balanced. To a winemaker, **BALANCE** is achieved when each of a wine's components is in equilibrium, and no one component overpowers the other attributes of the wine. The components in the equation can include acid, sugar, alcohol, fruit, tannins, bitterness, and oak. Additional factors, like body, serving temperature, and certainly how the wine tastes when paired with food, can also play a role in how a wine's balance is perceived. Not all wines include each of the components listed above, and some components may be more important than others for a particular wine style. For example, high acid is a critical component of grape varieties like Sauvignon Blanc or in wine styles like sparkling wine, whereas high tannins are significant in varieties like Cabernet Sauvignon. Winemakers are trained formally or informally to achieve the goal of a balanced wine.

To a consumer, balance may mean nothing more than enjoying a wine that tastes good, and a consumer's ability to evaluate balance may be limited by his or her experience with different varietals, blends, countries, and appellations.

Often disagreements about a wine's quality center on its balance, and the opinions of winemakers and consumers can differ widely. To one person, a highly acidic traditional style Sauvignon Blanc may seem too acidic. Yet another may find that it is perfectly balanced when served with the right food. Some might think a Cabernet Sauvignon is oaky, overripe, and flabby, while others would consider it juicy and flavorful.

White Wine

White wine balance is often a simple equation because the wine has fewer components than red wine. Dry wines do not have sugar and many rarely have oak contact. Tannins are usually a minor or nonexistent component because the wine is fermented without skins. Acid and alcohol have the most critical impact on balance. Balance means acids are not sharp and unpleasant, and alcohol is not flat and flabby. In certain varietals and styles of wine, like Sauvignon Blanc from the Loire or New Zealand, acid appears to dominate, but when paired with food the wine is likely to seem less acidic because of the richness of the food.

In off-dry, semisweet, and sweet wines, sugar plays a large part in the balance equation. Residual sugar counterbalances acidity. If you want a wine to taste drier, you want the acidity level to be higher than the sugar or alcohol levels. If you want the wine to be sweeter, you want a higher level of sugar to acid.

White wines that undergo oak fermentation and oak aging have a different balance equation. In addition to acid and alcohol, oak plays a more dynamic role. Oak flavors and body from the barrels can complement the fruit and acid in the wine, and even a touch of residual sugar can bring out its fruitiness.

Red Wine

Acid, alcohol, fruit, oak, and tannins are possible components of a balanced red wine. Red wine balance is frequently more complex than in white wine because of the presence of additional components, like tannins and oak.

Some red wines can produce such powerful tannins that they throw off the harmonious balance and make a wine unpleasant to drink when young. Micro-oxygenation and extended aging can soften the tannins, resulting in a balanced wine. Suppler, softer, early-drinking red wines minimize tannins by reducing the maceration time, which eliminates tannic harshness and achieves balance.

Wines made from grapes allowed to have extensive hang time in the vineyard generate high sugar levels and potential alcohol and ripe fruit flavors, but lower acidity. This style of wine is very popular in today's culture because of its intense fruit flavors, but it also has high alcohol. To achieve balance, the winemaker adds acid back into the wine.

As the number of countries where grapes grow expands, winemaking practices evolve, and the population of wine consumers grows, wine styles are becoming so varied that perceptions of balance are changing. It is ultimately up to you to decide whether or not you consider a wine to be balanced.

WINE FAULTS

When you first begin drinking wine, it can be difficult to tell if you do not like a wine because of its characteristics or because there is something wrong with it. As an example, if you are most familiar with the ripe, fruity, oaky style of California wine and taste an acidic, vegetal, unoaked wine from Europe, you may think there is something wrong with the European wine. In all likelihood the wine is fine; it is just different from what you usually drink. As you learn to evaluate wine, you can develop an awareness of not only appropriate style characteristics from different parts of the world but also when faults are present.

Wine faults are offensive colors, aromas, and flavors that develop from bad grapes, improper winemaking, poor bottling procedures, or improper storage and handling after the wine is bottled. Each individual has his or her own tolerance for wine faults. One person may immediately reject a wine with a fault, while another may decide that it is inconsequential or even adds to the wine's overall appeal. Only the person consuming the wine can decide whether or not to continue drinking it. Being able to identify and explain a wine's faults gives you greater confidence when rejecting a wine than saying, "I don't like it" or "It isn't any good." In a restaurant, you should reject a flawed wine and request a replacement.

Faults You Can See

It is fairly easy to recognize visible faults because sight is one of our most developed senses. Cloudy wine, bubbles, and color are readily apparent when wine is poured into a glass. Following are wine faults you can see:

Cloudy Wine Cloudy wine can occur if wine is poorly made or if it has not been stabilized. Most likely, either yeast or proteins are visible. Although it is rare, cloudiness indicates that something went wrong in the winemaking process. Generally, cloudy wine should be returned.

Tartrate Crystals Tartrate crystals are caused by excessive potassium bitartrate in wine. The tartaric acid that crystallizes leaves a residue that looks like shards of broken glass or rock candy. If you rub your finger on the bottom of a cork and it feels rough, you are feeling the crystals. They also fall out of solution and settle to the bottom of the bottle as wine ages, even when it is young. Virtually all wines produced in the United States are cold stabilized, which prevents the formation of tartrate crystals. However, many Old World winemakers do not use this process, so European wines commonly form tartaric crystals, which are not harmful. To most professionals, they are not a fault, and many consider this condition a quality characteristic. There is no reason to return the wine.

Bubbles in Table Wine Bubbles are a key component of sparkling wines and are fairly common in lighter style nonsparkling white wines and some lighter-style red wines. As an example, Lambrusco and Beaujolais commonly have a little SPRITZ. However, most dry table wines should not have bubbles. Bubbly dry wine is most likely going through an undesirable secondary fermentation caused by surviving yeast cells and unfermented sugar remaining in the wine. If you open a dry, still wine, like Chardonnay or Cabernet Sauvignon, that has bubbles, it is almost surely flawed.

Orange or Brown Wine Small amounts of oxygen in wine are beneficial; however, excessive amounts are detrimental to wine quality. Over time, too much oxygen in white wine turns it orange, or even brown. Red wines can absorb higher oxygen levels than white wines but also can turn brown. Not only is the color unpleasant, but heavily OXIDIZED wine gives off flat, stale aromas. Very old wines may also turn color, but most table wines of a recent vintage should not. Although many people consider oxidized wines undrinkable, others enjoy this style of wine. Some fortified wines, like sherry, are purposefully oxidized.

Faults You Can Smell

Several faults can show up when you smell wine. Some of these clearly are unacceptable, and the wine should be returned. However, one person may find some aromas objectionable, while another may believe that they add to the wine's complex character. It is important to be able to identify these faults and to determine your personal preferences. Faults you can smell are:

Corked Wine Cork taint, or corky or corked wine, is the first wine fault you should learn to identify. Corked wine has a distinctive musty, muddy, or wet newspaper aroma. Pour yourself a small amount of wine, then smell and taste it before serving your guests. Ask yourself, "Is the wine corked?" A corked wine indicates that the cork has been infected with TCA (see Chapter 5, TCA-Corked Wine, page 89), which gives the wine the tainted aroma. It is the cork that is damaged, not the wine. A second bottle of the exact same wine will probably be fine. Corked wine should always be returned or exchanged for another bottle.

Brettanomyces BRETTANOMYCES, or Brett, is a yeast infection that gives wine a barnyardlike, leathery, sweaty-saddle odor. Probably no other characteristic in this category causes more controversy. When present in a small amount in red wine, many, particularly Europeans, consider Brett an acceptable aroma that adds another layer to the wine's bouquet. However, most New World wine professionals regard it as a major fault and will reject the infected wine.

Oxidation As mentioned previously, oxidized wines are off-color. They may also have a flat, sherrylike smell. Oxidation can start when the grapes are being transported from the field, occur during any phase of winemaking, and continue after the wine is bottled. It is most commonly noticed in white wines but also develops in red wines. In different phases of the winemaking process, exposure to oxygen can be controlled by the addition of SO_2.

Acetic Acid Acetic acid, commonly called volatile acidity (VA), produces a vinegar scent in the wine. VA is formed when oxygen comes in contact with alcohol. Although the wine may be undrinkable, more fully developed it might become an excellent base for salad dressing.

Acetaldehyde Acetaldehyde forms from yeast and acetic acid bacteria when oxygen reacts with ethanol. At low levels it can provide pleasant fruit aromas, but at high levels it gives the wine a strong, dry sherrylike aroma.

Ethyl Acetate Ethyl acetate, a common ester, is formed by the interaction of acetic acid and ethanol, the most common alcohol in wine. At low and moderate levels, ethyl acetate contributes to the desirable fruity aromas in wine. At high levels, ethyl acetate makes the wine smell like nail polish remover, a distinct flaw.

Excessive Sulfur Dioxide Sulfur dioxide is added to virtually all wines, but when it is overused it can cause an unpleasant burnt-match aroma. It is most often detected in sweet wines.

Hydrogen Sulfide and Mercaptans Hydrogen sulfide, which gives wine the smell of rotten eggs, develops due to yeast and nitrogen deficiency during fermentation. It can normally be corrected by the addition of diammonium phosphate at this early stage of winemaking. If not corrected during fermentation, aerating the wine through racking should eliminate the problem. The interaction of hydrogen sulfide and alcohol creates mercaptans, which give wine a skunky, cooked-cabbage odor. Adding copper sulfite during fining or earlier may help dissipate the smell, but it will not be completely eliminated.

Faults You Can Taste

Appearance and aroma offer the first clues that a wine might taste bad. Bitterness and metallic tastes, however, cannot be detected by checking a wine's look or smell. These faults can only be discovered by tasting the wine. They are:

Excessive Bitterness Excessive bitterness is caused by too many phenolics (flavor compounds) or the use of dirty oak barrels. Very tannic wines are sometimes misinterpreted as overly bitter, when in reality they are astringent and not bitter. Whereas tannic wines soften with time and oxidation, bitter wines remain bitter, and the fault can only be disguised by adding sugar or serving the wine with rich, fatty foods, like a grilled steak.

Metallic Taste An excessive metallic taste can occur when certain minerals are absorbed from the soil or from fertilizers and pesticides, but it is found most often when wine is contaminated by contact with metal. A small amount of metallic taste from minerals can add complexity, but when caused by poor vineyard techniques or winemaking, the wine should be returned.

WINE AND FOOD PAIRING BASICS

Drinking wine by itself can be a pleasurable experience, but consuming it with food makes both more enjoyable and satisfying. Historically, wine has been a dining beverage. In Europe, where wine and food at the table is a daily experience, entire cuisines have been built around the wines of individual regions.

Wine and food combinations can be daunting. How do you start? Do certain wines go better with certain foods and vice versa? In this chapter you have learned how to taste wine and have developed an understanding of the components of balance. Now you will look not only at the characteristics of wine but also at the components of food that make each more enjoyable. This chapter contains guidelines for matching food and wine. Specific foods to serve with each varietal and with regional wines are covered in the chapters that follow.

Pairing Wine with Food or Food with Wine?

When matching food and wine, which comes first, the food or the wine? If you have a great bottle of wine that you want to try, your question has been answered. You search for a recipe or menu item that pairs with the wine. Alternatively, if you have a craving for a particular dish or food, then you also have made a choice. You seek out a wine that goes with it. In restaurants most diners typically select menu items first, then the wine. It is usually easier to choose a key ingredient, like beef, chicken, or fish, and select the wine based on that choice. However, many professionals find it easier to select the wine first because the diner can change the taste and balance of the dish to match the wine by adding seasonings, particularly salt, lemon juice, or vinegar (acid).

Rules of Pairing

At one end of the wine-pairing spectrum, some subscribe to the rule of no rules. They believe that you should pick what you like, both wine and food, without considering how they interact. At the other end of the spectrum, some seek out the perfect match, delving into the complex characteristics of the food and its preparation as well as the wine. Either approach can work, but in the middle ground, understanding some simple guidelines can help you to more fully enjoy selecting foods and wines that enhance one another.

The most traditional rule is red wine with red meat (beef, lamb, game), and white wine with white meat (chicken, fish, seafood). Another is serving rich wines with rich foods and light wines with light foods. This may provide a good starting point, but with experience, other factors will become equally important and will allow you to break some of the standard rules.

The key factors that should be considered when combining food and wine are:

Wines should complement the foods being served. Here Italian wines accompany a buffet lunch in Piedmont, Italy.
BILL WHITING

- acidity level of the wine
- alcohol level of the wine
- oak and tannins in the wine
- sweetness of the wine or food
- seasoning of the food
- cooking methods

Acidity

The acid level of a wine has the most profound effect on food and wine pairing. Acid balances out the richness, fattiness, and saltiness of many foods but is also compatible with acidic foods. A wine with distinct acidity generally goes with a broad range of foods.

High-acid white wines, like sparkling wines and varietals such as Sauvignon Blanc, Riesling, and Pinot Grigio, work well with salty foods, fried foods (usually seasoned with salt), and rich foods. All of these wines cut through the salty/briny taste of fresh shellfish, such as oysters and clams; fish coated with a rich beurre blanc sauce; smoked fish; and especially almost anything deep-fried. They are also excellent matches for lighter poultry dishes and salads with acidic dressings.

High-acid red wines pair with a wide range of foods. Italian varietals like Nebbiolo, Barbera, and Sangiovese are excellent food pairing wines. As an example, Sangiovese (Chianti) goes well with tomatoes and tomato-based sauces and complements hard cheeses that contain a bit of saltiness, like Parmigiano cheese. Nebbiolo and Barbera are commonly served with rich pasta dishes and game. High-acid wines also contrast well with the fattiness and richness of a variety of foods, including red meats.

Alcohol

Alcohol in wine contributes its own richness to a meal. Generally, the higher the alcohol level, the richer the wine, and the richer the food needs to be. High-alcohol wines work best with rich or high-fat foods and sauces. The flavors of light dishes are lost to the power of a high-alcohol wine. Lower-alcohol wines lack richness and require more delicate foods.

A high-alcohol red wine, like New World Cabernet Sauvignon, Syrah, or Zinfandel, cries out for a rich, fatty piece of meat. A high-alcohol Chardonnay is perfect for rich braised chicken or duck thighs. A poached chicken breast is overwhelmed by such a wine, whereas a lighter, more acidic wine complements the delicacy of the chicken and its cooking method.

Oak and Tannins

Although they are separate components of wine, oak and tannins complement the same kinds of foods. Tannins provide an astringent quality that works best with grilled foods, especially high-fat red meat, like a well-marbled steak. Oak contributes nutty, toasted, and smoky flavors. Again, grilled foods, especially red meats, pair with oaky red wines, as do roasted meats and rich braised dishes. Smoked meats work with oak-aged red wine that has a smoky character. Oak-aged white wines generally have lower tannins and pair better with rich cream or butter sauces on poultry or fish.

Sweetness

Whether a sweet wine is paired with an appetizer, entrée, or dessert, the wine must almost always be sweeter than the food. If the food is sweeter than the wine, it makes the wine taste sour and unpleasant. Much German food has a sweet or sweet and sour component and is a perfect companion to an off-dry Riesling, such as a German Kabinett or Spätlese. Many appetizers and main courses are accented by tropical fruits or sauces made from dried fruits, and off-dry wines, like Riesling or Chenin Blanc, are good choices for these dishes. Sweet-style wines also help to counter the spiciness of Asian cuisines. A very sweet dessert wine is a perfect accompaniment to a sweet dessert.

Seasoning

Seasoning impacts the balance of food and wine pairings. There are no general guidelines for seasoning. Each must be dealt with based on its individual characteristics.

Salt is the number one seasoning. It can enhance a dish by bringing out the underlying flavors of the ingredients and can inhibit bitterness, but it can cripple a wine, stripping it of its flavors, if too much is used in food preparation. Sparkling wine is an example of a wine that stands up well to salty foods.

Black pepper helps perk up dishes and the wines served along with them when used in moderate amounts. In heavier amounts, it goes with rich, peppery wines, like Zinfandel, but overpowers subtler wines. The amount of black pepper added to a dish can determine whether a peppery Zinfandel or a softer Merlot should be served.

Herbs can play either a minor or significant role in food and wine pairing. Subtle use of seasonings, like thyme, oregano, or marjoram, can bring out similar characteristics in the wine. How the herbs in the food interact with flavors in the wine affects the style of wine that is selected. Many herbs work well with white wines; fewer work as well with red wines because the reds are too strong and overpowering for the herbs. Rosemary is a significant exception. Because of its intense aromatic characteristics, it can stand up to red wine. A range of red wines, from light Pinot Noir to rich Cabernet Sauvignon and Brunello, make the perfect partners for lamb and beef preparations that include rosemary.

Spices, like chili, cinnamon, cumin, and cloves, complement many wines, but sweeter styles of flavorful white wines, like Riesling, or pungent wines, like Gewürztraminer, work best. The spiciness in many Asian dishes can have a huge impact on wine selection. If the dish is mild, a range of wines might work with light white wines accompanying poultry, fish, or vegetable dishes. Spicier dishes call for a sweeter wine that can counterbalance the heat of the food. If a dish is very spicy, beer might be a better choice than wine.

Other seasonings, like garlic, soy sauce, and mustard, have their own distinct characteristics, so they must be carefully paired with wines that enhance their unique flavors. A grilled or roasted leg of lamb with slivers of garlic throughout, for example, requires a strong red wine, like Syrah or a Grenache-based Southern Rhône blend.

Cooking Methods

Knowledge of basic cooking methods helps in matching food and wine. Poaching, steaming, sautéing, grilling, roasting (baking), braising, and frying are the principal techniques of food preparation.

Steaming and poaching are most commonly used with more delicately flavored foods, including poultry, fish, and vegetables. The food and preparation call for a light, delicate, low-alcohol wine, usually white, like Sauvignon Blanc, Riesling, Pinot Grigio, or Chenin Blanc.

When foods are sautéed, grilled, or roasted, the richness and texture of the food become the basis for selecting a wine. White wines that would be appropriate with lighter foods, like poultry and fish, include Chardonnay, Viognier, Pinot Gris, and Marsanne. With richer foods, like steaks, chops, and fatty fish, red wine choices include Pinot Noir, Cabernet Sauvignon, Merlot, and even Syrah.

Braising produces moderately flavored to richly flavored dishes. This cooking method removes the hard edges found with some foods and creates savoriness (umami) that works with a range of both white and red wines. Usually wine is a component of the braising liquid, and the same wine or something similar will pair well with the dish. For white meat braises try Italian varietals, like Arneis or Gavi, and light reds, like Pinot Noir. For red meat braises the standard Cabernet Sauvignon and Merlot work, but many other reds can also be used, including Zinfandel, Cabernet Franc, and Barbera.

RULES FOR PAIRING FOOD AND WINE

1 If unsure of which wine to match with a certain food, pick something that is more acidic. It will pair with many different foods.

2 Serve low-alcohol wines (less than 14 percent) with delicate foods and high-alcohol wines (greater than 14 percent) with rich foods.

3 Minimize the use of salt in dishes to be paired with wine. Too much can get in the way of the pairing.

4 Sweet wine must always be sweeter than the food that it is served with. Dry or even off-dry wines will not be successful with dessert if the wine is less sweet than the dessert.

The combination of fat and salt found in most fried foods is a perfect complement to many wines, especially those with high acidity. French fries or any deep-fried vegetables or cheeses work with tangy sparkling wines as appetizers. French fries are also a common accompaniment to a thick rare steak served with a rich Cabernet Sauvignon. Fried calamari or other seafood or fish goes with a range of lighter high-acid wines, like Pinot Grigio, Sauvignon Blanc, Verdejo, or Muscadet. Pan-fried white or red meats, dredged or battered, like fried chicken or chicken-fried steak, can work with a range of white or red wines, for example, Chardonnay or Merlot.

SUMMARY

The process of examining a wine encompasses more than just tasting. You must take the time to look at the wine, enjoying its color; smell the wine to appreciate its many aromas; and finally taste the wine to explore all of the different flavors and tactile sensations it has to offer. In addition, recognizing a wine's balance can increase the pleasure you get from drinking it.

As you prepare to serve a wine, consider whether it will be consumed by itself or with food. Pairing wine with food requires careful consideration of several factors, including acidity, alcohol, tannins, and oak. The sweetness of the wine or the food, seasonings in the dish, and cooking methods used can also affect your impressions of a wine.

7 Chardonnay

Chardonnay is the most widely recognized white grape variety in the world today. It is produced in virtually every major wine-producing country and its popularity continues to grow. Chardonnay grapes originated in Burgundy, France, where they were used to make the best white wine in the world in the 1950s and 1960s, and many believe that is still the case. New World producers from wine regions such as California and Australia helped to drive the popularity of Chardonnay, as producers went to Burgundy in search of grapevine cuttings to start their vineyards. They planted the cuttings in their new homes, where Chardonnay had immediate success.

Chardonnay is best known as a dry table wine, but the grape is also used for high-quality sparkling wines throughout the world. Like any grape variety, Chardonnay can be made into a sweet dessert-style wine, but more aromatic varieties are generally preferred for dessert wines.

A high-density vineyard in Côte de Beaune, France, is covered in frost during the winter months.

THE MOST POPULAR WINE IN THE WORLD

Producers find Chardonnay to be reliable, easy to grow, and adaptive to a range of climates. It grows virtually everywhere in a broad range of conditions. It is also highly malleable, so winemakers can use different winemaking techniques to produce a variety of unique styles of wine.

In addition to these strengths, New World producers and consumers have also helped bring about Chardonnay's tremendous popularity. In the New World, winemakers developed a style of Chardonnay that was quite different from Old World styles. Because much of the New World has a warmer climate than the Old World, New World Chardonnay generally achieves a higher degree of ripeness. In addition, malolactic fermentation (MLF) is used extensively to create a richer style of wine, and new French oak is used in fermentation and aging to develop distinctive oak aromas in the wine.

From the consumer's standpoint, Chardonnay is popular in part because it appeals to novice drinkers. First, because it is pronounced the way it is written, inexperienced buyers are not intimidated by ordering it. In addition, many inexpensive Chardonnays, like other lower-priced, entry-level wines, have a touch of sugar, which suits the palates of consumers who are new to wine drinking. And because it is available in a range of styles, Chardonnay is a wine that offers new options as the consumer becomes a more sophisticated wine drinker.

There is some concern that Chardonnay has become the thousand-pound gorilla of white wine in the United States, preventing other white varietals from entering or expanding in the marketplace. If everyone is drinking Chardonnay, wine retailers and restaurateurs are less likely to build an inventory of lesser-known grape varietals or blends, and if other wines do not sell, winemakers are less likely to produce them. Hopefully this trend will change because while Chardonnay has a lot to offer, increasing the number of other varietals and blends creates the greatest number of choices for the consumer.

Table 7.1 Top 10 Chardonnay-Producing Countries

1	United States	111,500 acres/47,712 ha
2	France	88,125 acres/35,252 ha
3	Australia	42,125 acres/16,855 ha
4	Italy	29,500 acres/11,800 ha
5	Moldova	12,500 acres/6,000 ha
6	South Africa	12,500 acres/5,983ha
7	Chile	11,000 acres/4,402 ha
8	Slovenia	8,875 acres/3,565 ha
9	Argentina	8,500 acres/3,373 ha
10	New Zealand	6,125 acres/2,449 ha

Source: chardonnay-du-monde.com

REGIONS AND APPELLATIONS

There are about three dozen countries where Chardonnay is grown, with approximately 346,000 acres/140,000 hectares planted to Chardonnay worldwide. Table 7.1 lists the top 10 Chardonnay-producing countries.

Other countries that produce Chardonnay include Spain, Bulgaria, Romania, Israel, Portugal, Hungary, Greece, China, Uruguay, Canada, and Austria.

Within each country, the grape is grown in numerous appellations, each with a distinct character. Prominent appellations are highlighted in the following section. (See Appendix 2 and Appendix 3 for maps of Chardonnay-producing countries and appellations, pages 380–447.)

The United States

In the United States close to 90 percent of Chardonnay grapes are planted in California. Some of the better vineyards outside California are located in Washington and Oregon in the Northwest, New York and Virginia in the East, and Texas in the South. In California Chardonnay is grown throughout the state, from Mendocino County in the north to the Central Coast and San Joaquin Valley (Central Valley) to Temecula in the south.

Table 7.2 Prominent Chardonnay Appellations

COUNTRY	REGION/APPELLATION	SUBAPPELLATION	VILLAGE OR SUBAPPELLATION
France	Burgundy	Chablis	
		Côte de Beaune	Montrachet Meursault Corton-Charlemagne
		Mâconnais	Pouilly-Fuissé
	Champagne	Côte des Blancs	
United States	California	Napa Valley	Carneros
		Sonoma County	Carneros Russian River Valley Sonoma Coast
		Santa Barbara County	Santa Rita Hills
		Mendocino County	Anderson Valley
	Washington State	Columbia Valley	
	East Coast	New York	Long Island
		Virginia	
Australia	South Australia	Clare Valley	
	New South Wales	Hunter Valley	
	Victoria	Yarra Valley	
Italy	Friuli Trentino-Alto Adige Veneto		
South Africa	Stellenbosch		
New Zealand	North Island	Auckland Hawkes Bay	

California has achieved the notable status of producing more Chardonnay than any other location in the world and it is by far the most widely consumed wine varietal in the United States. Napa Valley, Sonoma County, and more recently Santa Barbara County produce some of the most desirable Chardonnays in this country, many made in a rich, heavily oaked style. Chardonnay throughout the United States is labeled by grape variety, not by region.

France

France, the second largest producer, is highly regarded for Chardonnay from the **APPELLATION D'ORIGINE CONTRÔLÉES** (AOCs) of Burgundy and Champagne, but the grape is widely planted in other parts of the country as well. In AOCs it is identified and sold by the appellation, not by grape variety. In Burgundy it must be 100 percent Chardonnay by law. Chardonnay grown in Champagne, with very few exceptions, is made into sparkling wine.

Of the other regions in France, Languedoc-Roussillon, which runs along the Mediterranean, produces the most Chardonnay. This much warmer region yields Chardonnay that is similar to moderately priced California Chardonnays. The classification of Chardonnay from the Languedoc-Roussillon is **VIN DE PAYS** (VDP), a lower classification than AOC. VDP wines are easier for wine novices to purchase and understand because they are labeled by grape variety rather than by appellation, and they are widely sold at affordable prices in the international market. Other notable appellations where Chardonnay is produced include Anjou and Touraine in the Loire, Ardèche, Jura, and Savoy.

Jean-Jacques Vincent, owner of Château-Fuissé in Burgundy, France, is a prominent producer of Pouilly-Fuissé, a 100% Chardonnay wine.

Australia

The third major producer of Chardonnay is Australia, where it is the most popular and most extensively planted white variety. It is exported throughout the world, particularly to Britain and the United States. Australian Chardonnay is noted for its rich, buttery New World style. Australian plantings include Hunter Valley, Clare Valley, Yarra Valley, Eden Valley, and Adelaide Hills in the eastern part of the country; Margaret River in the west; and the island of Tasmania.

Other Countries

Italy, Moldova, South Africa, and Chile also plant substantial amounts of Chardonnay. In Italy Chardonnay vineyards are replacing some indigenous vineyards. The best Italian Chardonnay is grown in the northeast in the regions of Friuli, Trentino-Alto Adige, and the Veneto, but it is also planted throughout much of the rest of the country, including Piedmont, Tuscany, and Sicily. Moldova, a part of the former Soviet Union, is making great strides with grape and wine quality, and South Africa and Chile are producing internationally regarded Chardonnays.

Other Chardonnay-producing countries of note are Spain, New Zealand, and Argentina. As in Italy, Chardonnay is not indigenous to Spain, but is grown there and used along with both international varieties and indigenous Spanish grapes in a wide range of blended wines. Recently Chardonnay has been approved for use in making Cava, Spain's sparkling wine. New Zealand has also become recognized for its outstanding Chardonnay. Although noted more for its Sauvignon Blanc and Pinot Noir, Chardonnay is the most widely planted grape variety in New Zealand. Because of its dramatically changing geography from the North Island to the South Island, the wine styles can range from richer, riper, fruitier styles on the North Island to leaner, crisper styles on the cooler South Island. In Argentina Chardonnay is used as the base for most sparkling wines and also makes moderately priced table wines for the international market. Although their wines are not generally available in the American market yet, Asian countries are starting to grow wine grapes, and Chardonnay is a leading variety. China is becoming a huge grape-growing country and has the population to support wines produced there.

Table 7.2 includes an abbreviated list of countries that produce highly regarded Chardonnays along with their most prominent appellations and subappellations.

CHARDONNAY: THE GRAPE

Chardonnay is a neutral, nonaromatic grape that takes on the characteristics of its terroir and, perhaps more so, the winemaker's influence. The grape's greatest flaw is that it buds early, making it susceptible to spring frost that can damage or kill the crop. Chardonnay typically produces high yields, but when better-quality wines are desired, the vines can be managed using dense planting to control crop quality. In warm to hot climates, Chardonnay loses its acidity quickly in the final stages of ripening, making it a candidate for acidification. Chardonnay grapes and vines are not overly susceptible to diseases and molds, and in the winery Chardonnay responds well to winemaking techniques such as MLF, sur lies aging, barrel fermentation, and barrel aging.

The Chardonnay grape struggles in cold appellations, like Champagne and Chablis, in France, and can do well in hot appellations, like the San Joaquin Valley (Central Valley) in California. However, the greatest success with Chardonnay occurs in moderately cool climates—like Burgundy, parts of Sonoma County, Carneros in Napa Valley, Hunter Valley in Australia, and Northern Italy—all regions where ripening is slow and flavors have a long time to develop

IN THE VINEYARD

Chardonnay is produced in a wide range of geographical regions, and a number of viticultural factors in each region impact the resulting wines. Climate, soil, and water access play a role in the quality and characteristics of Chardonnay wines. The favored vineyard-management techniques for Chardonnay have tended to follow Old World traditions, but many producers are embracing New World techniques as well.

Climate

The best Chardonnay wine comes from grapes planted in cool-climate regions (those with an average mean temperature below 61°F/16°C between April and October) with good sun exposure and adequate rainfall, although it also grows in warm or even hot climates with success. Chardonnay grown in cold northerly latitudes can be successful if it is grown on sloped vineyards facing southeast to southwest. Two benchmark appellations, Chablis and Champagne, are located at latitudes where temperatures are so cold that Chardonnay barely ripens.

Chardonnay vineyards are planted at elevations ranging from sea level to 1,500 feet or more, but vineyards planted at lower, warmer elevations must have a moderating influence. For example, Carneros sits close to sea level in Napa Valley and Sonoma County, warm locations. However, daily fog off the Pacific Ocean keeps the temperatures cool enough for the grapes to thrive. This is an example of a cooling maritime influence. A mountain range like the one found in Sonoma Coast or the hilly plateau called Morvan west of the Côte d'Or in Burgundy can also moderate temperatures that might otherwise be too cold.

Some of the best Chardonnays in California come from wineries in higher, cooler elevations, such as Chateau Montelena, Mayacamas, and Kistler in the North Coast. The most famous Chardonnay appellation in Burgundy, Côte de Beaune, is also cool, with the better vineyards at a moderately higher elevation. Chardonnay does well in cooler climates because it matures early and yields grapes that maintain good acid levels and ripen at lower sugar levels. Cool climate wines with higher acidity and lower alcohol are leaner, steely, and edgy, and more likely to reflect their terroir. In unusually cool and wet years in Europe, it can be very difficult to achieve fully ripened grapes. In this case, wines will be chaptalized.

Chardonnay that comes from warmer climates typically makes a very different style of wine than Chardonnay from cool climates. In hot weather, Chardonnay can overripen and lose the acidity needed to provide balance. The wines are fruitier, softer, rounder, less acidic, and higher in alcohol. If overly ripe, the wines made from these grapes may be referred to as flabby or having no structure, which means that little acid is left in the wine. Acid—tartaric, citric, or malic—is commonly added when Chardonnay has ripened too much and has lost its natural acidity. Hot climates, such as California's San Joaquin Valley (Central Valley), Australia's Hunter Valley, and Languedoc-Roussillon in the south of France, produce riper, fruitier grapes that are generally used for box, jug, or lower-priced bottled wines.

Soil

Chardonnay is planted in diverse soils around the world. The grapes of the Côte de Beaune come from moderately sloped, well-drained vineyards with soil composed of limestone and clay. In Chablis the soil is chalky limestone and clay, and in Champagne it is chalk and limestone. Limestone is a common factor in these great French Chardonnay appellations. Limestone, in tandem with climate, produce wines higher in mineral content and higher in acidity that tend to be more austere, or steely, than New World wines. But Chardonnay is also grown successfully in soils ranging from clay to stony gravel to volcanic and sand around the world.

Water Access

In Europe, appellation laws dictate which appellations and varieties can be irrigated. These laws prohibit Chardonnay from being irrigated in most appellations, particularly Burgundy. Although Old World Chardonnay vines are not irrigated, they generally receive enough rainfall or have access to adequate moisture through their root systems. Chardonnay responds well in nonirrigated situations in other cooler or wetter climates of Northern Europe as well. Most appellations in high-production countries outside Europe, including the United States, Australia, Chile, and South Africa, use irrigation for Chardonnay.

Vineyard Management

Vineyard-management techniques have varied between Old World and New World practices. Many New World producers have reverted to Old World clones and density, but still use some advanced canopy-management techniques. The kinds of vineyard-management techniques used at the time of planting, the standards and requirements of a particular appellation, and the particular characteristics of a vineyard's terroir can influence the wine.

Clone selection is important and growers pay tremendous attention to using clones for new plantings that will increase quality or productivity. In the United States and elsewhere, Dijon clones from Burgundy are quite desirable. Chardonnay requires trellising, but adapts to a variety of training systems from Guyot to vertical shoot positioning to Scott Henry. The selection depends on the climate and country where it is planted, the exposure to the sun, and the growing philosophy of the producer. Cane pruning is most commonly used. Canopy-management techniques for Chardonnay include leaf pulling to maximize sun exposure and green harvest to ensure flavor development of the remaining bunches and to obtain the desired yield.

Vineyard density of Chardonnay has changed markedly in the United States and other New World countries. In Burgundy it was historically common to plant high-density Chardonnay vineyards with approximately 4,000 plants per acre/10,000 plants per hectare. In the New World, by contrast, it was more common to plant less densely with as few as 1,000 vines per acre/2,500 vines per hectare. The vines were spaced farther apart to accommodate tractors used in the vineyards. But as vineyards are replanted, many New World producers are switching to higher-density vineyards similar to those in Burgundy, a change that requires new vineyard equipment that functions well with the denser

Newly planted Chardonnay vineyard in Marlborough, New Zealand

planting. Increased vines per acre do not mean increased crop size; crop volume remains the same because there are fewer bunches (and weight) on each vine.

The Chardonnay vine is fairly prodigious and is perfectly capable of yielding a crop of five tons or more per acre/approximately 75 hectoliters per hectare, but at this volume the grapes begin to lose character and are likely destined for the bulk wine category. Some of the best Chardonnay vineyards limit output to as little as two or three tons per acre/approximately 30 to 45 hectoliters per hectare to achieve higher-quality grapes.

IN THE WINERY

The real nature of Chardonnay wine develops in the winery. Chardonnay is the most manipulated grape because it has such neutral flavors. As a result, winemakers have a tremendous impact on how the finished wines taste. There are five key techniques a winemaker can use to influence Chardonnay's aromas and flavors:

- maceration
- fermentation vessel
- MLF
- sur lies aging
- barrel aging

Using (or not using) one or more of these techniques, Chardonnay wine styles can range from an unoaked, steely mineral, and quite acidic wine from Chablis to a heavily oaked, buttery MLF-influenced wine made with ripe fruit from Napa Valley. Intensity and complexity in flavors and even tannins can be enhanced using the techniques of maceration and sur lies aging. Winemakers use these methods with other varieties, but the Chardonnay grape has a reputation for its ability to respond to the full array of these winemaking techniques to create unique wine styles.

Maceration

The winemaker decides whether to let the skins macerate with the juice for as little as several hours to as long as a few days before or during fermentation as a way to add flavors, body, and tannins. Although white varieties are not typically macerated, some winemakers, mostly New World, believe maceration gives the finished Chardonnay more richness and body, similar to a red wine.

Fermentation

Most New World Chardonnay is fermented in stainless steel or oak barrels. In the Old World stainless steel, concrete, ceramic, or old oak tanks and oak barrels are used. The selection of the fermentation vessel affects the outcome of the finished wine. Chardonnay must in stainless steel is kept in temperature-controlled conditions to slow fermentation, and the airtight tank maintains a bright, fresh fruit character in the wine. Chardonnay fermented in concrete and ceramic loses some of the bright, fresh fruit flavors but gains more complexity because it has contact with oxygen. Oak casks are porous, so the wine has continuous contact with air (micro-oxidation), which allows it to develop even greater complexity and become more balanced. Some believe Chardonnays made in older oak tanks are the most balanced of all.

Many producers, particularly in California, but also in Australia, Burgundy, and elsewhere, ferment the highest-quality grapes in French oak barrels. Barrel fermentation, particularly in new oak, is part of a process that creates a very distinctive style of Chardonnay with powerful oak-dominant aromas and flavors. French oak is used most often, but some producers like the overt character of American oak.

Malolactic Fermentation

MLF is a technique that is widely used to make Chardonnay. The process of adding lactic acid bacteria changes aromas, flavors, and complexity by converting a crisp, high-acid wine into a softer, rounder, buttery wine. Not all Chardonnays undergo MLF, as there are some producers and winemakers who prefer to maintain the bright, crisp character of the grape. Many producers use MLF for only part of each batch of wine to maintain some of the fresh fruit character in a softer, creamier style, while others opt for 100 percent MLF to obtain a very rich, buttery character that complements the richness created by significant barrel contact.

Sur Lies Aging

Sur lies aging is another common technique used to impart flavor and complexity to Chardonnay's neutral character. The process occurs in stainless steel tanks or in oak barrels. The lees (grape sediment and dead yeast cells) are stirred by hand (called *bâtonnage* in France) or mechanically frequently to reincorporate the solids with the liquid. The breakdown of the dead yeast cells, known as **AUTOLYSIS**, develops a nutty character that is absorbed into the wine by stirring the lees.

Barrel Aging

Chardonnay has the strength to stand up to changes that occur in the barrel and is the most commonly barrel-aged white wine. Whereas less than 20 percent of Chardonnay, usually only the best grapes, is fermented in oak, more than 80 percent of Chardonnay is aged in oak or placed in contact with some type of oak alternative. Barrel contact creates a distinctive characteristic ranging from subtle to overt. Older oak or a light toast on the barrel develops milder character, showing as vanilla and toasty aromas in

Chardonnay is the most commonly barrel-aged white wine.

some Chardonnays. New oak with a heavier toast develops a very powerful style with smoky aromas. In addition because oak barrels are porous and lose as much as 8 percent of their moisture through evaporation, the wine becomes more concentrated.

Some Chardonnay producers only put their wines in new French oak barrels in an attempt to replicate a Burgundian style, which is generally preferred by American connoisseurs. Whereas Burgundian styles tend to be well integrated, some California examples can be too one-dimensional, with the oak dominating. Other producers prefer Chardonnay aged in older oak barrels because they believe that the oak flavors are less powerful and other components are more integrated, creating a more balanced wine.

Chardonnay that shows oak character and costs less than $10 has probably not been aged in an oak barrel. Low-cost wines obtain their oak character from an oak-barrel alternative, such as oak chips, oak dust, or oak staves. These can be added during fermentation or during the tank- or barrel-aging period. In 2007, France passed a law allowing for the use of oak chips in wine, which you can expect to see in the lower-priced Chardonnay produced for the export market.

Blending

Although Chardonnay is generally perceived as unblended, that is not always the case. Why is Chardonnay blended? Cost. Chardonnay is an expensive wine because it is made from the costliest white grape variety and is often fermented or aged in expensive oak barrels. Blending less expensive varieties with Chardonnay brings down the net cost of a bottle of wine. Blending can also add a missing characteristic, like acidity, fruit, or sweetness to the wine, to create a more balanced wine. In California Chenin Blanc is often used as a blending variety for less expensive Chardonnays. In Australia Chardonnay is blended with Sémillon, and in Northern Italy with Pinot Blanc or Pinot Grigio.

Each country or specific appellation within a country has its own requirements for blending Chardonnay. By law, white Burgundies are 100 percent Chardonnay. In New World countries blending is allowed. U.S. law requires a bottle labeled Chardonnay to contain at least 75 percent Chardonnay but does not require the label to indicate the other varieties in the wine. Australian law states that 85 percent of a varietally labeled wine be of that variety and the label must indicate what other varieties are in the wine.

So how are you to know which varieties are in your bottle of Chardonnay? For New World wines, use price as a broad guideline as well as the information provided on the label. A low-priced bottle of California Chardonnay is almost guaranteed to be closer to the 75 percent minimum required than the possible 100 percent. On the other hand, an expensive bottle of Chardonnay will most likely be close to, and may even be, 100 percent Chardonnay. Also, several producers identify the varieties that they have blended into the wine or state the percentage of Chardonnay in the bottle on the back label.

CHARDONNAY: THE WINE

Color

The color of Chardonnay is largely dependent on the fermentation and aging techniques used and on the age of the bottled wine. If Chardonnay is fermented and aged in stainless steel, the color will be pale yellow. If the wine has contact with wood, it begins to turn deeper yellow. With longer barrel contact, the color becomes almost golden. As Chardonnay ages, oxidation also contributes to turning the wine a deep golden color.

Aromas and Flavors

The aromas of Chardonnay wine can be quite different from the aromas of the Chardonnay grape. Chardonnay grapes have aromas of citrus (limes, grapefruit, lemons) and fresh tree fruits (apples, pears). The wine may have aromas similar to the grape, but can also have melon, floral, buttery, vanilla, nutty, toasty, and smoky aromas. Table 7.3 lists aromas that might be found in Chardonnay.

The aromas and flavors of the grape can evolve dramatically due to the winemaker's influence and the amount of time the wine ages in the barrel. MLF imparts buttery or butterscotch characteristics and sur lies aging gives Chardonnay a nutty flavor. Oak probably makes the greatest contribution to the aromas and flavors of Chardonnay. New oak imparts a robust vanilla characteristic and heavily toasted oak gives smokiness to the wine. Fermenting Chardonnay in older oak barrels and larger oak casks diminishes the oak aromas and flavors, which develops a subtler, more complex style of wine.

Table 7.3 Chardonnay Aromas

FROM THE GRAPE	FROM WINEMAKING	FROM BOTTLE AGING
Apples	Baked bread	Figs
Apricots	Burnt caramel	Honey
Floral	Butterscotch	Prunes
Gooseberries	Buttery	Tobacco
Grapefruit	Charred	
Lemons	Cheesy	
Limes	Coconuts	
Melons	Creamy	
Peaches	Leesy	
Pears	Nutty	
Pineapples	Smoky	
Straw	Toasty	
Tropical fruits	Vanilla	

Wine Styles

Chardonnay falls into three general styles—light, integrated, and bold and powerful—that are made using traditional or modern winemaking methods (see Table 7.4). Light-bodied Chardonnay emphasizes fruit freshness and grape flavors. The grapes are typically harvested at fairly low sugar levels of 22 to 23 percent, and the wine is generally not put through MLF, or at most partial MLF to maintain its fresh acidity. It is fermented in an inert vessel, frequently stainless steel. This style is most prominent among European Chardonnays, but there is a rekindling of interest in this style in other appellations throughout the world, even in California. Because it lacks barrel contact the cost of this wine to the consumer is generally moderate.

The integrated style is most reflective of a Burgundian Chardonnay. Grapes that are a bit riper are preferred for this style, and all of the techniques that can be used in making Chardonnay are employed. Sur lies aging is almost always

Table 7.4 Chardonnay Wine Styles

	LIGHT	INTEGRATED	BOLD AND POWERFUL
WINEMAKING STYLE	Traditional	Traditional	Modern
WINEMAKING METHODS	Low harvest sugar No oak No MLF No or some sur lies aging	Low to moderate harvest sugar Partial new and/or old oak Partial MLF Sur lies aging	Moderate to high harvest sugar New oak fermentation New oak aging Full MLF Sur lies aging
WINE CHARACTERISTICS	Emphasis on grape Bright, crisp, high acid Moderate alcohol	Balance of grape and winemaking characteristics Moderate alcohol	Ripe fruit Extensive barrel and MLF aromas High alcohol
COST	$-$$	$$-$$$$	$$$-$$$$

used. MLF can be partial or full, and barrel fermentation and barrel aging combine new and older oak barrels. Following traditional methods, larger, older oak casks are used, adding to the subtlety of the wine. Burgundian wines from the best vineyards are still the most expensive Chardonnay wines in the world, but the style is also copied around the world, and prices for the wine can be quite variable depending on the country, appellation, fruit quality, and reputation of the producer.

The bold and powerful style is best exemplified by Chardonnays made in prominent New World appellations, like Napa Valley and Sonoma County. Grapes are riper because they come from warmer climates, and are sometimes allowed extra hang time before harvesting. Great care is taken to select only the best fruit, which is then typically barrel fermented, receives 100 percent MLF, and is aged primarily, if not entirely, in new oak barrels, which may even have a moderately heavy toast. The result is wines that are rich and high in alcohol, with a strong oak character. Because of these factors, bold and powerful Chardonnays are usually quite expensive.

ENJOYING CHARDONNAY

Buying Chardonnay

Now that you know something about Chardonnay, how do you put your knowledge to work to enjoy or serve it? Price and label can provide you with much information about a wine. What is the difference between a less than $10 bottle and a $30 or more bottle of California Chardonnay? The inexpensive bottle most likely comes from a broader appellation, such as California, and was made from less expensive grapes. The wine likely had only partial MLF and no oak barrel contact or contact only with oak chips, and will be a lighter yellow color and lower in alcohol. The higher-priced wine is probably from a more distinct appellation, was made from more expensive grapes, has undergone more extensive MLF, and has been aged in oak barrels, probably new. The wine will likely have a more golden color and higher alcohol content and will be full-bodied.

WHAT IS A BURGUNDIAN-STYLE CHARDONNAY?

People in the trade—winemakers, retailers, and sommeliers—speak of Chardonnay made in the Burgundian style, and most agree that the best Chardonnays still come from Burgundy. The reference refers to wines that are made from the great Grand Cru and Premier Cru vineyards of the Côte de Beaune, headed by Montrachet, Meursault, and Corton-Charlemagne. Within this appellation each producer or winemaker follows a distinct methodology to make Chardonnay. These "best" Chardonnays are fermented and aged in oak, a percentage of the wine in new oak barrels, and the remainder in larger oak casks. Partial MLF is typically used as is aging and bâtonnage on the finer lees. These techniques allow the fruit to show through, and they are best with at least two years of aging. Balance is the key to a fine Burgundian Chardonnay.

Examples of regional Chardonnay, each pro-
duced in a different style, from left: Chablis,
France; Australia; Côte de Beaune, France;
Sonoma County, California

TOM ZASADZINSKI/CAL POLY POMONA

If you see the term *Reserve* on the label of an expensive Chardonnay, the wine is likely to have been made from high-quality fruit and heavily manipulated, and to have undergone new oak contact and full MLF, making it riper and higher in alcohol.

If the alcohol level is 13.5 percent or lower, the wine will generally have good acidity with fresh fruit character. This indicates moderate or slight MLF, barrel aging, or both, and the wine will match well with a range of foods. High-alcohol Chardonnay, at 14 percent and above, usually has ripe fruit flavors and aromas, softness, and a strong influence of MLF and oak with broad and robust flavors.

The more specific the appellation, whether in Burgundy, California, Australia, or elsewhere, the better the wine is likely to be. As you drink Chardonnays from different regions, you will become familiar with the unique style of each appellation.

Storing and Serving Chardonnay

Most American Chardonnays should be consumed within two to three years of the vintage date. The best-quality Burgundies can be stored for a decade or more. The standard temperature for serving white wine is 50°F/13°C. However, if you are opening a bottle of full-bodied Chardonnay, you might want to serve it slightly warmer, at around 55°F/15°C, to allow the aromas to stand out.

Chardonnay is best served in a Burgundy-style glass with a fairly large bowl that is moderately open at the top. Because Chardonnay is considered a non-aromatic variety, the larger bowl adds oxygen to the wine and allows the aromas to develop when the wine is swirled.

Chardonnay and Food Pairings

To pair Chardonnay with food, first look at the cuisine of the country and appellation where your wine was made. What do Burgundians, Australians, or Italians eat when they serve Chardonnay? A region's cuisine complements the wine produced in that region. Second, match a component of the wine with the same component in the food, like a buttery Chardonnay with a buttery beurre blanc sauce. Third, consider the Chardonnay's style.

Simple preparations, like roast chicken or a sauté of chicken, fish, or veal with fresh herbs, go well with lighter Chardonnays because they are moderate to low in fat. Chablis, because of its high acidity, goes well with shellfish, particularly oysters, clams, and mussels. Lighter-style Chardonnays work well with egg dishes, like quiches and omelets, vegetable dishes from white mushrooms to zucchini to tomatoes, and rice and lighter pasta dishes, particularly risottos from Northern Italy. Light cream sauces also pair with light Chardonnays. Avoid serving light Chardonnay with sweet foods, which will make the wine seem very sour, and with rich foods that might overpower the light character of the wine.

Integrated Chardonnays have increased richness and a buttery character, making them a better pairing with richer fish and seafood. A poached chicken breast or fillet of sole would be overpowered by this style of wine, but richer fish, like turbot and halibut, and seafood, such as scallops, lobster, or shrimp, should pair well. This style of wine also complements fish and white meats served in a sauce. Other foods and dishes that go with integrated Chardonnays are cream soups and a wide range of butter sauces. Something as simple as macaroni and cheese can be terrific with an integrated Chardonnay, and any dish with nuts is also a good choice.

Powerful, oak-laden wines are best by themselves. This style of Chardonnay is truly wonderful when you smell and sip, focusing on the distinct character of the wine with nothing else to get in the way. If you do want to match a powerful Chardonnay with food, look for rich, powerful foods that can stand up to the oak and full MLF. This might be the perfect style to serve with charcoal-grilled fish, such as tuna or swordfish, or even a well-marbled and juicy steak. Smoked dishes also work.

SUMMARY

From bright, fresh, and clean to complex and rich, Chardonnay is a balanced wine with a range of attributes that make it the most sought-after white wine. Opening a bottle of Chardonnay from a different country or region each day or week is a great way to explore different variations of the same varietal. Within each appellation, a wine from one producer to the next can also be quite distinct. The difference between a white Burgundy and a Chardonnay from Friuli is astounding, and both are so dissimilar from an Australian Chardonnay that you may think you are tasting three very different varietals. When pairing each style of Chardonnay with food, think of the foods that are typically eaten in the area where the grapes are grown. You will have the opportunity to learn not only about wine but also about the diverse cuisines and cultures of these regions.

TOP 10 FOODS TO PAIR WITH CHARDONNAY

1 Oysters, clams, or mussels with Chablis
2 Roast chicken with a light, unoaked Chardonnay
3 Braised veal or pork in a cream sauce with an integrated Chardonnay
4 Risotto alla Milanese or primavera with an unoaked Chardonnay from Northern Italy
5 Rich seafood (lobster, shrimp, scallops) with an integrated Chardonnay
6 Pasta with ham, spinach, and cream with a light or integrated Chardonnay
7 Grilled marinated pork chops or grilled tuna with a barrel-aged Chardonnay
8 Brie with a rich, barrel-aged Chardonnay
9 Smoked fish (salmon, tuna, sturgeon) with an oaky Chardonnay
10 Roasted mixed nuts with a barrel-aged Chardonnay

Sauvignon Blanc

Sauvignon Blanc has never achieved the popularity of Chardonnay; nevertheless, it is highly regarded throughout the world for its penetrating crispness, clean flavors, and ability to pair so well with a wide variety of foods. It is an uncomplicated varietal and in a blind tasting its high acidity level makes it the easiest varietal to pick out. New wine drinkers often do not like Sauvignon Blanc because its acidity makes it seem tart and sour. It is unfortunate that most people try wine for the first time without food, particularly in the case of Sauvignon Blanc. If Sauvignon Blanc is sampled with almost any food, the perception of it changes dramatically. Until you have developed a palate for Sauvignon Blanc on its own, it is a wine that cries out for food. For those who have acquired a taste for this varietal, it also makes a refreshing aperitif and has the added benefit of being low in alcohol, generally below 13.5 percent.

REGIONS AND APPELLATIONS

Sauvignon Blanc is grown worldwide, with total plantings of approximately 125,000 acres/50,000 hectares. Approximately 50,000 acres/20,000 hectares of that is planted in France, followed by New Zealand with 22,500 acres/9,000 hectares in production, and the United States with a bit more than 15,000 acres/6,000 hectares in production. Australia, Chile, Argentina, South Africa, Italy, Spain, Slovenia, and Austria also produce wines for the international marketplace. Central and Eastern European countries have extensive plantings of Sauvignon Blanc, but the wines are not yet receiving much international recognition. (See Appendix 2 and Appendix 3 for maps of Sauvignon Blanc–producing countries and appellations, pages 380–447.)

APPELLATION CONFUSION

As you learn about French wines, appellation names can be quite confusing because they often sound similar. Two examples: Pouilly-Fumé is 100 percent Sauvignon Blanc and comes from the Loire; whereas, Pouilly-Fuissé is 100 percent Chardonnay and comes from the Mâcon in Burgundy. The appellation Pouilly-Fumé is adjacent to the appellation Pouilly-sur-Loire. Pouilly-sur-Loire, the wine, is made from the Chasselas grape and is inferior to the wines of Pouilly-Fumé.

France

Sauvignon Blanc originated in the Loire and Bordeaux regions of France. Within the Loire the two most renowned appellations are Sancerre and Pouilly-Fumé. Other significant appellations in the northeastern part of the Loire, also known as the Upper Loire, are Menetou-Salon, Quincy, and Reuilly. The most distinguishing factor about wines produced in these appellations is that, like Chardonnay in Burgundy, Sauvignon Blanc by law must be 100 percent of the variety. In Touraine and Anjou, appellations in the central part of the Loire, blending of Sauvignon Blanc with Chenin Blanc is allowed.

The white wines of Bordeaux have always played second fiddle to the red wines of Bordeaux; nevertheless, Bordeaux's Sauvignon Blanc comes with a great pedigree. Sauvignon Blanc in Bordeaux is produced in a very different style from the Sancerre and Pouilly-Fumé wines of the Loire. Unlike the 100 percent varietal wines of the Loire, in Bordeaux Sauvignon Blanc is blended with Sémillon and perhaps a touch of Muscadelle, an indigenous variety not grown much outside Western France. Winemakers also use barrel fermentation and aging to create the characteristic Bordeaux white wine. White wines of Bordeaux are classified under four appellations:

Bordeaux Bordeaux Blanc is the generic white wine produced in the general Bordeaux appellation. The general white wines of Bordeaux can also be labeled Bordeaux Supérieur.

Pessac-Léognan Based on expert evaluation and selling price, the white wines of Pessac-Léognan, actually a subregion of Graves, are considered the best. This very small appellation with relatively small production produces classic blends that include Sauvignon Blanc and Sémillon.

Graves These wines follow closely in quality and style the wines of Pessac-Léognan and have the same characteristics.

Entre-Deux-Mers A large, centrally located appellation that produces a broad range of Sauvignon Blanc styles, including wines like those from Pessac-Léognan and Graves, as well as wines that contain a higher ratio of Sauvignon Blanc, or are even 100 percent Sauvignon Blanc, which have no oak contact.

Other appellations that produce Sauvignon Blanc include the South-West, Languedoc, and Provence. Even Burgundy and the Rhône produce small amounts of Sauvignon Blanc, but these are generally not found outside of the region where they are produced.

Bordeaux Label

New Zealand

The most talked-about Sauvignon Blancs today are those from New Zealand. For years New Zealand had no wine pedigree, then suddenly in the 1980s New Zealand Sauvignon Blanc captivated the world with its clean, grassy, citrus, unoaked style. New Zealand Sauvignon Blanc is now considered a benchmark style along with French Sauvignon Blanc, and is considered a model of high-tech viniculture. The high-acid grapes are grown in a cool climate and rapidly machine harvested, then fermented in temperature-controlled tanks with minimal

A newly planted Sauvignon Blanc vineyard in Marlborough, New Zealand

oxidation. In the northern section of the South Island the largest and most prominent plantings are in Marlborough, with smaller plantings found in Nelson and Canterbury. On the North Island the Martinborough area, at the southern end, creates wine styles that resemble the cool-climate wines of the South Island. In the central and northern regions, the Sauvignon Blanc produced in Hawkes Bay, Gisborne, and Auckland has a riper, richer style. (It is important to remember that in New Zealand, which is in the Southern Hemisphere, the farther north you go the warmer the climate becomes.)

The United States and Canada

In North America, California produces a range of styles echoing those of the Loire, Bordeaux, and New Zealand as well as a style that is more like Chardonnay than Sauvignon Blanc because it receives full malolactic fermentation (MLF) and is fermented and/or aged in new oak barrels. Sauvignon Blanc production is highly dispersed throughout the northern, central, and coastal regions of the state. The highest-producing counties in California, as recorded by the *California Grape Harvest Report*, are Sonoma County followed by San Joaquin (Central), Napa, Lake, and Sacramento Valleys. The wines of Sonoma, Napa, and Lake all produce a variety of Sauvignon Blanc styles that are typically identified by their regional or local appellation. In Sonoma County, Dry Creek Valley makes some of the better cool-climate examples, while Alexander Valley yields Sauvignon Blancs from a warmer climate. The Central Coast region, including cool areas as far south as Santa Barbara County, produces some exceptional Sauvignon Blancs as well. The newest appellation to appear on the Sauvignon Blanc wine scene is Lake County, which sits north of Napa County and east of Mendocino County and Northern Sonoma County.

In North America, well-regarded Sauvignon Blanc is also produced with notable plantings in the states of Washington, Texas (Lubbock and the high plains), and New York (Long Island) as well as in the Okanagan Valley in British Columbia.

Other Countries

Italy, Austria, Slovenia, and Spain in Western Europe are generating some enticing Sauvignon Blancs. In Italy the northeastern regions of Friuli and Trentino-Alto Adige are producing world-class Sauvignon Blancs. Within Friuli, the appellations of the Collio and Colli Orientali del Friuli are making wines that are either variety-specific or blended. The Collio is adjacent to the wine-producing region of the Brda hills of Slovenia. The wines from this area of Italy and Slovenia share the same characteristics, and it is not unusual for vineyard owners, Italian and Slovenian alike, to maintain vineyards in both countries. During harvest truckloads of grapes move back and forth over the border, headed from vineyards in one country to producers' wineries in the other. In Alto Adige the wines are also variety-specific or blended with other cool-climate grapes of the region.

In Spain the most prominent Sauvignon Blanc appellation is in the north-central region of Rueda in Castile and Leon. Here it is commonly blended with the indigenous variety, Verdejo, which makes a light, fruity wine. Austria, better known for Grüner-Veltliner and Riesling, also produces some excellent clean, fruity, and dry varietal Sauvignon Blancs. The best and largest quantities come

Table 8.1 Prominent Sauvignon Blanc Appellations

COUNTRY	REGION/APPELLATION	SUBAPPELLATION	VILLAGE OR SUBAPPELLATION
France	Loire	Sancerre Pouilly-Fumé	
	Bordeaux	Graves Pessac-Léognan Entre-Deux-Mers Sauternes (sweet)	
New Zealand	North Island	Martinborough Hawkes Bay	
	South Island	Marlborough Nelson	
U.S.	California	Sonoma County	Dry Creek Valley Alexander Valley
		Mendocino County	Anderson Valley
		Lake County	Clear Lake
		Napa Valley Santa Barbara County	
	Washington State	Columbia Valley	
	New York	Long Island	
	Virginia		
Australia	South Australia	Adelaide Hills Clare Valley	
	New South Wales	Hunter Valley	
	Victoria	Yarra Valley	
Italy	Friuli	Collio Colli Orientali del Friuli	
	Alto Adige		
Slovenia		Brda	
South Africa		Stellenbosch Paarl	
Spain	Castile and Leon	Rueda	
Chile	Aconcagua	Casablanca	
	Central Valley	Curicó Valley	
Austria		Styria	

from the appellation of Styria in Southeastern Austria, adjacent to Slovenia. All of these European countries are producing world-class wines that are often hard to find in the marketplace, but it is worth the effort required to hunt them down in order to experience their unique characteristics.

South of the equator, Australia is producing Sauvignon Blancs in the cooler appellations of South Australia, like Adelaide Hills and Clare Valley, Hunter Valley in New South Wales, and Margaret River in Western Australia. The appellations of Casablanca, Curicó Valley, and Valparaiso in Chile and the Paarl and Stellenbosch in South Africa are making some tasty Sauvignon Blancs at very affordable prices. Most of the wines from these New World countries emulate New Zealand or Loire style, by being fermented in stainless steel with no oak contact.

Table 8.1 includes an abbreviated list of countries that produce highly regarded Sauvignon Blancs along with prominent appellations and subappellations.

SAUVIGNON BLANC: THE GRAPE

Sauvignon Blanc is the antithesis of Chardonnay. While Chardonnay is considered nonaromatic, Sauvignon Blanc is considered one of the most aromatic grape varieties. Chardonnay is round, broad, and luscious; whereas, Sauvignon Blanc has an aggressive laserlike crispness with a penetrating aroma and mouthfeel. Sauvignon Blanc produces a wine that in most styles is dry, crisp, and refreshing, like lemonade or limeade without the sugar.

In cool climates Sauvignon Blanc produces a traditional crisp, acidic style wine. In moderately warm climates it results in an alternate style that is riper, less acidic, and rounder with ripe fruit aromas and flavors. It is a grape that grows in almost as many places as Chardonnay, but the total amount of the global harvest is about one-eighth that of Chardonnay. A midseason-ripening grape, Sauvignon Blanc typically reaches the maturity and ripeness needed to make a wine with the desired mix of acidity, aromas, and flavors sooner than other varieties. Sauvignon Blanc vines are capable of very high yields, so it is quite easy to grow with minimal complications in the right locations. Because Sauvignon Blanc is so prolific, vineyard-management techniques are needed to control its growth.

The grapes are bright grassy green in color with conical clusters. Depending on a number of factors in the vineyard and when the grapes are harvested, the breadth of aromas of the grape ranges from a green vegetal character at one end of the spectrum to tropical fruit flavors at the other end. The grape's most distinctive component is its naturally high acidity—a trademark of the variety.

IN THE VINEYARD

While much of what is done to influence Chardonnay occurs during winemaking, most of what is done to influence Sauvignon Blanc takes place in the vineyard. Climate, soil, vineyard-management techniques, and harvest date play a dynamic role in the outcome of the wine.

Climate

The climates of the great Sauvignon Blanc appellations, where the best examples are grown, are typically cool. Sauvignon Blanc also grows well in warmer climates, but it loses some of its crisp, acidic, grassy, and citrus qualities.

The Upper Loire has a continental climate reflected by its harsh winters; long, sunny summer days; and rains that can come at the wrong time of year, late summer and early fall. Because the vineyards of the Upper Loire are so far north, the best vineyards face south to maximize potential sun exposure.

Bordeaux and Marlborough have maritime-influenced climates with cold winter temperatures, rains that typically come early in the season, and hot, dry summer and fall days. Although California Sauvignon Blanc typically grows in a warmer climate than those of France and Marlborough, the vineyards' proximity to the Pacific Ocean creates temperature differences of as much as 50°F/28°C between day and night. Cool nighttime temperatures benefit the development of the classic Sauvignon Blanc characteristics of crispness, acidity, and citrus and grassy flavors. In order to avoid California's warm valley temperatures, which would potentially result in overripe grapes, some of the better vineyards may be planted closer to the ocean or at higher elevations that can also provide benefits such as lower fertility and better drainage.

Soil

Sauvignon Blanc flourishes in soils as far ranging as the limestone of the Loire and the gravel of Bordeaux and Marlborough. But it needs well-drained soil, and therefore does not do as well in silt or pure clay because such soils hold too much moisture.

The soils in France are influenced by their respective mountain ranges. The ancient runoff from the Massif Central onto the vineyards of the Upper Loire and from the Pyrenees onto the Bordeaux vineyards has had tremendous influence on soil content. Depending on the specific site, soils in the Upper Loire can be a combination of limestone, gravel, and flint, which are considered key factors in the mineral characteristics of the wines from Sancerre and Pouilly-Fumé. The soils of Bordeaux are more gravelly with pebbles and stones intermixed with silt and light clay. These alluvial soils produce vines with more vigor and higher yields, and the wines have softer, rounder acids than the racy, bracing wines typical of the best vineyards of the Loire. Marlborough, with the largest production area of Sauvignon Blanc in New Zealand, contains a mix of different soil types ranging from stony to sandy. California and other North American appellations also plant on a great range of soil types, including gravel, clay, loam, and sand.

Water Access

Water issues for Sauvignon Blanc are similar to those of other grape varieties. In the Old World Sauvignon Blanc is typically not irrigated; whereas, in New World vineyards, which are often more arid, irrigation is permitted. Water is one key

TERROIR VS. SUN
Old World producers tie the distinctiveness of their Sauvignon Blanc to the terroir, including the climate and soil where the grapes are grown. New World producers pay greater attention to the amount of sun (Winkler-Amerine Heat Summation Scale) and use different styles of vineyard management to achieve the desired grape character and make the desired styles of wine.

factor that changes the character of a wine each vintage. Overwatered vines can overproduce, resulting in wines that are thin, watery, and lacking flavor.

Vineyard Management

Sauvignon Blanc has the potential to be a prolific producer (10 to 12 tons per acre/175 to 210 hectoliters per hectare), but at those levels the juice is watery, neutral, and thin, losing much of its varietal character of bright acid and classic flavors. At moderate levels of five to six tons per acre/80 to 105 hectoliters per hectare Sauvignon Blanc grapes develop greater concentration. To keep production at moderate levels, vineyard managers use clones, trellising styles, and canopy-management techniques.

Sauvignon Blanc cuttings, propagated in a nursery, are ready for planting.

When planting new vineyards, growers select rootstock that helps to control vigor and reduce foliage to enhance a vine's sun exposure, which improves ripening. Trellising systems are selected based on past successful outcomes to achieve the grape volume or quality desired. In France Guyot and double Guyot are prevalent. In California and other New World countries vertical shoot positioning (VSP) is most common, but double canopy systems, such as Geneva double curtain or Scott Henry, are used to control vigor. Canopy-management techniques must be aggressive to keep yields moderate and include pruning, shoot thinning, leaf pulling, and dropping fruit in a green harvest.

Finally, harvest timing is also important, as it determines ripeness and therefore has an impact on grape quality. Sauvignon Blanc harvest usually occurs at sugar levels of 21.5° to 23.5° Brix, which retains the grapes' natural acidity. If a fruitier, rounder, possibly barrel-aged wine is the goal, harvest will not take place until the sugar is as high as 25° to 26° Brix.

IN THE WINERY

The historic rule in winemaking has been that aromatic grapes, such as Sauvignon Blanc, should be treated simply; whereas, nonaromatic grapes, such as Chardonnay, should be subjected to heavier treatments, including maceration, barrel fermentation, barrel aging, MLF, sur lies aging, and blending. Following this tradition, wine experts have always said that the wine produced from Sauvignon Blanc grapes comes from the vineyard. With advances in winemaking technology and the expansion of regions where Sauvignon Blanc is grown and produced, however, there has been a corresponding increase in the range of techniques used in the winery to change the grapes' characteristics.

All of the techniques used with Chardonnay are now used to make some Sauvignon Blancs. Most Sauvignon Blancs do not undergo oak contact during either fermentation or aging, and the use of the other techniques is far more limited than with Chardonnay. However, individual winemakers and producers are deciding to break with the traditions of their appellations to create innovative styles. So it is important to be aware of the appellations where winemaking techniques are likely to be used, which techniques are used, and how they impact the wine's style.

Maceration

Maceration has become a fairly common practice in Sauvignon Blanc production. A number of appellations around the world, including those in France, macerate Sauvignon Blanc. Many California winemakers refrain from using this technique, preferring to keep the wines lighter in style. On the other hand, New Zealand producers identify the skin contact that occurs with machine harvesting and partial crushing in the field as a technique that plays a role in their distinctive style.

Maceration increases the flavors and body of the wine, adding more depth and creating fruitier flavors. However, maceration can reduce the life of a wine and can also lead to an increase in phenolics, causing bitterness and higher pH, which produces a less racy quality in the wine.

Fermentation

Most Sauvignon Blanc is fermented in an inert vessel, usually in temperature-controlled stainless steel, but in the Old World sometimes lined concrete tanks or old oak vats are used. Stainless steel maintains the focus on the classic Sauvignon Blanc characteristics because it adds no flavors and can prevent oxygen from coming in contact with the wine. Fermentation normally occurs at cooler temperatures, allowing the wine to achieve more fresh fruit flavors. Nontemperature-controlled vessels normally have warmer fermentation temperatures, sometimes achieving more mineral flavors.

Historically, barrel fermentation has seldom been used with Sauvignon Blanc outside of Bordeaux. However, individual winemakers in a few locations throughout the world, including the Loire, California, and New Zealand, have tested and used barrel fermentation. It is done more often with Sauvignon Blanc–Sémillon blends than with unblended Sauvignon Blanc because the richness of the Sémillon makes the blend a better candidate for barrel fermentation. The best vintages from top producers in Pessac-Léognan and Graves are fermented in oak, and in California some winemakers like the complexity barrel fermentation grants to the wine.

Sauvignon Blanc is typically fermented in temperature-controlled stainless steel tanks like these at Cloudy Bay Vineyards in Marlborough, New Zealand.

Malolactic Fermentation

Because of its high acidity, Sauvignon Blanc would seem a good candidate for MLF, which softens acidity. But the high-acid, crisp style of wine that comes from the Sauvignon Blanc grape is what makes it distinctive. Therefore, most Sauvignon Blancs are not subjected to MLF or receive only partial MLF. The wines from Old World producers are less likely to use MLF to any great degree, and even in the New World MLF, in most cases, is restrained to ensure that the wine retains its bright, crisp flavors.

Sur Lies Aging

Lees contact through aging and stirring the lees is a common practice in Sauvignon Blanc production, particularly for the riper styles. As with Chardonnay, Sauvignon Blanc is left in contact with the lees to build greater complexity in the wine. Less expensive wines are typically separated from the lees and racked early to move the wine to market as quickly as possible. Better quality wines may receive lees contact to develop flavors. In Pessac-Léognan and Graves lees aging is used in conjunction with barrel maturation.

Barrel Aging

When Sauvignon Blanc is barrel aged, its color deepens, rounded fruit flavors develop, and vanilla flavors are absorbed from the wood, which creates a richer wine. Although barrel aging is seldom used, some producers in Pessac-Léognan and Graves, in California, and on the North Island of New Zealand as well as a smattering of producers elsewhere, choose to incorporate barrel aging in Sauvignon Blanc production. When barrel aging is used, it typically takes place in old oak barrels and lasts only from four to 12 months, which moderates the impact of oak on the wine. In some cases, producers create wines with prolonged new-oak-aging that are reminiscent of Chardonnay rather than Sauvignon Blanc.

In Pessac-Léognan and Graves wines may be aged in oak barrels for up to 12 months, and occasionally new oak gets used with grapes from the best vineyards in the best vintages. In California, barrel aging became popular after Robert Mondavi used oak aging in his popular Fumé Blanc wine in the 1970s (see Mondavi and Sauvignon Blanc, at left), and it is now an accepted style along with nonoak-aged wines.

Blending

Most Sauvignon Blanc wine is varietal. However, it is typical in Bordeaux to blend Sauvignon Blanc with Sémillon and the local Bordeaux variety, Muscadelle. Some believe that Sémillon takes the hard edge off of Sauvignon Blanc's acidity, increases complexity and aging ability, and makes a better style of wine.

By appellation law in the Upper Loire, where Sancerre and Pouilly-Fumé are produced, wines must be 100 percent Sauvignon Blanc. In New Zealand and California Sauvignon Blanc labeled as varietal wine may in fact be blended. Most examples from New Zealand and California are made entirely of Sauvignon Blanc, but some are blended with a few percentage points of Sémillon or other varieties, which balance the acidity in the wine.

Stylized Wines

Sauvignon Blanc is often blended with Sémillon to produce an outstanding dessert wine. Sémillon's thin skin, which is susceptible to the **BOTRYTIS CINEREA** mold, and Sauvignon Blanc's high level of acidity make the two grapes a good

---

MONDAVI AND SAUVIGNON BLANC

The history of Sauvignon Blanc in the United States dates back to the 1870s when producers brought the grape to Northern California from France. In the 1930s Sauvignon Blanc was one of the first wines labeled with its grape variety name, but as late as the 1950s and 1960s the American Sauvignon produced and sold most frequently was a sweet table wine with little distinctive character. In 1968 Robert Mondavi decided to re-craft the wine to produce Sauvignon Blanc, convinced that it would become a core varietal at his new Napa Valley winery. He took the name Blanc Fumé, as Sauvignon Blanc was called in the Pouilly-Fumé appellation of the Loire, inverted it to Fumé Blanc, and launched his new wine. His risk paid off, and today both names, Sauvignon Blanc and Fumé Blanc, are used to identify the popular wine in the United States.

combination for this style of wine. The most famous wines of this style are the dessert wines of Sauternes and Barsacs from Bordeaux. Sauvignon Blanc can also make a wonderful sweet dessert wine when unblended. (Dessert style wines will be discussed in greater depth in Chapter 18, Dessert Wines, page 292.)

SAUVIGNON BLANC:
THE WINE

Color

The color of Sauvignon Blanc normally ranges from almost colorless to pale yellow, and some Sauvignon Blancs may even have a greenish tinge. Put Sauvignon Blanc in touch with oak, however, and the wine becomes darker yellow, deepening as the wine ages.

Aromas and Flavors

Most Sauvignon Blancs are youthful, crisp, acidic wines with aromas and flavors ranging from green, grassy, and herbaceous to mineral and citrus, to melon and tropical fruits. Sauvignon Blancs that are most different from the norm are those that have been oak aged and/or blended with Sémillon. The aromas of Sauvignon Blanc come from the grape and are influenced by vineyard-management and winemaking techniques. Soils impart flavors, most commonly minerality, to the wine. Table 8.2 lists aromas that might be found in Sauvignon Blanc.

Vegetal aromas of green peppers (capsicum) and asparagus come from methoxypyrazine, a chemical compound found in Sauvignon Blanc grapes that develops when they are grown in cool climates like the Loire or New Zealand. When the grapes in the vineyard have a heavy canopy, they possess more vegetal and herbal aromas, but if the trellising system that was selected gives the grapes more sunlight and the canopy is thinned, the grapes will have a much fruitier character. The longer the grapes are allowed to hang on the vine for additional ripening, the more the aromas of melons, peaches, and figs develop. If the winemaker uses low-temperature fermentation, as they do in many New World regions, the wine acquires a more tropical fruit character. The distinctive soils of the Upper Loire showcase the soils' mineral content if the grapes are

Table 8.2 Sauvignon Blanc Aromas

STYLE	AROMAS
Lean unoaked	Apples
	Asparagus
	Cat pee
	Citrus
	Cut grass
	Gooseberries
	Grapefruit
	Green olives
	Green peppers (Capsicum)
	Kiwis
	Lemons (or zest)
	Limes (or zest)
	Minerals
	Musk
Ripe fruit	Figs
	Melons
	Nectarines
	Peaches
	White peaches
Low-temperature fermentation	Bananas
	Figs
	Guavas
	Passionfruits
	Pineapples
	Tropical fruits
Malolactic fermentation	Butter
Oak	Smoke
	Vanilla
Botrytis	Almonds
	Apricots
	Honey
	Oranges
	Pineapples

Table 8.3 Sauvignon Blanc Wine Styles

	LOIRE	BORDEAUX	NEW ZEALAND	CALIFORNIA OAK
TYPICAL VARIETAL BLEND	100% Sauvignon Blanc	Blended with Sémillon and Muscadelle	100% Sauvignon Blanc or blended with a touch of Sémillon	Mostly Sauvignon Blanc with Sémillon
ACID	High acid	Softer acids	High acid	Round, softer acids
ALCOHOL	11%–13%	12%–13.5%	12%–13.5%	12.5%–14.5%
AGING	No oak	Oak	No oak	Oak
AROMAS/FLAVORS	Minerals, citrus, grassy	Earthy, nutty	Melons, tropical fruits, green peppers (capsicum)	Buttery, toasty, smoky
BODY	Light	Rich, viscous	Medium	Round, full
COLOR	Pale	Yellow	Light straw	Honey

fermented at moderate temperatures. When MLF and oak play a role in the winemaking, Sauvignon Blanc, like Chardonnay, develops the aromas of butter, vanilla, and smoke.

"Cat pee on a gooseberry bush" has been used as a descriptor for a New Zealand Sauvignon Blanc, and indeed cat pee and gooseberries are two of the more unusual aromas found in Sauvignon Blanc. Gooseberries, a small green berry, are a part of the same fruit family as currants. They are little grown in the United States but are plentiful in New Zealand and Australia. When picked unripe, they have a green, herbal character, and when picked ripe, they have a spicy, musky aroma. Cat pee, cat box, or *pee pee du chat*, as the French call it, is an aroma that many find characteristic of Sauvignon Blanc. Some find this aroma pleasing, while others find it offensive.

Wine Styles

Traditionally, each producing country and its appellations had its own unique styles. However, because of globalization and easier access to new winemaking techniques, today producers all over the world are experimenting with style variations not typical of their appellations. Oak aging and fermentation have been tried in the Loire and New Zealand, for example, and blending has been employed in every appellation, except those in which it is illegal. So in order to understand Sauvignon Blanc wines, it is essential to know not only the style of a particular appellation, but also the styles made by various individual producers (see Table 8.3).

Sauvignon Blanc from the Loire, or Loire style, is very crisp with bright flavors of grass and minerals; whereas a wine from Bordeaux, or Graves style, is likely to be softer and rounder because it has been blended with Sémillon and oak aged. Sauvignon Blanc from the South Island of New Zealand has fresh citrus, melon, and herb flavors and is not oak aged, while wine from the North Island has riper fruit character and more likely has been oak aged.

One Producer, Four Styles of Sauvignon Blanc

Fred Brander in Santa Barbara County, California, crafts a number of different Sauvignon Blancs that reflect wine styles from around the world.

Vintage Sauvignon Blanc, Santa Ynez Valley

This New World style wine is made from 95 to 100 percent Sauvignon Blanc that is cold fermented in stainless steel tanks and blended with a small percentage of Sauvignon Blanc that has been barrel fermented. It does not undergo MLF. The result is a crisp, fruity wine that is ready for immediate drinking.

Cuvée Nicolas

This Graves, or Pessac-Léognan, style wine is a blend of approximately 75 percent Sauvignon Blanc and 25 percent Sémillon. The grapes are cold soaked for 24 hours before fermentation, then barrel fermented, and barrel aged with lees contact for two to three months. This is a more full-bodied wine with medium acidity.

Au Naturel

This is a Loire style wine with 100 percent Sauvignon Blanc that receives no MLF and has no oak contact. The grapes are cold soaked for 24 hours and then fermented in stainless steel tanks at slightly warmer temperatures than the Santa Ynez Valley wine. This creates a wine with distinct mineral and lime aromas.

Cuvée Natalie

This is a nontraditional style developed by the winemaker that is a blend of Sauvignon Blanc, Riesling, Pinot Gris, and Pinot Blanc. It is fermented in stainless steel, does not undergo MLF, has no oak contact, and is bottled early, in late winter. The additional varieties give it floral aromas and add a touch of sweetness.

Sauvignon Blancs from California are often more difficult to characterize. Some are like fresh, crisp fruit in a bottle. Others, made in a Loire or New Zealand style from 100 percent Sauvignon Blanc, are not oak aged and have more depth of flavor with herbs over fruit. Others are made in a more integrated Graves style, using Sémillon for blending and then oak aging the wine. Finally, there are a few new barrel-fermented and barrel-aged wines that are very similar to Chardonnays.

ENJOYING
SAUVIGNON BLANC

Unoaked Sauvignon Blanc is a straightforward, youthful wine that has strong supporters and equally strong detractors. Its supporters love its crisp freshness and herbal and fruit aromas as well as its refreshing palate-cleansing ability when matched with a wide range of foods. Detractors find it too sour to be pleasant, or one-dimensional and uninteresting.

With a few exceptions, Sauvignon Blanc is a young, fresh wine that should be consumed within about two years of the vintage. Wines that will be held for a period of time are usually sealed with cork to allow air exchange while they age. Because Sauvignon Blanc does not require aging, it is a good candidate for a screw-cap closure to avoid the risk of trichloroanisole TCA. New Zealand producers led the conversion from cork-finished Sauvignon Blanc wines to those with screw-cap closures.

Sauvignon Blanc is a tart and bracing wine by itself but goes well with many foods. Your first experience sampling Sauvignon Blanc should be as an accompaniment to food. A creamy American goat cheese would be a good complement to your first Sauvignon Blanc.

Buying Sauvignon Blanc

Sauvignon Blanc is a great value wine. Many wonderful brands cost less than $15 a bottle and, with a few exceptions, you can buy the best Sauvignon Blancs for $30 or less in a retail wine shop. Your first decision when you are buying Sauvignon Blanc is to determine what style you want to try, and knowing where the wine is produced will give you a good idea of its style.

Selling price and alcohol level are also good style indicators. Lower-priced wines are likely to be straightforward Sauvignon Blancs that have been fermented in stainless steel with bright, fresh fruit or herbal aromas and flavors. As price increases, it indicates that the wines may be coming from a distinctive appellation, like Sancerre or Pouilly-Fumé, or there has probably been some blending with Sémillon as well as oak aging. Higher alcohol levels indicate riper grapes with flavors of melons, peaches, and figs, and high-alcohol Sauvignon Blancs are more likely to have been oak aged.

Examples of regional Sauvignon Blanc, each produced in a different style, from left: Santa Barbara, California; Marlborough, New Zealand; Loire, France; Sonoma County, California

Storing and Serving Sauvignon Blanc

Sauvignon Blanc is not a wine that is normally cellared and most should be drunk within the first two years of release. Sauvignon Blancs that have a significant amount of Sémillon in the blend and have been oak aged can be kept longer. The best example of wines that can be cellared for five to six years, and in some cases 10 years or more, are those from Pessac-Léognan and Graves.

In order to showcase the light, bright freshness of Sauvignon Blanc, you should serve it well chilled. Cooler temperatures, 45°F to 50°F/7°C to 10°C, heighten the acidity in the wine. If the wine is riper, rounder, and barrel aged, serve it at 55°F to 60°F/13°C to 16°C to draw out the more complex character of the wine.

Sauvignon Blanc does not require the same large-bowl glasses as Chardonnay. Because Sauvignon Blanc is aromatic, even a small glass with a closed rim works well.

Sauvignon Blanc and Food Pairings

Sauvignon Blanc's crisp acidity makes it a great food-pairing wine. Traditional styles of Sauvignon Blanc are excellent accompaniments to classic French and other European dishes, and also complement modern and other non-European ethnic cuisines, but oak-driven Sauvignon Blancs are best with the same foods that pair with Chardonnay.

Two foods—chèvre and oysters—go particularly well with traditional Sauvignon Blanc styles. Chèvre, a French goat cheese, is a typical food of the Loire. What makes chèvre so good with Sauvignon Blanc? The goat's milk is highly acidic; therefore, its cheese is also highly acidic but creamy. Sauvignon Blanc plays off the creaminess and tartness of the cheese, making an incredibly enjoyable taste experience. Chèvre works well by itself, in salads, omelets, or pastas, or used as a stuffing in poultry dishes.

Oysters, clams, or mussels from France, New Zealand, the coasts of the United States, or any other coastal area that produces Sauvignon Blanc pair well with the local Sauvignon Blanc. They can be served fresh on the half shell (sliders) or cooked in a variety of preparations, such as Oysters Rockefeller, linguini with clams, or grilled mussels.

Most things vegetal and herbal go with Sauvignon Blanc, so it is an excellent wine with many vegetarian dishes. The wine matches with simple steamed or sautéed vegetables, pasta or cheese, or soups. Tomatoes served fresh or made into a sauce also work well.

Sauvignon Blanc is an excellent accompaniment to most fish and seafood, as well as white-fleshed meats, like poultry, pork, or veal. Successful preparation styles range from simple steamed, poached, sautéed, or roasted dishes to complex dishes made with moderately rich cream and butter sauces or even hearty fish stews. Sauvignon Blanc also complements moderately spicy Asian, Latin, and Mediterranean dishes. You will find that Sauvignon Blanc makes almost every food it is paired with taste better.

What does not go with Sauvignon Blanc? Anything that is sweet: some sweet and spicy Asian dishes, foods that are too piquant, and red meat. If you are serving an oak-driven Sauvignon Blanc, you might want to consider some of the pairings listed for Chardonnay on page 127.

TOP 10 FOODS TO PAIR WITH SAUVIGNON BLANC

1 Chèvre (goat cheese)
2 Fresh oysters (sliders)
3 Sushi
4 Sautéed scallops in a lime butter sauce (beurre blanc)
5 Fresh tomatoes with mozzarella and basil
6 Almost any type of salad: chicken, seafood, vegetarian, green
7 Eggplant Parmesan
8 Roast pork
9 Moderately spicy Mexican food (not too hot)
10 Fried chicken

SUMMARY

Of all of the major wines, Sauvignon Blanc is perhaps the easiest to understand—the grapes are in the glass. There is not much interference from the vineyard or the winery. Growing and winemaking techniques are reasonably straightforward and produce a range of interesting styles. There are no hidden nuances to Sauvignon Blanc; it is all right in front of you. No other wine is as refreshing or cleans your palate so well. Sauvignon Blanc is ideal with food, but its lively, fresh components make it an excellent wine for almost any occasion.

Riesling

Riesling is perhaps the most misunderstood varietal wine produced, particularly in the United States. Many Americans believe that it is a sweet, innocuous wine that only unsophisticated consumers drink. In fact if you talk to some of the world's leading wine experts, you find that their favorite varietal is not Cabernet Sauvignon or Pinot Noir or Chardonnay but Riesling. Why is this so? It is an aromatic grape variety with beautiful and diverse aromas. The wine results in layers of aromas and flavors, racy acidity, and a range of tastes from bone dry to unctuously sweet, and best of all, it is a wine that goes with a broad range of foods.

Along the Mosel River in Germany, Riesling is planted on some the world's steepest slopes.

Riesling originated from an indigenous wild vine in Germany and was first produced as a wine sometime before 1500. In the Old World it largely remains a variety grown close to its German home, extending only to Alsace in France, Austria, and Northern Italy. The climates in Southern France, Spain, Central and Southern Italy, and other Mediterranean countries are generally too warm to produce classic Riesling.

Riesling's importance and popularity peaked a little over 100 years ago when it commanded the highest price of any varietal, including Cabernet Sauvignon and Chardonnay. However, because of wars, difficult growing conditions, and the desire for quantity over quality, Riesling lost much of its luster. Over the last 20 years, Riesling has been making a comeback with a dramatic increase in the quality of the wine from a number of regions around the world. In 2009 it was the fastest-growing varietal wine by sales volume in the United States, according to A.C. Nielsen, exceeding the pace of growth in Chardonnay and Pinot Noir.

REGIONS AND APPELLATIONS

From its origins in Germany, Riesling spread to other German-speaking regions and countries, most notably Alsace, Austria, and Northern Italy. It then found its way to Central and Eastern Europe as well as Australia, North America, and South Africa, most likely carried by German immigrants. More recently it has been planted in South America and New Zealand. (See Appendix 2 and Appendix 3 for maps of Riesling-producing countries and appellations, pages 380–447).

Germany

German Riesling is the benchmark against which all other Rieslings are compared. Germany produces the most Riesling worldwide and has the most diverse styles. Germany's wine regions consist of 13 appellations concentrated in the southwestern part of the country. Riesling is the most widely cultivated variety in Germany, accounting for almost 25 percent of the planted vineyard area, and is the only German varietal sold widely throughout the world. German Riesling sold internationally is generally classified in the top two quality levels, Qualitätswein (QbA) and Prädikatswein (formerly QmP). (See Appendix 2, German Wine Classifications, page 400.) The most highly praised Riesling appellations, which regularly produce the highest classifications of Riesling, are the Mosel-Saar-Ruwer, Rheingau, and the Pfalz with the premier vineyards planted on the steep southern slopes facing the Rhine, Mosel, Saar, and Ruwer rivers. Mechanical harvesting is never used in the vineyards of the Mosel and Rheingau: The slopes are too steep, ranging from 25 to 65 degrees—as steep as any expert ski slope in the world.

Other notable Riesling appellations are the Rheinhessen and Nahe, which have noticeably flatter topography but more diverse soils and microclimates than the Mosel and Rheingau regions. Depending on the objective of the pro-

Trimbach winery and vineyards in France produce an Alsatian style of Riesling.

ducer, wines usually are made for lower classifications but higher-classification wines are produced as well. Rheinhessen is most well known for being the largest producer of the lightly regarded, but widely sold, white blend, Liebfraumilch, which only has a very small amount of Riesling in it.

Alsace

Alsace is located in the northeast corner of France bordering the German wine region Baden. The vineyards are distributed along a thin strip of land sandwiched between the Rhine River in the east and the Vosges Mountains in the west, and between the Bas-Rhin in the north and the Haut-Rhin in the south. Riesling is planted primarily in the Haut-Rhin at cooler, higher elevations on the western side of the valley on hilly, sheltered sites climbing toward the Vosges Mountains. Most Alsatian Riesling is made in a drier, higher-alcohol, fuller-bodied style than German Riesling, but table wines may still have some residual sugar, depending on the producer.

Australia

Riesling is produced in a number of Australian regions in sizable quantities. Australia's climate ranges from moderately warm to hot, which is generally not conducive to growing Riesling. However, the Australians have found microclimates that are cool or that have large day-night temperature variations, which allow the grapes to maintain desirable levels of acidity. The style is quite consistent throughout the country, with the most highly regarded Riesling appellations in Clare, Eden, and Barossa Valleys in the broader South Australia appellation.

North America

Riesling is planted in widely dispersed appellations throughout North America. Unlike other varieties, the best Riesling is not produced in California but comes instead from lesser-known appellations. The state of Washington is the largest producer of Riesling in North America, much of it from one vintner, Chateau Ste. Michelle, a leading proponent and supporter of Riesling production and the largest producer of Riesling in the world. The vineyard sites are almost all located in the Columbia or Yakima Valleys, which have brutal winters that suit Riesling just fine. The Willamette Valley in Oregon is waking to the idea that it may have a perfect climate for producing cool-climate Riesling, as is the state of Idaho.

Some of the best Riesling-producing regions in the United States are the Finger Lakes appellation in North Central New York and the Leelanau and Old Mission Peninsulas in Northern Michigan, as well as pockets in Ohio and Virginia, and even parts of Western Colorado. Grapevines in both New York and

Michigan benefit from the lake effect. Although the lake effect works most of the time, both regions can experience severe freezes that can damage the crop or the vines themselves.

The Finger Lakes appellation has achieved tremendous notoriety for its Riesling over the last few years. Very similar to appellations in Germany, Finger Lakes producers attain only three or four great vintages out of 10, but when they are great, the wines are exceptional.

The grapevines of Canada lie in the same general latitude as the Finger Lakes and Northern Michigan. They have the same harsh winters and also benefit from the lake effect. In Eastern Canada Riesling grapes are planted along Lakes Ontario and Erie, and in Western Canada they are grown in the Okanagan Valley in Eastern British Columbia.

California produces a few excellent Rieslings, but Riesling production is dwarfed by every other important variety, red and white. This is largely due to the climate, which is usually too warm to grow Riesling. The best Rieslings come

Table 9.1 Prominent Riesling Appellations

COUNTRY	REGION/APPELLATION	SUBAPPELLATION	VILLAGE OR SUBAPPELLATION
Germany	Mosel-Saar-Ruwer Rheingau Pfalz Nahe Rheinhessen		
France	Alsace	Haut-Rhin	
Australia	South Australia	Barossa Valley Clare Valley Eden Valley	
	Victoria Western Australia Tasmania		
Austria	Wachau Kamptal Styria		
Italy	Alto Adige Friuli		
United States	New York		Finger Lakes
	Michigan		Old Mission Peninsula Leelanau Peninsula
	Washington	Columbia Valley	Yakima Valley
	Oregon		Willamette Valley
	California	Mendocino County	Anderson Valley
		Monterey County	
Canada	Ontario	Lake Erie	
	British Columbia	Okanagan Valley	

from the coolest appellations, like the Anderson Valley, Potter Valley, and Cole Ranch in Mendocino County, or from microclimates within Napa Valley, Sonoma County, and Monterey County.

Other Countries

Austrian Riesling is produced in smaller quantities than Austria's leading white variety, Grüner-Veltliner. The Wachau is the prime production area for Riesling, but the variety is also planted in Kamptal and Styria. Small amounts of Riesling grown in the northerly Neusiedlersee appellation yield a highly prized, unctuously sweet dessert wine.

Alto Adige and Friuli appellations in Northern Italy produce Riesling in a style that is similar to Austrian Riesling.

Table 9.1 includes an abbreviated list of countries that produce highly regarded Rieslings along with their most prominent appellations and subappellations.

RIESLING: THE GRAPE

Riesling is an aromatic variety. The Riesling grape is most noted for its striking acidity followed by the depth and length of its aromas and flavors. It is a grape that is capable of producing stellar wines in a broad range from bone dry to successively higher gradations of sweetness. The grape is unique in that it can achieve high sugar levels without losing acidity.

Riesling grapes are most distinctive when grown in cool climates but produce good wine that maintains the variety's characteristics in warmer climates as well. Riesling can accept extended ripening—it can be one of the first varieties harvested, but with good sun in cool weather it continues to ripen into late October and early November. Riesling bunches are compact, making them susceptible to the mold *Botrytis cinerea*, which may not necessarily be a bad thing. Winemakers desire *Botrytis* when making dessert wine because it develops an unctuous, honeylike character in the resulting wine (see Chapter 18, Botrytized Wines, page 293).

One of the grape's most notable characteristics is its inability to tolerate new oak, which strips it of its distinctive character. Riesling may be fermented in old oak casks, but it is almost never fermented or aged in new oak barrels. Riesling is rarely blended with other varieties. The character of the grape is so strong no other grape can enhance the finished wine.

IN THE VINEYARD

Riesling develops in the vineyard; once harvested, little manipulation occurs in the winery. Old World producers believe that no other variety is impacted as significantly by the soil as Riesling, and growing conditions and vineyard-management techniques are critical components of Riesling production in the New World.

Climate

Riesling performs best in cool or even cold climates, but also does well in moderate climates in some circumstances and may have warm-climate potential with a different aroma profile if picked early. Riesling is an ideal choice for planting at the northernmost and southernmost latitudes where it is possible to grow *vinifera* grapes. Many of the world's most outstanding Rieslings grow at the edge of these acceptable latitudes. Riesling, one of the heartiest *vinifera* vines, is able to survive brutal winters because the wood at the base of the vines is very thick and resistant to frost. Bud break and flowering come late in the spring, which is beneficial in cool climates because the buds are not damaged by early frosts. In warmer climates Riesling tends to ripen early, but in cool climates, like that of Germany, it may not ripen until quite late, even into November.

Factors that contribute to quality Riesling in a cool climate include sloped vineyards that enhance sun exposure and facilitate good drainage, minimal fall rain, and a strong heat source. Grapes growing in cool climates cannot withstand heavy fall rains because it is too late in the season for the grapes to recover from excessive moisture and there is not enough heat to ripen the grapes. Sloped vineyards help alleviate this problem because they naturally prevent standing water and ensure good drainage. In the Northern Hemisphere, vines are planted on south-facing slopes to maximize afternoon sun exposure and increase the heat the vines receive, which helps to ripen the grapes. Vineyard managers further intensify the heat from the sunlight that hits sloped vineyards by using canopy-management techniques and unusual practices, like overlaying slate on the ground around the vines. The lake effect warms cool-climate grapes in vineyards located next to a lake or river.

Riesling grown in warmer climates is usually planted on alluvial soils requiring irrigation, implementation of vineyard-management techniques, and an earlier harvest to protect acidity levels. Vineyard sites in warmer regions do well in areas with large temperature variations between day and night that allows the grapes to maintain good acidity levels.

Riesling grapes are acidic and aromatic.

Soil

Riesling is noted for reflecting the composition of the soils where it is planted. Riesling adapts to most soil types, but the most famous Riesling vineyards grow in the blue and red slate soils of the Mosel-Saar-Ruwer. The steep slopes are composed of old, low-fertility soils covered with slate that maintains heat, affords terrific drainage, and is believed to transmit its characteristics into the wines. Although this stone is common to the region, the sloped vineyards have been covered with additional slate brought in from other locations. The slate maintains heat from the sun to help warm the grapes, particularly toward the end of ripening.

In the other German appellations of the Rheingau, Rheinhessen, Pfalz, and Nahe, you find a more diverse set of

In the Mosel-Saar-Ruwer district Riesling vines are tied to tall poles to obtain as much sunlight as possible.

soils that consist of slate as well as calcareous silt, quartzite, clay, sand, and loam. The flatter appellations of the Rheinhessen, Pfalz, and Nahe typically have more fertile soils, allowing grapes to ripen more easily and producing larger yields. Typically this generates less-distinctive wines than those made from grapes grown on the steep slopes of the Mosel-Saar-Ruwer and Rheingau.

Alsatian soils, which are quite different from slate-based German soils, consist of a mix of clay, chalk, limestone, and sandstone that changes as elevation changes. Chalk and limestone are characteristic on higher slopes, which are considered better than the clay-based soils located on lower slopes.

Soils in the New World tend to be a combination of clay, sand, loam, calcium, limestone, and granite. Unlike the Old World where producers believe that soil type causes aroma and taste differences, other factors such as day-night temperature changes, irrigation management, and sun exposure play a more conspicuous role in the New World.

If you do not believe that soil influences the aromas and flavors of wine, sample Riesling from a good Mosel producer side-by-side with a Riesling from a Washington State producer. Although the wines will have underlying similarities, they will be starkly different. The wine from the grapes grown in the Mosel's steep slopes, covered by blue and red slate, have absorbed the mineral characteristics of the slate, whereas the sandy soils of Eastern Washington emphasize the fruit characteristics of the grape.

Water Access

Just like other varieties, Riesling needs moisture but not too much. The key for Riesling is good drainage. The best Riesling vineyards throughout the world have well-drained soils regardless of the climate. With excellent drainage, as they have in the Mosel, heavy rains are less of a problem.

Vineyard Management

Creating great Riesling can be about terroir, as in the Mosel. However, site selection, trellising, irrigation management, canopy management, determining crop levels, and selecting a harvest date all play a significant role in making a superior wine.

Trellising is unique on the steep slopes of the Mosel and Rheingau. Each individual vine is tied to a pole that can be as much as eight feet high, allowing the vines to obtain maximum sun exposure.

In other locations in Germany, as well as elsewhere in Old World and New World appellations, growers use standard trellising systems. The grower needs to understand exposure and shading before making a selection. Canopy-management techniques of pruning, leaf pulling, and crop thinning are used throughout the world to improve crop quality and control yield.

On steep slopes all pruning, thinning, trellising, and harvesting must be done by hand, which tremendously increases production costs. However, in reasonably flat vineyards vines can be trained for mechanical pruning and harvesting.

Riesling is capable of generating a quality crop at higher yields than other comparable varieties. Why? Riesling has higher levels of extract (the solids in wine) and acidity than other varieties. This combination allows for greater concentration of aromas and flavors that are not diluted even at higher yields. Yields vary because of vintage weather conditions, but range from three to five tons per acre/45 to 75 hectoliters per hectare in the better vineyards in Germany and Alsace and up to six to eight tons per acre/90 to 120 hectoliters per hectare in lesser vineyards. In the New World, five to six tons per acre/75 to 90 hectoliters per hectare is standard in a normal vintage.

Harvest

While for most varieties the length of time between bud break and harvest can be up to eight months, for Riesling it can be even longer, particularly for late-harvest grapes. In the Mosel, for example, harvest can last easily into October and sometimes even November for very ripe, even overripe Beerenauslese (BA) and Trockenbeerenauslese (TBA) grapes.

A German Riesling hand-harvest can take as long as five or six weeks from first picking to last, most unusual when compared with other varieties. Riesling is picked only as the grapes ripen, not during one short period of time. Different vineyards ripen at different levels. It is not unusual in some of the best vineyards producing grapes for BA and TBA wines for pickers to go through the vineyard three or four times, each time only picking grapes that have achieved the desired ripeness and leaving the less ripe bunches for a later harvest.

As they are harvested, grapes are selected for the type of Riesling to be produced. In premier German vineyards where grapes are handpicked, they are separated by ripeness into batches that will be fermented into different Prädikatswein wines. Grapes harvested later in the season that have higher sugar levels are reserved for higher-quality Prädikatswein wines. If perfect conditions exist, the grapes in certain vineyards may remain on the vine in hopes of harvesting them frozen to make an Eiswein.

IN THE WINERY

Making Riesling is simpler than making other white varietals because, unlike Chardonnay and even some versions of Sauvignon Blanc, it is not manipulated. Riesling's style is determined by ripeness of fruit, fermentation temperature, length of fermentation, and sweetness in the finished wine.

Fermentation

If handpicked, grapes typically arrive at the winery separated into lots by ripeness and sugar level. The grapes are then fermented by lot generally in stainless steel fermenters, or in some cases in the Old World, in large oak casks or tanks that are old enough to no longer influence the wine with oak character.

As with other white varieties, fermentation occurs at cool temperatures, 60°F/16°C or below, to maintain the fruit and terroir characteristics in the wine.

In warm climates, like those found in some appellations in Australia or California, tartaric acid may be added during fermentation if the grapes arrived at the winery with high sugar levels and low acidity. Acidification is definitely not the standard in cool-climate appellations and is illegal in Germany and France for classified wines.

Winemakers can employ a few methods to arrive at a sweet wine. In the past German winemakers fermented wine dry and then added unfermented Riesling juice, called Süssreserve, to the wine just before bottling to achieve the desired level of sweetness and also to dilute the alcohol level. Although this method is still used for lower-quality German wines, Riesling winemakers more commonly stop fermentation artificially at the desired level of sweetness by using a centrifuge or by cold-chilling the wine, which causes the yeast to fall out of solution. The wine is then racked and the yeast is left behind.

Winemaking techniques used with other varieties are almost never applied to Riesling. It is rarely macerated with the skins, and malolactic fermentation (MLF) is never performed because softening its strong acidity would create an uncharacteristic style. Sur lies aging would add undesirable flavors to Riesling.

Blending

Although permitted by law in some appellations, Riesling is rarely blended. In Alsace Riesling must be 100 percent varietal by law. Other countries allow blending: 85 percent is the requirement in Germany and Australia, 75 percent in the United States. However, most Riesling is 100 percent varietal because no other grape adds quality or complexity to Riesling. A producer may decide to blend different Riesling lots to achieve a desired style or sweetness level, and occasionally Riesling is added to other varieties to give them fruitiness or acidity.

Stylized Wines

In addition to a wide range of table wines, Riesling transforms into some of the finest dessert wines in the world. In Germany sweet wines are classified as BA, TBA, and Eiswein. In other parts of the world, Riesling is made into late-harvest wine and ice wine.

Riesling also yields wonderful sparkling wine—lively and crisp with beautiful aromas. In Germany it is known as Sekt and is made from Riesling and other white varieties. Riesling sparkling wine is also produced in Australia.

RIESLING: THE WINE

One of Riesling's outstanding characteristics is its variable level of sweetness. Unlike most other varietals, sweetness can be a desirable attribute in Riesling. It adds an additional layer of complexity that balances its acidity and pairs with a much broader range of foods.

Color

Riesling is pale to light yellow with even a hint of green, and because of its concentration and high acidity, it can maintain this pale color without deterioration for several years. Riesling is a varietal that resists color change by oxidation. Riper grapes can turn Riesling to darker shades of yellow, and *Botrytis cinerea* can also affect color, giving the wine a more golden hue.

Aromas and Flavors

Riesling certainly fits in the category of aromatic wines. Its intensity may not be as great as that of Muscat or Gewürztraminer, but the breadth of possible aromas one finds in Riesling is incomparable. Riesling's aromas include the fresh tree-fruit scents of apples, peaches, nectarines, and possibly pears. Rieslings are noted for their significant mineral scents and even more distinctive petroleum aromas, which some may find off-putting. Aromas do not evolve much during the winemaking process, though oxidation can gradually change some of the aromas in the wine. Table 9.2 lists aromas that might be found in Riesling.

Table 9.2 Riesling Aromas

CHEMICAL	FLORAL	FRUITY	OTHER
Diesel	Carnations	Apples	Earthy
Kerosene	Ginger	Apricots	Honey
Petroleum	Jasmine	Citrus	Minerals
	Orange blossoms	Citrus peels	Slate
	Roses	Lemons	Spicy
		Limes	Steely
		Melons	
		Nectarines	
		Peaches	
		Pears	
		Tangerines	

Wine Styles

Riesling can be dry or sweet, or any gradation in between. The level of sweetness is the defining characteristic in determining the style of a Riesling table wine (see Table 9.3).

The International Riesling Foundation (IRF) has recently developed a set of standard styles to help consumers better select a bottle of Riesling. There are four different styles in the IRF's system: dry, medium-dry, medium-sweet, and sweet (see International Riesling Foundation Sweetness Scale, page 152). The Germans have their own notable classification system, but the IRF style guidelines are consumer-focused and based on the sweetness of a wine. It is not only the residual sugar that determines sweetness level, but also the level of acidity in a wine. Higher acidity reduces the perception of sweetness in wine. Wines with high acidity can make a wine with high sugar levels taste drier. Dry Rieslings can range from wines with no residual sugar to those that have some residual sugar. Increasing residual sugar can change a Riesling's style classification from dry to medium-dry, medium-sweet, or even sweet, but the style classification will also depend, to a lesser degree, on the acidity level in the wine.

Wines that commonly fit into the dry classification are German QbA trocken, Classic, and Selection wines, Prädikatswein Kabinett wines, and Alsatian, Australian, Austrian, Northern Italian, and some North American Rieslings. Wines that are commonly found in higher sweetness classifications are German QbA halbtrocken wine, Prädikatswein Kabinett to Auslese wines, a few Alsatian wines, particularly Vendange Tardive and Séléction de Grains Nobles, and an array of wines from North American appellations.

Overlaying the IRF style scale based on sweetness are the special characteristics of each producing country or appellations within a country. Germany has its own unique character, as does Alsace, Austria, Northern Italy, Australia, the Western United States, the Eastern United States, and Canadian appellations. Climate, soils, and winemaking can influence flavors, leaving wines within the same sweetness level with different profiles.

In Germany wines from the steeply sloped vineyards of the Mosel-Saar-Ruwer, and Rheingau and the flatter Pfalz are best recognized for maintaining their acidity at all sweetness levels, which results in balanced wines, a richness of flavor that develops with aging, and a youthful freshness at any age. These wines develop distinguished mineral and petroleum characteristics.

Though in close geographic proximity, Alsatian Riesling is quite different from German Riesling, primarily because the macroclimate in Alsatian appellations produces riper grapes with more potential alcohol, in turn producing full-bodied and dry wines. Austrian wines generally echo the dry Rieslings of Alsace with their clean and concentrated aromas. In years when the weather allows for extra ripening, late-harvest grapes are made into special wines classified as Vendange Tardive and Séléction de Grains Nobles in Alsace. Vendange Tardive are made from very ripe grapes and can range from dry to sweet. Séléction de Grains Nobles wines, riper still, are invariably sweet.

International Riesling Foundation Sweetness Scale

Table 9.3 Riesling Wine Styles

	DRY	MEDIUM-DRY	MEDIUM-SWEET	SWEET
COUNTRY/APPELLATION	Germany, Alsace, United States, Austria, Italy, Australia	Germany, United States	Germany, United States	Germany, Canada, United States, Alsace
VARIETAL	100%	100%	100%	100%
ALCOHOL	13% or more	11%–13%	9%–11%	7%–9%
ACID	High, searing	Balanced	Balanced	Balanced
AROMAS/FLAVORS	Floral, tree fruits, minerals, petroleum, limes	Stone fruits, minerals	Stone fruits, minerals	Dried apricots, honey
BODY	Light to medium, delicate	Round	Rich, full	Heavy, thick
COLOR	Light yellow	Light straw	Pale yellow	Yellow

The typical Australian style is dry and full-bodied, more similar to Alsatian than to German Riesling. In almost all cases they are bone-dry. Even when they have some residual sugar, they have such high acidity that they appear to be dry.

American Riesling wines, particularly those from Washington and California, are generally riper, softer, and fuller, many of them with a peachy character. Made in all sweetness levels, they sometimes lack the high acidity found in European and Australian Rieslings. Riesling grown in the East and Midwest is produced in both dry and sweet styles and can reflect a more European style, with higher acidity in some vintages and a softer, Western American style in other vintages.

ENJOYING RIESLING

Riesling is a varietal that provides different sweetness-level styles and appellation characteristics that can match many different tastes. New consumers often start with a medium-dry or medium-sweet Riesling and then expand their palates to accept drier styles or ultrasweet dessert wines.

Buying Riesling

Riesling is easy to buy because labels in both the Old World and New World identify the wine by variety, not appellation. Even in Alsace, wines are identified by grape variety as opposed to appellation, which is the norm in France for Appellation d'Origine Contrôlée (AOC) wines. Many producers make multiple styles of Riesling at different sweetness levels, but these differences may not be readily indicated on the label. How can you tell how sweet a Riesling will be?

1   Determine if the wine you are considering follows the IRF style guidelines. The IRF label is very helpful if it is printed on the back label. However, because the guidelines are quite new, and because of the individualist nature of some producers, many producers may not have adopted the guidelines.

2   If the wine is German, follow the German classification system (see Appendix 2, page 400). The labels on both QbA and Prädikatswein wines may state whether they are trocken (dry) or halbtrocken (half-dry). Classic and Selection QbA wines are almost always dry. Prädikatswein wines from Kabinett to TBA, although based on grape ripeness, typically increase in wine sweetness as you ascend the scale.

3   Look at alcohol level to provide an indication of sweetness. As alcohol level decreases, residual sugar increases. A Riesling at 11 percent alcohol or lower will almost always have distinguishable sweetness. If a bottle of Riesling is high in alcohol, 13 percent or above, it will likely be dry. Many sweet-style German Rieslings drop to 7 or 8 percent alcohol.

4   Consider the origin of the wine. Wines from Australia, Austria, Italy, and the Okanagan Valley in Canada are almost always dry. Wines from Germany, Alsace, and the United States have a range from dry to sweet.

5   If you like wines from a particular appellation, taste wines from different producers within that appellation. Each producer maintains the same styles

Examples of regional Riesling, each produced in a different style, from left: Michigan; Mosel, Germany; Alsace, France; South Australia

TOM ZASADZINSKI/CAL POLY POMONA

from vintage to vintage. This is particularly helpful with European wines. For example, an Alsatian Trimbach Riesling is likely to be bone-dry and laserlike on the palate; whereas, a Hugel Riesling is likely to be richer and a touch sweeter.

6 Check the back label, which may state the residual sugar. Riesling is dry when it has less than 1 percent sugar, medium-dry with 1 to 2 percent sugar, medium-sweet with 2 to 4 percent sugar, and sweet with more than 4 percent sugar.

7 When the label simply reads "Riesling," with no other assisting information, it may be necessary to ask the sommelier, server, or retailer to determine how dry or sweet the Riesling is.

While Rieslings from larger producers in California and Washington are widely available, Rieslings from other North American regions are more regional in their distribution. A selection of European and Australian producers may best be found in specialty wine shops. Late-harvest or dessert wines from European and New World countries are made in smaller batches under select conditions and are generally only found in fine wine shops or directly from the producer.

Pricing for Riesling is generally modest in comparison with similar-quality wines from other varieties. Increasing sweetness is considered a sign of increasing quality, particularly with German Riesling. Prices increase accordingly. Ultra-sweet Rieslings can be some of the most expensive wines you can purchase.

Storing and Serving Riesling

Riesling's high acidity and high extract make it a wine with great aging potential. Some Rieslings, particularly those of the Mosel and Rheingau, age longer than any other white wine and compare in aging ability to many red varietals. High-quality Rieslings are known to age two or three decades or even longer, and German Riesling is probably the most age worthy of the white wine varietals. We have sampled German Rieslings 30 years old that taste as fresh as a newly vinted bottle. However, not all Rieslings can be aged; those with high acidity (low pH concentration) and sweetness age best. It is also beneficial to know a producer's track record for aged wines before attempting to age a Riesling.

Riesling should be stored like any other wine. A temperature of 57°F/14°C at 70 percent humidity is optimum. The coolest, darkest interior closet is the best alternative to a temperature-controlled cellar or wine storage cooler.

The classic Riesling glass is smaller than a typical Burgundy or Bordeaux glass, but if you are lacking an array of varietal wineglasses, an all-purpose glass works well or you can use a larger Bordeaux glass.

Riesling and Food Pairings

Because of its many styles, Riesling matches well with many different kinds of food. However, the style of Riesling you choose must be carefully selected to complement the food you are serving. Drier styles are natural matches with many foods. The sweeter the wine, the more likely it is that it should be enjoyed by itself or that it will require more specific foods.

Sweetness, acidity, alcohol level, and concentration all play a role in making a good match:

– Riesling's acidity refreshes and cleans the palate and provides a complement to rich foods or high-fat foods.
– Sweeter styles go well with hot-and-spicy and sweet-and-sour foods and, therefore, complement a wide range of ethnic dishes.
– Low-alcohol styles work particularly well with hot-and-spicy foods.
– Lighter Rieslings—those with less concentration—pair well with delicate foods.
– Rieslings with high concentration have enough depth to work with a broad range of foods. They go well with rich foods and can even be served with red meats.
– Medium-dry Rieslings are ideal by themselves as an aperitif.

TOP 10 FOODS TO PAIR WITH RIESLING

1 Quiche Lorraine (dry)
2 Tandoori chicken (medium-sweet)
3 Thanksgiving dinner (medium-dry)
4 Picnic of fruit, cheese, and cured meats (dry to medium-sweet)
5 Platter of sausages, pâtés, and sauerkraut (dry or medium-dry)
6 Braised duck with fruit (medium-dry)
7 Coq au Riesling (dry)
8 Poached or steamed halibut (medium-dry)
9 Strong cheeses like goat or Roquefort (sweet)
10 Smoked salmon (medium-dry)

SUMMARY

The beauty of Riesling starts with the grape itself. Its pristine aromas and flavors, which are a true reflection of its origin, are maintained because it is not fermented or aged in oak barrels. The combination of acidity, concentration of aromas and flavors, and sweetness levels make for an exciting, complex wine that rivals many reds. Its range of styles makes it spectacular and versatile when paired with different foods; no other wine seems to work with such a varied range of cuisines. As an additional benefit, Riesling provides flavor at low alcohol levels.

Other White *Vinifera* Grape Varieties

We have just discussed the three major international noble white varieties that are most frequently found in the marketplace around the world. However, beyond these important three, there are dozens of others that produce wonderful wines. Some are quite common and you may already be familiar with them. Others you may be less familiar with. With the exception of Muscat, plantings of these grapes are generally restricted to climate-specific regions, and some have distinctive aroma and flavor profiles that appeal to select consumers.

This chapter identifies seven *vinifera* varietals that are generally available in restaurants and retail stores. We provide a summary of the key elements of each variety. In addition we list 32 notable indigenous *vinifera* varieties that are most typically found in their respective countries of origin, but also may be available in the United States.

Seven Common White *Vinifera* Varietals

Chenin Blanc

Gewürztraminer

Muscat

Pinot Blanc

Pinot Gris/Pinot Grigio

Sémillon

Viognier

CHENIN BLANC

Chenin Blanc emerged during the ninth century in the Loire in France, where some of the most notable examples of Chenin Blanc wines in the world are still produced today. The grape contributes to a range of styles from bone-dry to unctuously sweet dessert wines. It also makes outstanding sparkling wines because of its high acidity. Chenin Blanc is planted around the world in significant amounts because producers like its high yields and high acidity. However, its reputation has been damaged outside of the Loire because it is primarily used for making generic jug wines or is blended with other varieties, like Chardonnay.

Countries That Produce Chenin Blanc

The most noteworthy Chenin Blanc grapes and wines come from the Loire, followed by South Africa. In California a few producers are making Chenin Blanc into a quality varietal wine.

Seven white varietal wines commonly found in restaurants and retail stores, from left: Muscat, Alsace, France; Pinot Grigio, Friuli, Italy; Pinot Blanc, Alsace, France; Sémillon, Napa Valley, California; Viognier, McLaren Vale, Australia; Gewürztraminer, Anderson Valley, Mendocino County, California; Chenin Blanc, South Africa
TOM ZASADZINSKI/CAL POLY POMONA

The Loire

In the Loire, Chenin Blanc is commonly called Pineau de la Loire. It is the most widely planted white variety in the region, exceeding Sauvignon Blanc. The most notable vineyards are located in the Touraine and Anjou. The most famous Chenin Blanc appellation is Vouvray, which is in the Touraine on the northern side of the Loire River. Montlouis, another excellent appellation, sits across the river to the south. Savennières, Saumur, and Coteaux du Layon, known for their stunning *Botrytis*-infected sweet wines, are located in Anjou.

The Loire, which is in the north of France, has a very cool climate. This causes ripening to be difficult, although it becomes a bit easier in the southern and western appellations. The wines of Vouvray and Montlouis are typically crisp and low in alcohol, whereas the wines of Savennières, Saumur, and Coteaux du Layon are ripe, rich, and higher in alcohol.

South Africa

Chenin Blanc has been planted in South Africa since the seventeenth century, and was commonly known there as *Steen*. Although that name is still used, it is more frequently called Chenin Blanc today. It was once the most widely planted white variety in the country, but now it is outstripped by Chardonnay and Sauvignon Blanc. It was originally used to make high-quality wines, then later for diluted, blended, quaffing wines. Recently it has recaptured its luster as a high-quality grape capable of making outstanding wines in the hands of skilled winemakers.

The United States

In California most Chenin Blanc is blended into generic white wines or is used as a blending grape with other varieties, making up part of the unlisted 25 percent allowed on varietal labels. Chenin Blanc brings acidity or fruit character to less-expensive varietal wines that need those characteristics. Clarksburg in the San Joaquin Valley (Central Valley) is the dominant region for growing quality Chenin Blanc. Most California producers of varietal Chenin Blanc use Clarksburg grapes and display the Clarksburg appellation on the bottle.

Characteristics

The Grape

Chenin Blanc is a difficult grape to cultivate because it buds early, making it susceptible to frost damage, and ripens late, putting it at risk of inclement weather. It performs best in a cool climate where it maintains its vibrant acidity. Perhaps more than any other grape variety, quality depends on low yields, which allow the fruit aromas and flavors of the grapes to develop. Yields in California may be up to four times higher than those in the best vineyards in the Loire, leaving no doubt that California's fruit quality is lower.

The Wine

Chenin Blanc is high in acid, which makes it an outstanding choice for sparkling, table, or sweet wine. It is made in a wide range of fermentation vessels from stainless steel to concrete tanks to barrels. Some are barrel aged but rarely in new barriques. Most high-quality Chenin Blanc wines are 100 percent varietal although in some places it is blended with Chardonnay or Sauvignon Blanc. Typically Chenin Blancs are unoaked, but old wooden tanks are still used in the Loire and the occasional New World producer uses small oak barrels.

The aromas of Chenin Blanc differ from region to region. In the Loire the classic aromas are honey and wet straw with tree or stone fruit character. In South Africa they are dominated by an array of tree fruit aromas, as well as tropical aromas. In California it tends to show simple fruit flavors because its generally high yields produce neutral and dilute wines.

Because Chenin Blanc from the Loire has the ability to ripen in any given vintage, a variety of styles can be made. When grapes do not ripen fully, a frequent occurrence in the Loire, the wine is likely to be made into a dry sparkling wine, called crémant, or a slightly sparkling wine, called pétillant. In riper vintages a medium-sweet style, called demi-sec, and a sweet style, called moelleux, are produced. These wines, particularly the demi-sec and moelleux, have great aging ability because of their high acidity. Many of the best do not achieve full maturity for 10 years or more and many can age for several decades.

Chenin Blanc is the most widely planted white grape variety in France's Loire wine region.

In South Africa Chenin Blanc styles mirror those of the Loire. It is made into sparkling wines and table wines ranging from dry to sweet. The wines are either unoaked or barrel aged.

In the United States, where Chenin Blanc fully ripens, it lacks the acidity and intensity of the Chenin Blancs from the Loire and is almost always produced in an off-dry to medium-sweet wine when it is made into a varietal wine. A very few producers make a sweet, late-harvest style.

Enjoying Chenin Blanc

As with most white wines, style plays a key role in determining food matches. In the Loire, where the wine has high acidity, it goes naturally with the cheeses of that region, particularly goat cheese. It also matches well with lighter chicken dishes, freshwater fish, and vegetarian dishes. Sweeter styles work well with salads as well as lightly spiced Indian or Asian foods. Crémant de Loire can replace any sparkling wine or Champagne and be served with the same types of food.

GEWÜRZTRAMINER

Gewürztraminer is quite a change if you have become accustomed to Chardonnay, Sauvignon Blanc, and Pinot Grigio. It shows an intensity of aromas and flavors found in few other white varieties. It is not a varietal that you want to drink every day, but it provides a refreshing change. It matches with a wide range of food styles and specialty foods. It is a variety that is not dominant anywhere, but core supporters refuse to let it be replaced entirely by Chardonnay or Sauvignon Blanc in the marketplace.

Countries That Produce Gewürztraminer

Gewürztraminer has its French home in Alsace, but it is widely produced in relatively small quantities throughout the world. In Western Europe it is also made in Germany, Austria, and Northern Italy as well as Central and Eastern European countries, particularly in Moldova and Bulgaria.

In the New World, California, Australia, and New Zealand make notable Gewürztraminer on a small scale with strong support from growers, winemakers, and consumers. Gewürztraminer grows best in cool areas of the United States. In California, Mendocino County is considered the best climate for the grape and several producers there grow and make excellent Gewürztraminer.

Characteristics

The Grape

Gewürztraminer is a grape with color, aroma, and body. Unlike most white grape varieties, the grape is pink and produces a wine that is golden in color. Its perfumed aroma is perhaps the most distinctive of any variety. The first half of the grape's name, *Gewürz*, means "aromatic" or "spice."

Gewürztraminer is a difficult grape to grow. It buds early, making it susceptible to frost damage, is more likely to incur viral diseases than other varieties, and loses its acidity rapidly during the ripening phase. Finding the balance between flavor development and acidity is problematic for any grower. This grape performs best in a cool climate.

The Wine

A well-made Gewürztraminer starts with ripe grapes from a long, cool growing season, which allows them to maintain an acceptable level of acidity. This environment creates a wine with moderately high alcohol but with acidity that borders on the low end. Caution must be taken during vinification because Gewürztraminer has a tendency to oxidize. The wine at its best has a thick, deep golden color; has perfumed aromas of heavily scented roses, lychee fruit, and minerals; and is full-bodied with substantial strength and concentration. Gewürztraminer is at the opposite end of the spectrum from high-acid wines, such as Sauvignon Blanc and Chenin Blanc, but it should carry at least moderate acidity.

Although Gewürztraminer is not the most widely planted variety in Alsace, it is one of the cherished grapes of the region because of its distinctiveness. Wines are commonly made dry, but grapes can ripen in good vintages, allowing for late picking to create Vendange Tardive or even the ultrasweet Sélection de Grains Nobles.

In the Pfalz and Baden in Germany and in Austria, wines are made in similar styles to those made in Alsace, both dry and sweet. In the Alto Adige of Northern Italy producers make wines typically in a dry style. The largest producer of Gewürztraminer in the United States, Fetzer Vineyards, located in Mendocino County, makes an off-dry style.

Gewürztraminer's exotic character is either loved or hated. Those who lack an appreciation for the wine are likely to note its weight, high viscosity, heavy floral aromas and flavors, and high alcohol, which is usually above 13 percent. Detractors dislike its low acidity and potential bitterness. Fans talk about a wine you can taste and feel—its pungent aromas, lingering finish, and ability to go with foods that other varietals cannot stand up to.

Enjoying Gewürztraminer

Although not suited to everyone's taste, Gewürztraminer works exceptionally well with an interesting array of foods. Drier styles complement richer foods, like those found in Alsace: Alsatian tarte flambé (grilled pizza), smoked cheese, Muenster cheese, smoked fish, pâté, and rich duck and pork dishes. Sweeter styles work better with spicier foods, including dishes seasoned with ginger and pungent but not hot Asian and Indian cuisines. Gewürztraminer is also an interesting wine to sip by itself.

MUSCAT

Muscat is an ancient grape that may be the parent of all other *vinifera* varieties. It is considered by many to be the first grape variety made into wine. The Greeks introduced the grapes to the Romans, who spread it throughout Mediterranean Europe. Today Muscat is widely produced around the world.

Muscat has evolved into an extensive family of grapes with four major subvarieties, Muscat Blanc à Petits Grains, Muscat of Alexandria, Muscat Ottonel, and Muscat Hamburg, as well as many mutations of these varieties. No other variety, including the Pinot family, has as many different subvarieties as Muscat. This diversity has resulted in Muscat-based wines in every style from dry to sweet, sparkling to fortified. Table 10.1 highlights the major Muscat varieties and includes a very limited list of alternate names that are most commonly used. There are several more names that are unique to particular countries or regions.

Countries That Produce Muscat

The many varieties of Muscat are grown in a broad range of climates throughout Europe, the United States, Australia, and other countries.

Italy

In Italy Muscat is most widely grown and produced in Piedmont, but it is cultivated in most regions throughout the country. In Piedmont Muscat is made into two styles of sparkling wine both using Moscato Bianco: Asti (formerly known as Asti Spumante), a full-sparkling wine that can be dry or off-dry, and Moscato d'Asti, a semi-sparkling wine that is sweeter and lower in alcohol. In Southern Italy it makes naturally sweet wines or passito, an unctuous dried-grape wine that is redolent of the aroma and sweetness of raisins. These wines can be made from Muscat Blanc à Petits Grains or Alexandria (Zibibbo).

France

France is known for two styles of Muscat: the dry Muscat of Alsace and the sweet, fortified Muscat wines produced in vineyards planted along the Mediterranean Sea. Alsace has made dry Muscats seemingly forever from both Muscat Blanc à Petits Grains and Ottonel. Recently in outstanding vintages Alsatian winemakers have made a small amount of sweet botrytized Séléction de Grains Nobles.

In the South of France the most noted Muscat wines are those that are fortified using both Muscat Blanc à Petits Grains and Muscat of Alexandria. Muscat de Beaumes-de-Venise, Muscat de Frontignon, Muscat de Lunel, Muscat de Mireval, Muscat de Riversaltes, and Muscat de St.-Jean-de-Minervois are all examples of these wines (see Chapter 18, Fortified Wines, page 301).

The United States

Muscat is widely produced throughout California where it is mostly made into sweet styles of wine or blended into generic table wines. A small portion is

Table 10.1 The Muscat Family

MUSCAT BLANC À PETITS GRAINS

Muscat de Frontignan
Moscato Bianco

MUSCAT OF ALEXANDRIA

Muscat Romain
Zibibbo
Moscatel de Málaga

MUSCAT OTTONEL

Muskotály

MUSCAT HAMBURG

Black Muscat

used for varietal table wines. Much of the Muscat comes from the warm San Joaquin Valley (Central Valley), but it is also produced in the Central Coast appellation. Many of the wines mimic styles found in Italy, like Moscato d'Asti, and fortified wines, like the French Muscat de Beaumes-de-Venise and Muscat de Frontignan. Quady Winery in the San Joaquin Valley makes a number of sweet and fortified wines from the subvarieties Orange Muscat and Black Muscat (Muscat Hamburg).

Other Countries

Sweet and fortified Muscat-based wines are also produced in Spain, Portugal, Germany, and Greece. Ottonel is the dominant subvariety of Muscat used in the dry Muscat wines of Germany and Austria because of its resistance to cold temperatures. In Central and Eastern European countries, like Hungary, where Ottonel is called Muskotály, it is used for both dry and sweet wines. Australian Muscat comes from Northeast Victoria, where ultrasweet wines are made.

Characteristics

The Grape

The diversity of Muscat mutations and clones results in a wide range of grape characteristics. The sizes of the grapes of each subvariety are quite different, as exemplified by the small berries of Muscat Blanc à Petits Grains versus the large, oval Alexandria grapes. Color can be dramatically different as well, ranging from pale to rose to red and even to black for Hamburg. The grape grows successfully in different climates, but it tends to lack acidity, making relatively short-lived wines.

The most highly regarded subvariety is Muscat Blanc à Petits Grains. The small berries develop intense character and richness, making beautiful wines. Alexandria grapes are more likely to show up on the table as eating grapes or to be used for distillation, but they also are used in wines, particularly in fortified styles. Ottonel, the most neutral, is the primary Muscat variety used in cool climates usually to make dry wines. Hamburg, considered the lowest in quality of the Muscat varieties, is most commonly used as a black table grape. In California's San Joaquin Valley it is used to produce a fragrant dessert wine.

The Wine

Muscat has many faces, which are determined by the Muscat variety used, the terroir or country where it is grown, and how the wine is made. The characteristics of Muscat wines start with the grape's incredible perfume in all its subvarieties, except for Ottonel. To some, Muscat seems even more aromatic than Gewürztraminer. The wine can range in color from golden to pink to black and even brown for aged Austrian wines. Most Muscat is sweet. However, producers in Alsace and Italy make dry Muscats, and the occasional producer from other countries may experiment with a dry style.

Enjoying Muscat

Muscat is usually served as a sweet wine. To match the right food with this type of wine, consider the sweetness of the wine. It must always be sweeter than the food. In Italy Moscato d'Asti is served with light cookies, like biscotti, or perhaps a fruit salad.

In many parts of the world sweet, fortified Muscats are served as an aperitif before the meal, a less common practice in the United States. The wines also go with bittersweet chocolate or vanilla ice cream. True lovers of sweet, fortified Muscats prefer to drink them after a meal as a substitute for dessert.

The dry Muscats of Alsace go best with rich seafood, like scallops or smoked salmon, or lightly seasoned Asian foods that are not too spicy. In Alsace dry Muscat is frequently used as an aperitif.

PINOT GRIS/PINOT GRIGIO

Pinot Gris is a mutation of the Pinot Noir grape that today is known by both its French name, Pinot Gris, and its Italian name, Pinot Grigio. Both names are used interchangeably to identify the varietal, but they often denote very different styles of wine. Both styles are exceptionally food-friendly, and provide an easy-drinking alternative to Chardonnay or Sauvignon Blanc.

Countries That Produce
Pinot Gris/Pinot Grigio

The most well-known Pinot Gris/Pinot Grigio wines come from France and Italy. The grape is also cultivated throughout Europe and in several New World countries.

Europe

In France Pinot Gris is most widely grown in Alsace, which is considered its home region. Tokay d'Alsace is a common synonym for Pinot Gris that is still used in Alsace. The grape is also cultivated under the name Malvoisie in the Loire and Burgundy, where it was probably first grown in France.

Substantially more wine from Pinot Grigio is produced in Italy than from Pinot Gris in France. In Italy Pinot Grigio is cultivated primarily in the cool northeast in the general regions of Friuli and Trentino-Alto Adige. It is also grown in Lombardy, the Veneto, Emilia-Romagna, and as far south as Southern Tuscany.

In Germany and Austria the wine is called Ruländer or Grauburgunder, and in Central and Eastern Europe, it goes by a number of different names.

The New World

In the New World Pinot Gris/Pinot Grigio has taken hold in North America and is the fastest-growing wine in popularity in the United States. The most noted

Pinot Gris/Pinot Grigio comes from Oregon where it rivals Chardonnay as one of the two most widely planted white grapes.

California produces more Pinot Gris/Pinot Grigio grapes and wine than any place in the United States, and the amount of grapes under cultivation has increased more dramatically than any other variety in the last 10 years. In 1998 a little more than a thousand tons of Pinot Gris/Pinot Grigio were crushed. In 2007 almost 80,000 tons were crushed. The variety is also produced in several eastern states. The Okanagan Valley of British Columbia also makes outstanding wines from the variety.

In the Southern Hemisphere the most notable wines come from New Zealand where Pinot Gris/Pinot Grigio is grown on both the North and South Islands. It is also produced in Australia and South Africa.

Characteristics

The Grape

Pinot Gris/Pinot Grigio grapes are actually red in color, and if the skins are not immediately separated from the juice prior to fermentations, the wine will have a faint pink color. The grape performs differently in different climate conditions. In cool climates it maintains its acidity and crispness, but in warmer climates the wine develops more richness. In even warmer climates, like those found in California, it can lose its acidity and become flabby. The classic aromas of honey and flowers only develop in the richer styles.

The Wine

There are two distinct styles of wine made from this varietal: Alsatian Pinot Gris style and Italian Pinot Grigio style. Climate has a direct impact on which of these styles is produced. Although it is farther north, Alsace is warmer and drier than the mountainous vineyards of Northwest Italy. In Alsace Pinot Gris presents itself as a richer style of wine because the grapes become fully ripe, creating higher sugar content in concert with lower acid. The wine is fermented in stainless steel or oak tanks. It may or may not be aged in small oak barrels depending on the winemaker's preference. The finished wine is moderate in alcohol with balanced acid and has a round, full body with classic floral and honey aromas. The color is straw to golden. If aged in wood, there is a noticeable oak component in the wine's aroma and flavor.

Because of the cool temperature in Northwest Italy, Pinot Grigio is harvested earlier and has less ripeness and higher acidity than Alsace's Pinot Gris. Fermenting takes place in stainless steel, producing a light, fruity style with low to moderate alcohol and more neutral flavors. The finished wine is colorless to pale yellow with a citrus and green apple nose and is light and delicate on the palate. The wine is meant for early drinking.

Similar styles to those of Alsace and Northern Italy are found wherever the grape is planted and the wine is made. In the United States, Oregon is a leading producer of a wine called Pinot Gris that is made in a style that falls between the Alsatian style and the Northern Italian style.

Enjoying Pinot Gris/Pinot Grigio

When you select food to go with Pinot Gris or Pinot Grigio, it is important to determine the style of the wine. Pinot Grigio is a light quaffer that works with light foods. Pinot Gris is richer and works with richer foods. Pinot Grigio makes an excellent aperitif, particularly if it has a little spritz, which makes it a light, refreshing starter. It is a good complement to classic Northern Italian dishes, like frito misto, and it also works with simple grilled or sautéed fish, simple pasta dishes, and even fresh oysters and clams. The richer Pinot Gris goes well with fatty fish, like salmon; richer seafood, such as scallops; or pork and veal. It is also perfect with classic Alsatian dishes, like tarte flambé, or pork with braised potatoes, onions, and sauerkraut. In Oregon Pinot Gris is classically served with grilled salmon or veal.

PINOT BLANC

Pinot Blanc is not considered one of the great grape varieties, but it is widely planted in Northeast Italy and Alsace, as well as Germany and Austria. Lacking distinctive aromas and flavors, it is used in sparkling wine and for blending, where its acidity and body can play a beneficial role. Its ability to accept oak aging also makes it possible to create a wine similar to New World Chardonnay.

Countries That Produce Pinot Blanc

Pinot Blanc is produced widely in Old World countries and in some New World countries as well. In Europe Pinot Blanc is an important varietal in the Northern Italian regions of Friuli and Trentino-Alto Adige, where it is called Pinot Bianco. In France it makes a dry, vibrant table wine in Alsace or a very pleasant sparkling wine, called crémant, in Alsace and Burgundy. In Germany it goes by the name Weissburgunder. In Austria and Alsace it also goes by the name Klevener. The grape is also produced in Central and Eastern Europe.

In North America Pinot Blanc is cultivated in California as well as Oregon and the Okanagan Valley of British Columbia, Canada. In California it is planted most extensively in Monterey County. It is a key ingredient in many California sparkling wines and is made into varietal wines as well. Because it has the same body as many Chardonnays, in California it used to be called "poor man's Chardonnay."

In South America it is planted in Argentina and Uruguay.

Characteristics

The Grape

Pinot Blanc is a mutation of Pinot Gris, which itself is a mutation of Pinot Noir. It has similar characteristics to both these grapes but lacks their distinctive character. The grape adapts to the climates where it is grown and acts very much like the Chardonnay grape. In cooler climates the grape maintains higher acidity. In warmer climates it is likely to ripen more, creating a wine with more sugar and higher potential alcohol. It is a neutral grape with good acidity and body that is amenable to different winemaking techniques, which makes it a perfect vehicle for making sparkling wine.

The Wine

For table wines in Alsace, Pinot Blanc is typically blended with Auxerrois and made into a dry, rich style that is food-friendly. In Germany it is made in either a dry, unoaked style or a barrel-aged style that has a Chardonnay-like character. In Northern Italy it may be made in a light, acidic style or a barrel-aged style, or it may be blended with Chardonnay. In California it is likely to receive some combination of barrel fermentation, malolactic fermentation (MLF), sur lies aging, and barrel aging, making a wine that may be hard to tell apart from Chardonnay.

All of these countries also use Pinot Blanc as a principal ingredient in their sparkling wines. It provides the vibrancy and body that are valuable components of the wine.

Enjoying Pinot Blanc

The Italian and Alsatian syles of Pinot Blanc tend to be neutral, which allows them to pair well with many kinds of food. However, each style goes with a different type of food.

Italian Pinot Bianco is generally lighter and higher in acidity. It accompanies lighter dishes, including fish, poultry, pasta, eggs, and even salads. The richer Alsatian style works with richer foods, such as quiche Lorraine, a classic dish that combines pastry, eggs, cheese, onions, and bacon. It also works with rich-sauced poultry dishes or even roasted, braised, or grilled pork. In California much Pinot Blanc is made like a typical Chardonnay, with oak, and should be matched with foods that go with Chardonnay, such as richer fish and poultry dishes.

SÉMILLON

Sémillon has two strong advocate countries. In Bordeaux it is highly regarded and produces both rich, dry wines and some of the best sweet wines in the world when blended with Sauvignon Blanc. In Australia it makes some of the longest-lived dry white wines and is highly prized in blends with Sauvignon Blanc or Chardonnay.

Countries That Produce Sémillon

France and Australia are the most well-known producers of Sémillon. It is also produced in the United States, Chile, and South Africa, although it is less popular in these countries.

France

Bordeaux is credited as the primary growing region for Sémillon. It is also planted in the South-West region of France. In Bordeaux it is produced in two distinctly different styles: dry and sweet. In Pessac-Léognan, Graves, and Entre-Deux-Mers, it is made into a dry white wine that is most commonly blended with Sauvignon Blanc, with the possible addition of Muscadelle. Sémillon is also an ingredient in the broader, general white wine category of Bordeaux, Bordeaux Blanc, which typically includes these three varieties.

The other style is that of the internationally famous sweet wines, infected by *Botrytis cinerea*, of Sauternes and Barsac (see Chapter 18, Botrytized Wines, page 293). These wines are also blends of Sémillon and Sauvignon Blanc, but Sémillon is the dominant grape. In the South-West it also produces both dry and sweet wines.

Australia

Australia is the second country with significant production of Sémillon, but it is made in a different style than in Bordeaux. The most noted regions are Hunter Valley in Northern New South Wales and the Adelaide Hills of South Australia. It is also produced in the Barossa and Clare valleys in South Australia and in Margaret River in Western Australia. In Hunter Valley Sémillon is used for a dry 100 percent varietal wine. In other regions it is also made in a dry style but is more likely blended with either Sauvignon Blanc or, more recently, with Chardonnay under the name SemChard. Late-harvest and Sauternes-style botrytized wines are also made.

The United States

In the United States, Sémillon is not as widely cultivated as other varieties, but some producers value it for use in Sauvignon Blanc or Chardonnay blends. In the western part of the United States, it is grown in California, Washington, and Oregon. In California it is produced in the Livermore Valley, Napa Valley, and Sonoma County with a smattering of vineyards in other locations. Washington's Columbia Valley and Southern Oregon are the primary homes to the variety in those two states.

Characteristics

The Grape

Sémillon is a variety that responds differently depending on where it is grown, the growing conditions, and how it is made. The grape is yellow-gold and easy to grow with the potential for yielding a large crop. If overcropped, it produces a grape that is low in acid. Sémillon is the variety that is most susceptible to *Botrytis cinerea*, which has helped it to achieve its greatest fame as the primary grape used in making Sauternes.

The Wine

Most Sémillon is blended. The best of the blended wines are the rich, dry blend of Sémillon and Sauvignon Blanc found in Graves and Pessac-Léognan. These are hard to duplicate in other locations because of the combination of gravelly soils and Atlantic climate where the grapes are planted. In other Old World countries, Sémillon is used for blends in jug wines intended for local consumption because it is easy to grow and gives high yields. In the New World it is usually blended with Sauvignon Blanc or Chardonnay. SemChard, a blend of Sémillon and Chardonnay, has become a popular inexpensive wine produced in Australia and even California.

Hunter Valley Australian Sémillon and a limited number of other wines from around the world are unique because they are 100 percent varietal Sémillon. Hunter Valley is the most highly regarded region for making a dry style that requires a decade or more of aging before it reaches drinking maturity.

Enjoying Sémillon

Dry, medium-bodied styles of Sémillon that are typically blended with Sauvignon Blanc work with simple grilled or sautéed fish and seafood, and with lighter poultry dishes. If blended with Chardonnay, richer seafood and sauced fish dishes are good matches. Richer meats, like pork, also fit the bill.

To many the *Botrytis*-infected and late-harvest wines made from Sémillon are sublime by themselves. But Sauternes are the classic combination with foie gras and the salty blue-veined cheese Roquefort. Other blue-veined cheeses, like Maytag blue from the United States, also work. Dark chocolate is another possibility. Sauternes has even been served with rich beef, like prime rib. The richness of the wine pairs well with rich foods.

VIOGNIER

Viognier originated in the Rhône, France, where it produces some of the most expensive early-drinking white wines in the world in fairly small quantities. It achieved notoriety in California 15 to 20 years ago when it was planted extensively in the belief that it would become as popular as Chardonnay. It was meant to fill the gap between big, oaky California Chardonnays and lighter, high-acid Sauvignon Blancs, but never achieved widespread popularity. It is a variety with unique characteristics that produces wines in a number of countries and regions throughout the world, but nowhere is it a dominant grape.

Countries That Produce Viognier

Southern France, including the Rhône, Languedoc-Roussillon, and Provence, is home to Viognier. The most cherished appellations are the tiny Condrieu adjacent to the Côte-Rôtie in the Northern Rhône and the even tinier appellation Château Grillet within Condrieu. The largest French Viognier production area is in the Languedoc, which has dramatically increased its plantings over the last 15 to 20 years. It is also widely produced in South Australia and in California. On the East Coast of the United States, Virginia has become a surprisingly successful region for cultivating this grape. There are also a smattering of growers and producers in other wine regions of the world, including Italy, Chile, Argentina, and New Zealand.

Characteristics

The Grape

Viognier is a warm-climate grape. It is a difficult grape to grow, producing low yields that ripen easily but are low in acidity. Growers must maintain a delicate balance between maintaining acidity in the grapes and also ripening them enough to develop their rich, floral aromas. Harvesting before full ripeness produces good acidity but risks underdevelopment of the floral and fruit aromas that are its most desired characteristic. The loss of acidity can turn the wine from balanced to flabby quickly, creating a not-entirely-pleasant beverage even with its enticing aromas.

The Wine

Viognier is a wine with luscious dark yellow color, aromas of stone fruits and flowers, body from long ripening, and high-potential alcohol. The desired ripeness and richness of the wine causes low-acid levels. The aromas of the wine include jasmine, peaches, apricots, melons, and musk. Viognier is typically not allowed to go through MLF for fear of losing its delicate acidity. There are diverse views on how the wine should be aged, ranging from stainless steel with no oak to barrel aging and small barrique aging. Because the wine has the potential of losing its aromas quickly and lacks acidity, it does not typically age well. Viognier is sometimes blended in small amounts with Syrah in the Côte-Rôtie.

The wines of California and Australia are very different from those of the Rhône, developing not only richer flavors but higher alcohol levels as well. In California Viognier produces wines with rich stone fruit flavors from extended hang time and can easily achieve more than 15 percent alcohol. In California acidification can be used to compensate for low acids.

Enjoying Viognier

To some, Viognier is difficult to pair with food, but there are a variety of options and some foods match perfectly. If it is unoaked, serve it with lighter dishes like a roast chicken or pan-seared fish. If it is oak aged, treat it like Chardonnay. It has the same body so it goes well with richer foods, such as crab, lobster, and sauced dishes. It also pairs with lightly spiced Indian and Thai dishes. In the Languedoc it is commonly served with North African dishes. Two other possibilities are sushi and a broad range of cheeses from blue to Brie to Gruyère.

ADDITIONAL WHITE VINIFERA GRAPE VARIETIES

Some wines from the following varieties may be more difficult to find, and in some cases will only be found in a blended wine, but many are widely available. Often made from varieties native to the specific countries where they are grown, they offer unique tasting experiences, particularly with foods from the regions in which they are produced.

Albariño More than 90 percent of the production in Spain's Rías Baixas **DE-NOMINACIÓN DE ORIGEN** (DO) depends on the Galician grape, Albariño. It produces exquisite wines that are dense, tart, and aromatic, with low alcohol and high acids. The wine is often made from 100 percent Albariño, but in some DOs it is blended with more obscure local varieties. It is considered by many the perfect seafood wine, and in its local environs it is always served with pulpo (octopus).

Arneis This Italian variety is produced primarily in the Roero zone of Piedmont, where it is sold by its varietal name. It has become quite popular in recent decades as one of only two white varieties from Piedmont. It is made into a dry wine that has good acidity and a beautiful citrus and floral nose. Serve it with egg pastas, fish, and seafood.

In Rías Baixes, Spain, Albariño is grown on pergola trellises.

Assyrtico A Greek grape variety, Assyrtico, is grown on the Greek mainland and the island of Santorini. It is a high-acid grape with a fruit and mineral nose. It can be made into a wine on its own or blended with other varieties to create a softer, more rounded wine. Serve with light Greek foods, including fish and seafood.

Chasselas/Fendant The dominant white variety of Switzerland, it is produced primarily in the French-speaking region of the country, where many vineyards overlook Lake Geneva. It is also produced in Alsace and Pouilly-sur-Loire, France, where it is blended with Sauvignon Blanc. Today it has achieved greater status, producing wines specific to its terroir. Much of the Chasselas crop is used for excellent table wines. It is the perfect wine for a classic Swiss fondue.

Colombard/French Colombard A white variety originating in the Cognac region of France, Colombard was used alongside Ugni Blanc in the production of brandy. It was transplanted to California and became the most widely planted grape in the state. It is primarily grown in the vast Central Valley where, in the hot climate, it results in a grape that has little character but maintains good acidity and blends well with other varieties. It is an easy quaffer that would be delightful as a picnic wine.

Cortese Cortese is the most popular dry white wine of Piedmont, where it is best known by its place name, Gavi. It is also produced in Lombardy and the Veneto where it can be used in the blended wine, Bianco di Custoza. It is made in a dry style or with a little spritz that makes it a pleasant, easy-drinking wine that works well with regional Piemontese foods and with fish.

Ehrenfelser This **CROSS** of Riesling and Sylvaner was created in Germany. Researchers succeeded in developing a grape that can be grown in many climatic conditions but resulted in a wine that was generally too low in acidity. It is produced in the Pfalz and Rheinhessen, but it has also found a new home in the Okanagan Valley in British Columbia, where it makes luscious sweet wines, including ice wine. Serve it with a dessert of sweet cake or cookies.

Falanghina Falanghina is one of the native white varieties produced in the Southern Italian region of Campania. It is a dry, rich, flavorful wine that goes well with seafood and the local bufala mozzarella. It is excellent with pasta dishes or a salad composed of mozzarella, tomatoes, basil, and olive oil.

Fiano Another of the white varieties produced in the Southern Italian region of Campania, Fiano is a rich, spicy wine that has great aging potential. Fiano di Avellino is recognized with a **DENOMINAZIONE DI ORIGINE CONTROLLATA E GARANTITA** (DOCG). It is commonly served with seafood of the region, like seafood pizza or pasta with scallops.

Furmint A Hungarian grape variety that is best known as the dominant ingredient of Tokaji, the unctuous, botrytized sweet dessert wine that compares favorably with the best Sauternes. It is usually blended with Hárslevelü and Muscat Blanc à Petits Grains. It is also made into a dry style. When well made, it results in a full, rich, round complex wine with good acidity and aromas of pears and minerals. Hungarian dishes, like chicken dumplings, and both hard and soft cheeses work well. The sweet, botrytized Tokaji complements sweet cake or strudel.

Indigenous grapes are typically blended to create unique wines. Owner Francesca Spadafora and a winemaker blend white wine at Spadafora, Sicily, Italy.

Garganega A native Northern Italian variety, Garganega is considered one of Italy's most prolific grapes. When properly handled, it becomes an elegant, delicate, juicy wine. It is often blended with other varieties to give it needed body. Soave Classico DOCG represents Garganega at its best. It is commonly served with pastas, risottos, and polenta.

Greco/Greco Bianco Greco is grown throughout Southern Italy's appellations, including Campania, Apulia, and Calabria. The best wines are Greco di Tufo and the dried-grape sweet wine, called Greco di Bianco, from Calabria. The dry style goes with pastas, bufala mozzarella, tomatoes, fish, and seafood.

Grillo A white grape of Sicily, Grillo makes excellent dry wine and is also an ingredient in the fortified wine Marsala. Grillo can be rich in style with good structure and has an affinity for oak aging. In Sicily it is most frequently served with seafood.

Gros Manseng and Petit Manseng These two white grapes are from the relatively unknown South-West wine region of France. Gros Manseng is typically made dry in a bright, fresh style. Petit Manseng, when picked late, yields a fuller-bodied wine that stands up to oak aging and makes one of the great but little-known sweet wines of France. Drier wines work well with both cow's-milk and sheep's-milk cheeses. Duck confit (leg), fresh fish, if available, and salt cod dishes are commonly served in the Basque country.

Grüner-Veltliner Grüner-Veltliner is the most widely planted grape in Austria. It has become extremely popular internationally because of its affinity for food. It produces bright, fresh wines with citrus and stone fruit aromas and flavors as well as wines that have more weight and the ability to age. The wine works particularly well with fish and veal dishes.

Macabeo/Viura Macabeo is the most widely planted white grape in Northern Spain. In Rioja it is known as Viura and is the dominant white grape variety. In Penedès it is grown in large quantities for blending with Parellada and Xarel-lo to make Cava (see Chapter 17, Sparkling Wines, page 276). It contributes acidity to white wine blends and is added to red wines to soften them. Cava is a great wine to serve as a starter with deep-fried foods, like calamari.

Malvasia The Greeks brought Malvasia to Italy where it mutated into several subspecies that acquired unique traits as they adapted to their new habitats. Consequently, there are many permutations of wines made from Malvasia. In Italy it is blended with Trebbiano to add body and aroma. It produces a wine that has deep golden color and is full-bodied, crisp, and aromatic with good acidity. The Frascati **DENOMINAZIONE DI ORIGINE CONTROLLATA** (DOC) has the best reputation for making delicate, fragrant, and elegant Malvasia. It has also been transplanted to France, Spain, and Portugal where it is made into a range of sweet and fortified wines.

Marsanne Marsanne is planted in both the Northern and Southern Rhône as well as throughout Southern France. Small amounts are also produced in Italy and Spain. It has been transplanted to the United States where it is part of the Rhône-variety phenomenon in the Central Coast in California, and in Australia it is primarily produced in Victoria. The wine is made as a varietal in a full-bodied style and also blended with Viognier and Roussanne. In the Cen-

"Groovy"
Brundlemayer *
— salad
pancetta
vinaigrette
— spring rolls
— pork chops
(Weinerschnitzl)

tral Coast appellation it is served with fish and shellfish. In Southern France it works with pissaladière, the French pizza of olives, anchovies, and onions.

Moschofilero An aromatic Greek variety, Moschofilero is popular in Greece, most prominently in the Peloponnesos, and is starting to be seen in other countries. It is made into a varietal wine and also blended with other indigenous Greek white varieties. The wine is a natural partner of shellfish.

Müller-Thurgau Müller-Thurgau is a cross of Riesling and a type of Chasselas. It is the second most widely planted variety in Germany and is the dominant ingredient in generic blended wines, like Liebfraumilch. It has been transplanted around the world where it responds to the local climate. In higher, cooler climates it retains good acidity and can make a very drinkable wine. Freshwater trout pan-fried with potatoes and onions is a typical match for Müller-Thurgau.

Muscadet/Melon de Bourgogne Although commonly known as Muscadet, the grape is really Melon de Bourgogne. It is produced in the easternmost part of the Loire, bordering the Atlantic Ocean. It is a racy, acidic wine that goes perfectly with the local seafood and fish. Some consider it the perfect wine with oysters. The best come from a subappellation of the Loire, called Sèvre-et-Marne, which sits at the confluence of two rivers. The wine is never aged in wood but is usually left on its lees, which is noted on the label by the words "sur lies."

Parellada Parellada produces fine, high-quality wine in the poor soil and cool conditions of Catalonia, Spain, where it is cultivated in Penedès, Tarragon, and Lerida. When blended with Macabeo (Viura) and Xarel-lo, it adds freshness and a touch of citrus to Cava. It is also combined with Chardonnay and Sauvignon Blanc to make barrel-aged still white wines that are light, fresh, and floral with good acidity and low alcohol. Wood-fire grilled fish and seafood is a cooking style that matches well with this variety.

Gros Manseng and Petit Manseng are used for unique styles of dry and sweet white wines in Jurançon, France.

Prosecco A native grape of Italy, Prosecco performs well in the damp, cool climate of the Veneto and Friuli. The grapes are used primarily for light sparkling wine made using the Charmat method (see Chapter 17, Sparkling Wines, page 276). Prosecco is a delicate, fresh, and fruity aperitif with a straw yellow color. The best known DOC is Prosecco di Conegliano-Valdobbiadene. Serve it as you would with any sparkling wine. Salty, deep-fried foods, such as rice croquettes, are ideal. Lardo (cured pork fat) wrapped around a bread stick is a typical pairing.

Roussanne Roussanne is a white variety produced in the Northern and Southern Rhône regions as well as the Southern Mediterranean coast in Languedoc-Roussillon. It is blended with other white Rhône varieties and occasionally red Rhône varieties. It produces a rich wine potentially high in alcohol but with good acid balance and tree fruit flavors. Serve Roussanne with dishes that include garlic. A garlic-roasted chicken is perfect.

Sylvaner/Silvaner This is a widely planted grape in Alsace and Germany as well as other northern regions of Austria, Italy, and Switzerland and Central and Eastern Europe. Little is produced in the United States and other New World countries because the variety is quite neutral in flavor. Stuff a vol-au-vent, a buttery puff pastry, with a creamy seafood mixture, and garnish it with fresh herbs—it is a perfect match with Sylvaner.

Torrontés Torrontés is the name given to several different white grape varieties. The best known is the Torrontés produced in Argentina, where it is widely popular. It is a light, crisp, aromatic variety reminiscent of Muscat that is intended for early drinking. In Argentina Torrontés is served with cured meats, mild cheeses, and seafood.

Trebbiano Trebbiano (called Ugni Blanc in France) is the most widely cultivated and highest-volume white variety produced in Italy. It has mutated into many subspecies of varying quality. As a varietal wine, Trebbiano is pale, light-bodied, and lacks character. Because it is a neutral grape, it appears in numerous white wine blends, most unmemorable. However, when the best subspecies are blended with Malvasia or Chardonnay, the quality improves. Trebbiano, is one of the key grapes in Vin Santo, a dessert wine.

Verdejo Verdejo is the pride of Rueda, Spain, where the grape originated. Verdejo-based wines are known for their great character. They are full-bodied, fruity or herbaceous, with hints of pear, fennel, mint, citrus, grass, or apple. Verdejo is often blended with Sauvignon Blanc or Macabeo (Viura) to add complexity and reduce the potential for oxidation. The moderately alcoholic wines are very aromatic, are highly flavorful, and have good structure and acidity. Verdejo pairs with cured meats and the Spanish cheese manchego.

Head-pruned Verdejo, an indigenous grape, in the Rueda region of Spain

Verdicchio Verdicchio is an excellent although little-known nonaromatic white variety that makes wonderful wine in the Marche located in Central Italy on the Adriatic Sea. It is acidic and has green, herbaceous, and sour apple flavors that make it excellent with fish and seafood.

Vermentino Although its origins are probably Spanish, Vermentino is considered a variety indigenous to Sardinia, Italy. It is also planted extensively in the coastal regions of Tuscany and Liguria. Vermentino di Gallura, the only DOCG on Sardinia, is the area that shows off the best of the grape. It is a full-bodied, spicy wine with distinctive herbal qualities. As a coastal variety, it is one of the best wines to serve with seafood, particularly grilled squid.

Vernaccia Vernaccia produces a historic wine of Tuscany, Vernaccia di San Gimignano. It is a crisp, light wine with citrus and floral aromas that lacks body. For this reason it is blended with Chardonnay and/or Vermentino or Trebbiano. It is best as an aperitif wine with simple tomato bruschetta.

Xarel-lo A native of Catalonia, Xarel-lo adds body and alcohol to Cava. It is an aromatic grape that results in powerful and flavorful wines. It is primarily used in sparkling wine but can be found in still wines. Enjoy Cava with Spanish tapas.

SUMMARY

As you have discovered, you have many choices when selecting a white wine. The varieties described in this chapter provide alternatives to the major international varieties. Experimenting with these interesting wines can expand your horizons or open you to new tasting experiences. Lack of familiarity should not prevent you from trying different varietal wines. You may find that your ethnicity or culture is tied to one or more of these wines, and pairing them with foods of the same region can add to your knowledge of national or even regional cultures.

11 Cabernet Sauvignon

Cabernet Sauvignon, the most noble of red grapes, is the core ingredient of the best-known wine in the world, red Bordeaux, and it makes exceptional wines elsewhere, including the United States, Australia, and Italy. It is a grape that has become so popular that it has nudged out other varieties in vineyards throughout the world. It has replaced plantings of indigenous grapes in Italy and Spain and taken away vineyard space from Zinfandel in California and Shiraz in Australia. Newer appellations in South America, South Africa, and Central and Eastern Europe have all gravitated to Cabernet Sauvignon.

Amazingly, given its stature, it is more frequently blended than made into a 100 percent varietal wine. Growers consider Cabernet Sauvignon the perfect grape because it can grow an excellent crop under many conditions. Winemakers love it because of its fruit, tannins, and acid. Consumers like its depth of flavor and richness.

REGIONS AND APPELLATIONS

Although Cabernet Sauvignon is grown throughout the world, the discussion in this chapter highlights countries, appellations, or regions that demonstrate different methods and styles. (See Appendix 2 and Appendix 3 for maps of Cabernet Sauvignon–producing countries and appellations, pages 380–447.)

France

Cabernet Sauvignon is the absolute king in Bordeaux, although it is not the most widely planted red variety there. Bordeaux sits adjacent to the Atlantic Ocean and is influenced by three rivers: the Gironde, Garonne, and Dordogne.

Cabernet Sauvignon is dominant in all red wines produced west of the Gironde and Garonne, known as the left bank. East of the Gironde and Garonne, where Merlot is dominant, is known as the right bank. The leading Cabernet Sauvignon region is the Médoc. The most prominent village appellations, Margaux, Pauillac, St. Estèphe, and St. Julien, lie within the Haut-Médoc in the marginally higher southeastern portion of the Médoc adjacent to the Gironde. Lesser-quality Cabernet Sauvignon vineyards are located in the Bas-Médoc to the north and the Haut-Médoc to the west and south surrounding the esteemed village appellations. Below the city of Bordeaux lie Graves and Pessac-Léognan, which are also dominated by Cabernet Sauvignon and produce wines similar to those of the Médoc.

Cabernet Sauvignon is also produced in vineyards east of the Gironde in the Libournais where Merlot, and even Cabernet Franc, is more prevalent (see Chapter 12, Merlot, page 192). There are also substantial Cabernet Sauvignon vineyards in Entre-Deux-Mers between the Garonne and the Dordogne rivers. It is also produced in the various Côtes, including Premières Côtes de Blaye, Côtes de Bourg, Côtes de Francs, Côtes de Castillon, and the Premières Côtes de Bordeaux.

Cabernet Sauvignon is also cultivated in regions of the South-West. Bergerac in the Dordogne makes wines very similar to those in Bordeaux, as do the Côtes de Duras and the Côtes du Marmandais. From Madiran to the Pyrenees foothill regions of Jurançon and Irouléguy, Cabernet Sauvignon is used more frequently as a secondary blending grape with the dominant variety, Tannat.

Cabernet Sauvignon is also grown in the Central Loire regions of Anjou-Saumur and Touraine where it is also used as a blending grape, primarily with Cabernet Franc.

In Languedoc-Roussillon Cabernet Sauvignon is one of many varietal wines sold under the Vins de Pays d'Oc classification. The Languedoc is much sunnier and produces wines from riper grapes but at higher yields than those from Bordeaux, resulting in a wine that is less complex and is sold at moderate prices internationally.

The United States

Cabernet Sauvignon is grown in California from Mendocino County in the north to Temecula in the south. There are many great Cabernet Sauvignon appellations in California. Though other regions, including Sonoma County and Paso Robles, might disagree, Napa Valley is considered the best region by virtue of the value of Napa Valley Cabernet Sauvignon vineyards and the average selling price of their Cabernet Sauvignon wines. American Viticultural Areas (AVAs) within Napa Valley that produce Cabernet Sauvignon include Stags Leap District, Oakville, Rutherford, Mt. Veeder, Howell Mountain, Spring Mountain, Diamond Creek, and Calistoga. The vineyards on the valley floor, like Oakville, Rutherford, and Calistoga, are quite a bit warmer than the

A Cabernet Sauvignon vineyard at Rosenthal, the Malibu Estate along Southern California's coast

AVAs on the higher and steeper slopes of Mt. Veeder, Spring Mountain, Diamond Creek, and Howell Mountain, which typically produce a less-ripe style of wine.

Sonoma County is also a highly regarded California appellation. The AVAs Alexander Valley and Knights Valley in Northern Sonoma County are particularly recognized for Cabernet Sauvignon. Heading south in Sonoma County, Cabernet Sauvignon vineyards are found on the valley floor and hillsides in Dry Creek Valley, Russian River Valley, and Sonoma Valley.

In the Central Coast quite a number of Cabernet Sauvignon vineyards are planted from the Santa Cruz Mountains to Santa Barbara County. Cabernet Sauvignon performs particularly well in Carmel Valley in Monterey County and the warmer eastern side of Paso Robles. The Santa Cruz Mountains represent a cooler climate that produces wines more typical of a Bordeaux style.

The San Joaquin Valley (Central Valley) grows vast amounts of Cabernet Sauvignon grapes with high yields per acre. It produces most of the low- to modest-priced varietal and blended Cabernet Sauvignon wines bottled under the California appellation.

Outside California, some of the most dynamic Cabernet Sauvignon–based wines come from the Columbia and Yakima Valleys in Eastern Washington State. The valley floors produce riper grapes, but a range of excellent vineyards are found in the cooler Rattlesnake Hills; in Red Mountain with its steep southwest slope and temperature variation of more than 40°F/22°C from day to night; in Walla Walla, the oldest AVA in Washington; and in Horse Heaven Hills, which produces some of the best Cabernet Sauvignon in the United States.

Other warm climates producing Cabernet Sauvignon include Texas, Arizona, and even the state of Georgia. Cooler-climate Cabernet Sauvignon is produced in Southern Oregon and in New York on the North Fork of Long Island.

Italy

Cabernet Sauvignon was first planted in Piedmont in the early 19th century and is now cultivated throughout the country. It is used in the Trentino-Alto Adige, Friuli, and the Veneto in varietal wines and blends. In Piedmont, Angelo Gaja established one of the best Cabernet Sauvignon vineyards in Italy. Here he makes his famed Darmagi from a blend of Cabernet Sauvignon and the native Piedmont varieties, Nebbiolo and Barbera.

The best-known Italian Cabernet Sauvignon wines are modeled after Sassicaia and Ornellaia, which come from Bolgheri in the Maremma section of Tuscany on the Mediterranean coast. Sassicaia is an 85 percent/15 percent blend of Cabernet Sauvignon and Cabernet Franc. Ornellaia is a more traditional right bank Bordeaux blend of Merlot, Cabernet Franc, and Cabernet Sauvignon. It is now grown throughout Tuscany where it is commonly blended with Sangiovese to make Super Tuscan wines (see Chapter 12, Super Tuscans, page 194). Farther south Cabernet Sauvignon is planted from Umbria to Sicily where it is made as a varietal wine or a Bordeaux-style blend, or blended with indigenous varieties.

Australia

Cabernet Sauvignon is the second most widely planted red grape in Australia after Shiraz (Syrah). It is planted throughout most of the country's growing regions, highlighted by Coonawarra and Clare Valley in South Australia and Margaret River in Western Australia. Coonawarra is considered the top level of regions growing Cabernet Sauvignon in part because of special terra rossa soils (see Soil, page 183). Cabernet Sauvignon is also planted throughout Western Australia, other parts of South Australia, Victoria, and New South Wales.

South America

Cabernet Sauvignon is grown throughout South America in countries including Brazil, Uruguay, and Argentina, but Chile is by far the dominant producer on the continent both in quality and quantity. Cabernet Sauvignon, the leading variety of Chile, was first planted in the Maipo Valley in the 1800s. Today it is planted from the Elqui in the north to the Bío Bío in the south. Although vineyards are planted for more than 700 miles along the Andes mountain range, the best locations for the grapes are in the central part of the country starting from the more northerly Aconcagua east of Valparaíso moving south through the Central Valley's subregions, Maipo, Rapel, Maule, and Curicó Valleys. Rapel Valley's Cachapoal and Colchagua are, perhaps, two of the most important areas. In this moderately narrow band close to the base of the Andes, Cabernet Sauvignon is the dominant variety.

The success of Cabernet Sauvignon in Chile has led to its cultivation in Argentina and Uruguay. It is widely produced in Argentina in Mendoza, but from a quality standpoint it is overshadowed by the dominant Malbec. In Uruguay, a small wine-producing country, Cabernet Sauvignon is a major variety, but the country has yet to achieve much international recognition for its wines.

Eastern Europe

Little is written about Moldova, Romania, and Bulgaria along with other Eastern European countries because for so long they were part of the Communist bloc. Although these countries have historically made lots of wine, it was all consumed within Eastern Bloc countries, mostly Russia. High yields caused diluted wines, making them unmarketable to Western countries.

Since the breakup of the Communist bloc the wine industry has begun to change very slowly. Cabernet Sauvignon has been planted in Eastern Europe and is one of the dominant varieties in Moldova, Romania, and Bulgaria, but there are roadblocks to these wines entering the international marketplace. Quality lags because the local taste is for sweet red wines. Government regulations have failed to establish requirements that would improve quality and meet international standards. Foreign investment and improved winemaking technology have also been slow to arrive in the region.

Table 11.1 Prominent Cabernet Sauvignon Appellations

COUNTRY	REGION/APPELLATION	SUBAPPELLATION	VILLAGE OR SUBAPPELLATION
France	Bordeaux	Médoc-Haut-Médoc	Margaux Pauillac St. Estèphe St. Julien
		Graves Pessac-Léognan Entre-Deux-Mers	
	South-West	Dordogne	Bergerac
	Loire	Anjou-Saumur	
	Languedoc-Roussillon		
United States	California	Napa Valley	Stags Leap District Yountville Oakville Rutherford St. Helena Calistoga Mt. Veeder Spring Mountain Diamond Creek Howell Mountain
		Sonoma County	Alexander Valley Knights Valley Dry Creek Valley Sonoma Valley
		Mendocino County	
		Central Coast	Santa Cruz Mountains Monterey County Paso Robles
	Washington State	Columbia Valley	Yakima Valley Horse Heaven Hills Red Mountain Wahluke Slope Walla Walla Valley

Other Countries

Cabernet Sauvignon is also found in other major wine-producing countries from Spain to South Africa to New Zealand. In many cases, these countries are making exceptional wines. Cabernet Sauvignon is planted throughout Spain in relatively small amounts. It is blended with Tempranillo in many appellations, but has achieved particular stature in the Jean Leon wines of Penedès and the wines of Vega Sicilia in Ribera del Duero. Even so, Spanish Cabernet Sauvignon is a small player in the international market.

In both premiere regions of South Africa, Stellenbosch and Paarl, Cabernet Sauvignon is a dominant red variety. It is used to make varietal wines and excel-

Prominent Cabernet Sauvignon Appellations, continued

COUNTRY	REGION/APPELLATION	SUBAPPELLATION	VILLAGE OR SUBAPPELLATION
Italy	Tuscany	Maremma	Bolgheri
	Piedmont Trentino-Alto Adige Friuli Veneto Umbria Sicily		
Australia	South Australia	Coonawarra Barossa Valley McLaren Vale Clare Valley	
	Victoria		
	Western Australia	Margaret River	
Chile	Central Valley	Curicó Valley Maipo Valley Maule Valley	
		Rapel Valley	Colchagua Cachapoal
Argentina	Mendoza	Uco Valley	
South Africa		Paarl Stellenbosch	
Spain	Castile and Leon	Ribera del Duero	
	Catalonia	Penedès	
New Zealand	North Island	Hawkes Bay	

lent Bordeaux-style wines. With the overall quality of South African Cabernet Sauvignon wines improving, it is available more frequently in the United States and other countries, but general awareness of South African wines, including Cabernet Sauvignon, is quite low.

New Zealand is far better known for its Sauvignon Blanc, Pinot Noir, and Chardonnay, but on the North Island's Hawkes Bay, wineries produce some amazing Bordeaux blends, including those dominated by Cabernet Sauvignon. Because production is low they are not easy to find in the American marketplace.

Table 11.1 includes an abbreviated list of countries that produce highly regarded Cabernet Sauvignon along with their most prominent appellations and subappellations.

CABERNET SAUVIGNON: THE GRAPE

The Cabernet Sauvignon grape has been characterized as a grower's dream. It is easy to grow and adapts to various soils and climate conditions, and demand and prices for the harvested grapes are high.

Cabernet Sauvignon took root in Bordeaux and is relatively youthful, as *vinifera* varieties go. It is a cross of two other well-known varieties planted in the same region, Cabernet Franc and Sauvignon Blanc, that probably occurred naturally in the vineyards.

The vine produces small, dark grapes in small, tight clusters. The grape's color is closer to blue or black than red or purple. It is characterized as a skin wrapped around four seeds: a small berry with thick skins that has less pulp and, therefore, less juice than most grapes. Because of their small berries and high ratio of solids to juice, Cabernet Sauvignon grapes are extremely intense with a high concentration of phenolics, the color and flavor components of wine. The skins and seeds are high in tannins, which contribute to making Cabernet Sauvignon potentially a quite tannic wine.

In the growing cycle, Cabernet Sauvignon is the last of the major varieties to bud. This minimizes the risk of spring frosts that kill the buds, causing loss of the crop for the vintage. Conversely, late ripening opens the grapes to problems caused by rainy and cold fall weather before or during harvest.

One of the benefits of Cabernet Sauvignon is that it is generally disease- and rot-resistant. The vines are vigorous but tend to produce low yields. Newer variety and rootstock clones have increased productivity but require the use of more vineyard-management techniques.

IN THE VINEYARD

Cabernet Sauvignon can be planted wherever there is enough heat and sun to allow it to ripen. Whereas some varieties change dramatically in style when planted under different conditions, Cabernet Sauvignon retains its distinguishable characteristics under most circumstances. Climate, soil, sun, temperature, and human intervention all affect the quality of the harvested grapes.

Climate

Climate is the underlying factor that determines where Cabernet Sauvignon can be successfully cultivated. It is not normally planted in cool climates because the grapes will not fully ripen, and some northern countries, like Germany, do not even attempt to plant it for that reason. When planted in borderline climates, like the Loire in France, Trentino in Italy, or coastal regions of California, it is always questionable whether the grapes will achieve full ripeness. In these areas the vineyards need to squeeze out as much warmth as possible, otherwise the unripe grapes may develop an undesirable overly vegetal, herbal character that does not suit most Cabernet Sauvignon consumers.

One caveat concerning climate and ripeness has been the effect of global warming. Bordeaux is an excellent example of its impact. Historically, the region has occasional difficult vintages that are too cold to ripen Cabernet Sauvignon grapes. Recently, however, the area has experienced a string of years with warm temperatures and more sun, and it has produced riper grapes for several consecutive vintages.

Soil

Cabernet Sauvignon was originally planted, and achieved its earliest acclaim, on the gravelly soils of the left bank of Bordeaux. Gravel is ideal because it warms quickly in the sun and retains heat, a key factor that helps the grapes ripen fully in Bordeaux. Gravel also drains well. In warmer appellations Cabernet Sauvignon grows well in other kinds of soils as long as they provide good drainage. Because Cabernet Sauvignon is a vigorous producer, growers prefer planting it in low-fertility soils.

Another famous soil for Cabernet Sauvignon is the terra rossa of Coonawarra in Australia, which is a mix of sand and clay layered over limestone. The clay helps to retain moisture and the sand provides good drainage. The red color comes from the iron oxide in the soil. This soil mix gives minerality to the Cabernet Sauvignon grown there.

In California it is common to see Cabernet Sauvignon planted on alluvial soils like those found on the valley floor of Napa Valley. Usually composed of sandy limestone and often fertile, these soils work in environments like Napa Valley where there is ample sunlight and heat. The soil fertility does not seem to negatively impact grape quality in this area, as the region produces intense, ripe grapes. Off the valley floor, sloped vineyards at higher altitude have much poorer-quality soil and produce a leaner style of grape that is less fruit-driven.

Water Access

Although Cabernet Sauvignon thrives with heat wherever it is grown, it also requires continuous moisture to develop the intense color, aroma, and flavor characteristics that it is known for. In Bordeaux, where irrigation is not allowed, there is generally enough rain. In more arid climates, particularly California and Australia, Cabernet Sauvignon vineyards must be irrigated.

Vineyard Management

Rootstock for Cabernet Sauvignon is an important consideration. Almost all Cabernet Sauvignon rootstock is a phylloxera-resistant native American variety. Because it can be such a vigorous producer, grafting it to low-producing rootstock helps manage crop size. Cabernet Sauvignon rootstock can also be selected based on such characteristics as drought resistance and disease resistance.

Growers use several techniques to obtain the best possible Cabernet Sauvignon crop based on their local climate and soil. Trellising for Cabernet Sauvignon varies. In France and Italy the standard is cane-pruned Guyot, in which the vines are pruned close to the ground where they absorb the heat from the gravelly soils. In California and other New World countries, vertical shoot positioning (VSP) is commonly used, allowing vines to grow higher with less canopy to increase sun exposure on the grapes.

The biggest problem with Cabernet Sauvignon in the vineyard is potential mildew, which is controlled by spraying with sulfur. Mildew causes less difficulty in sunnier, drier climates but can be a serious problem in foggy areas.

In Bordeaux yields range from three to four tons per acre/50 to 65 hectoliters per hectare. The same is typical in California, but there can be a much wider variance. Some old high-quality vineyards may harvest fewer than three tons per acre, whereas high-yielding California San Joaquin Valley (Central Valley) fruit may reach a yield of six to eight tons per acre/100 to 130 hectoliters per hectare. These grapes end up in very different wines, both in terms of quality and retail price.

Harvest

With late bud break comes late ripening and late harvest. The benefit is that grapes are slow to ripen, allowing flexibility in the harvest schedule. However, growers face the challenge of deciding whether to pick the grapes before they have ripened to the desired level or risk exposing them to rain while waiting for more ripening.

If Cabernet Sauvignon is picked too early, the underripe grapes have a strong, green pyrazine character. The grapes will have an herbal or vegetal tone and will not have developed fruit flavors. Although some prefer retaining the typical herbal pyrazine characteristic, fully ripe grapes with rich fruit flavors are generally desired. Rain dilutes the flavors in grapes, so harvesting immediately after a storm reduces grape quality. However, leaving grapes on the vines following rain is risky if sunny, hot weather does not happen soon enough to dry them.

The grape quality determines, in some cases, how Cabernet Sauvignon is picked. High-quality grapes for more prestigious wines are hand-harvested, but by volume most Cabernet Sauvignon grapes are harvested mechanically. Historically, Cabernet Sauvignon has been picked at 23° to 24° Brix to achieve a potential alcohol of less than 14 percent. Some growers are allowing more hang time to obtain riper grapes, up to 27° Brix or more, resulting in a wine with potential alcohol of 15 percent or higher.

IN THE WINERY

Cabernet Sauvignon is a powerful grape that demands skilled winemaking to make the best wines. There has been a significant change in the style of Cabernet Sauvignon being produced by some countries and their winemakers over the last 25 years, but the basics of making Cabernet Sauvignon remain the same.

Cold Soak

From the receiving station or sorting table Cabernet Sauvignon grapes are conveyed through the crusher-destemmer to a tank where a cold soak begins. Some winemakers allow some stems to remain with the must for added complexity. Letting the grapes cold soak (macerate) before fermentation allows color, fruit flavors, and softer tannins to transfer to the juice. Temperature is kept to a desirably cool level (40°F/11°C) to prevent fermentation from starting.

The style of Cabernet Sauvignon to be produced determines the length of maceration both before and after fermentation. The longer the grapes soak, the more color, flavors, and tannins transfer to the juice. Darker, more age-worthy wines with heavier tannins undergo extended soaking that can last as long as two or three weeks. In the New World some producers shorten or eliminate maceration to create early-maturing wines in a softer, young-drinking style.

Cabernet Sauvignon is often cold soaked to bring out its colors and flavors and soften its tannins.
WINE INSTITUTE OF CALIFORNIA

Fermentation

Fermentation follows cold soaking in the same tank. Fermentation temperature for Cabernet Sauvignon is allowed to rise as high as 85°F/30°C, which initiates or speeds fermentation. The higher the temperature, the quicker fermentation takes place. For Cabernet Sauvignon converting all of the sugar to alcohol takes about seven to 10 days. If the winemaker wants fruitier flavors and less extraction, he or she uses lower fermentation temperatures, which extends fermentation time. During a Cabernet Sauvignon fermentation pump overs and punch downs are commonly used to continue extracting color and flavor components.

If the winemaker is producing an inexpensive or moderately priced wine that will not be barrel aged, oak alternatives, such as chips or powder, may be added during fermentation. Post-fermentation, the wine may go through additional maceration for as much as a week or two to extract more tannins. The wine is then pressed and readied for aging.

Malolactic Fermentation

Almost all red wines, including Cabernet Sauvignon, go through malolactic fermentation (MLF). This occurs naturally if the wine is put in oak barrels that have previously been used, which is typical of most Cabernet Sauvignon. If the wine does not go through barrel aging or new barrels are used, lactic acid bacteria are added to initiate the process. If acids become too soft, acidification is allowed in New World countries.

Barrel Aging

Among the red varieties, Cabernet Sauvignon is the most likely candidate for barrel aging because the strength and power of Cabernet Sauvignon holds up exceptionally well to oak. All high-quality wine that contains Cabernet Sauvignon is barrel aged. French oak is the preferred wood in the Old World, but in the New World, producers use both French oak, which is subtler, and American oak, which is more overt. Aging can last for up to 24 months, or even longer, in barrels that range from new to five or more years old depending on the winemaker's aging program for each wine. During this time the wine receives the additional benefit of micro-oxidation to soften the tannins.

At Château Reynier in Entre-Deux-Mers, France, wine is barrel aged in caves carved by the Romans.

Blending

Cabernet Sauvignon is a variety that is almost universally blended, mostly with Merlot and Cabernet Franc, but also with indigenous varieties. Typically each variety and lot is fermented and aged separately, leaving the blending until both fermentation and aging are complete.

In Bordeaux the classic red blend combines Cabernet Sauvignon with Merlot, possibly a smaller amount of Cabernet Franc, and maybe even smaller amounts of Petit Verdot and Malbec.

In Italy Cabernet Sauvignon can be blended with Sangiovese in Tuscany, Nebbiolo or Barbera in Piedmont, Corvina in the Veneto, Nero d'Avola in Sicily, Aglianico in Campania, or Carignano in Sardinia. In the Veneto and Tuscany, it may be combined with Merlot instead of indigenous varieties. In Spain it is most commonly blended with Tempranillo.

In the United States, Cabernet Sauvignon is primarily sold as a varietal wine. According to the Alcohol and Tobacco Tax and Trade Bureau (TTB), only 75 percent of the wine is required to be Cabernet Sauvignon. Up to 25 percent of other varieties may be added without disclosing which are included. Merlot and Cabernet Franc are commonly used. The classification **MERITAGE** is a classic example of a wine produced in the United States that is made only with Bordeaux varieties and is usually dominated by Cabernet Sauvignon (see Meritage Wines, at right).

In Australia, Cabernet Sauvignon is commonly blended with Shiraz or Grenache but is also combined with other classic Bordeaux varieties.

CABERNET SAUVIGNON: THE WINE

Cabernet Sauvignon produces a wine reflective of its grape characteristics. Most consumers expect a dark wine with intense aromas and flavors. The wine can be heavy, tannic, and brooding. It is potentially one of the longest-lived wines produced and possibly a candidate for long cellaring because of the high tannin level, acid, and richness contributed by the grape's high solids content. However, many California-style Cabernet Sauvignons may not age well for long periods because of overripeness and low acid.

Cabernet Sauvignon can also be made in a soft, early-drinking style, which is typical of many high-volume, value-priced wines. In California some of the better-known brands even leave a touch of sugar in the wine to enhance the fruitiness.

Color

Cabernet Sauvignon is generally concentrated in color because of the grape's dark, thick skins and high solids content. The length of skin contact determines the color; as with any red wine, the longer the maceration, the darker the wine. The color ranges from dark red to a deep bluish purple when young. As Cabernet Sauvignon ages, it can begin to lose some of its color, first fading around the edges, then turning a brick color, and eventually a little orange. These color changes can indicate more complexity in the wine or, in some cases, can mean that the wine is old and tired.

Aromas, Flavors, and Mouthfeel

The aromas and flavors of Cabernet Sauvignon are influenced by a number of factors. The soil and plants grown adjacent to a vineyard impart initial flavors to the grapes. The primary fruit aromas in Cabernet Sauvignon wines come from the grape and include black currants (cassis), blackberries, and black cherries. Climate has an impact on the level of ripening and therefore affects the dominant flavors. If the wine comes from grapes that are less than fully ripe with a lower Brix reading at harvest, herbal or even vegetal aromas may be noticeable. These include bell peppers, eucalyptus, green olives, and dried herbs. In riper wines a more pronounced jammy fruit character occurs. Even riper fruit shows plummy, overripe, or cooked-fruit qualities. If the wine is aged in oak barrels, another array of aromas and layer of depth can develop depending on the origin and age of the barrels. Vanilla, tobacco, coconut, cedar, and cigar box can be found in barrel-aged Cabernet Sauvignon. Table 11.2 lists aromas and mouthfeel that might be found in Cabernet Sauvignon.

MERITAGE WINES

In 1988 three California vintners founded an organization to highlight the qualities of blended American wines. They focused on creating a replica of Bordeaux wines using only Bordeaux varieties. Today, the Meritage Association represents over 200 producers from around the world, excluding Bordeaux, but most members are located in the United States.

The name *Meritage* is trademarked and can only be used by members of the Meritage Association. The name is a blend of merit and heritage, and is pronounced like heritage. Meritage is not certified as a class or type by the TTB and can only be listed on the label of a wine in the label areas where **PROPRIETARY NAMES** are allowed.

The key regulations for Meritage wines are:
- Only classic Bordeaux varieties can be used. The red varieties allowed are Cabernet Sauvignon, Merlot, Cabernet Franc, Malbec, Petit Verdot, Gros Verdot, St. Macaire, and Carmenère. The whites are Sauvignon Blanc, Sémillon, and Muscadelle.
- The blend must contain a minimum of two varieties.
- No single variety can make up more than 90 percent of the blend.

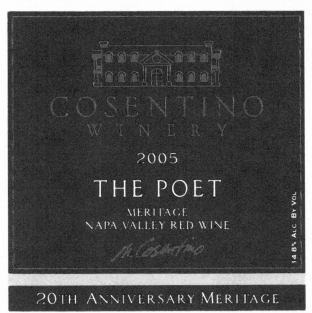

Cosentino is the first winery to create a Meritage label.

Table 11.2 Cabernet Sauvignon Aromas and Mouthfeel

	AROMAS		MOUTHFEEL
FRUIT/FLORAL	**HERBAL/VEGETAL**	**OTHER**	
Blackberries	Asparagus	Cedar	Astringent
Black cherries	Bell peppers	Cigar box	Hot
Black raspberries	Eucalyptus	Coconuts	Rich
Currants (cassis)	Green olives	Earth	Weighty
Plums	Herbs	Leather	
	Peppercorns	Oak	
		Pencil shavings	
		Smoke	
		Tar	
		Toast	
		Tobacco	
		Vanilla	

Blending with other varieties also affects the aromas, flavors, and mouthfeel. Merlot is added to provide additional fruit, soften the acidity, or provide texture. Cabernet Franc is added to lighten the color and body and add fruit flavors. It ripens earlier and adds an array of red fruit and herbal aromas, particularly in cool vintages. Likewise, indigenous varieties from Italy and Spain and Shiraz in Australia each add its own characteristics to Cabernet Sauvignon blends.

Cabernet Sauvignon with its thick skins and high ratio of solids to juice exhibits strong tannins, which produce an astringent mouthfeel. Extended maceration can make tannins extremely harsh, with an even more parched, puckery mouthfeel. Blending with less-tannic varieties, like Merlot, and aging can soften the hard edge of Cabernet Sauvignon tannins. The wine can also demonstrate a rich, full mouthfeel because of its concentration. In addition, extended grape ripening, which is common in New World wines, develops more sugar that gets converted to high alcohol, leaving a hot mouthfeel.

Wine Styles

For many years Bordeaux-style, Cabernet Sauvignon–dominant blends were favored by consumers. The style maintains a balance of black fruit, herbs, acid, tannins, and oak, and is made to be served with food and cellared for decades. They are typically lighter in color and higher in acidity than Cabernet Sauvignons from warmer New World appellations. These wines are moderate in alcohol (12 to 13 percent) and can be sharp and astringent when young. Following a few years of aging, the wines become balanced, demonstrating an increased array of aromas, softer tannins, and acidity that is complementary to many foods, earning these wines their international reputation. Terms used to describe this traditional style of wine are *complex, elegant, delicate,* and *austere.* To a local Bordelaise, or frequent consumer, the variety and regional character of the wines are easily identified and highly regarded (see Table 11.3).

A new Cabernet Sauvignon style has evolved over the last few decades in California and Australia, where grapes ripen more easily than in Bordeaux.

Table 11.3 Cabernet Sauvignon Wine Styles

	BORDEAUX STYLE (OLD WORLD)	CALIFORNIA STYLE (NEW WORLD)	BLENDED VARIATIONS
VARIETAL/BLEND	Always blended with Merlot, Cabernet Franc	Varietal or blended with Merlot or other varieties	Sangiovese, Shiraz, Tempranillo, or others
CLIMATE	Moderate	Warm to hot	Cool to warm
SOIL	Gravel	Gravel and other soils	Variable
ALCOHOL	12%–14%	14%–15% or higher	12%–15% or higher
ACID	High	Moderate to soft	Moderate to high
AROMAS/FLAVORS	Fresh fruits, herbals	Blackberry jam, plums, oak, vanilla	Cassis, tar, earth
COLOR	Light to medium	Medium to dark	Variable
BODY	Light to medium	Medium to heavy	Variable

California Cabernet Sauvignon is darker in color and higher in alcohol (14 to 15 percent or more). Riper grapes produce higher sugar and lower acid, creating deep, rich, jammy fruit, and fewer herbal notes. This riper, softer style of Cabernet Sauvignon requires less need to blend with Merlot or other varieties. The softer, fruit-forward wines have a sensation of sweetness from the riper fruit and higher alcohol. They are ready for drinking younger and do not require as long in the cellar. Terms often used to describe this style are *powerful, intense, concentrated, deep,* and *opulent.* These big wines are enjoyable to taste but work best with the richest foods. This style has become the preferred choice of American wine journalists and consumers from the United States and other New World countries. Even in Bordeaux many producers today are allowing their grapes to ripen longer to achieve wines with a similarly rich style and higher alcohol levels.

Variations on both basic styles have also evolved. In Italy a number of producers blend Cabernet Sauvignon with the more traditional Sangiovese, and it can now be included in a Chianti blend. Australian Cabernet Sauvignon, when not made into a varietal wine, is usually blended with varieties that are not in the classic Bordeaux mix, such as Shiraz and Grenache. In Spain Cabernet Sauvignon is commonly blended with Tempranillo.

Cabernet Sauvignon is also mass-produced in a generic style of wine for large-scale bottling. Increasing yields, simplifying winemaking, and eliminating barrel aging can make the wine quite in-

Examples of regional Cabernet Sauvignon, each produced in a different style, from left: Australia; Chile; Bordeaux, France; Napa Valley, California

TOM ZASADZINSKI/CAL POLY POMONA

expensive. It reduces the varietal character substantially but makes a very easy-drinking and affordable wine.

ENJOYING CABERNET SAUVIGNON

Buying Cabernet Sauvignon

Cabernet Sauvignon is the most salable red wine because it is familiar to most consumers and there is a cachet to buying and consuming it. Understanding the available styles can help guide you in purchasing Cabernet Sauvignon. For example, lighter, lower-alcohol wines are likely to be less expensive, while darker, riper, higher-alcohol styles are likely to be more expensive. The specificity of the appellation is also an excellent guide to a wine's potential quality and price. As an example, a Rutherford appellation is likely to be more expensive than a general Napa Valley appellation, which in turn is more expensive than a California appellation. The same holds true for wines from other domestic and foreign appellations.

When looking at a bottle of Cabernet Sauvignon, determine the appellation, verify the alcohol level, and then try to recall the regional style. A Cabernet Sauvignon–dominant Bordeaux wine is likely to be quite different from a varietal Cabernet Sauvignon from Napa Valley, Sonoma County, or Paso Robles, and will be different from a wine from Australia or Italy.

Proprietary wines made from Cabernet Sauvignon grapes are typically expensive, and it is sometimes difficult to determine what other grapes are in the bottle or the style of the wine. The back label may contain this information or your retailer or restaurant's sommelier or knowledgeable server can describe the wine's style.

Storing and Serving Cabernet Sauvignon

Cabernet Sauvignon should be kept at cellar temperature (55°F to 58°F/13°C to 15°C) or in the coolest, darkest location in a home, and like most red wines, should be served at cellar temperature or slightly above. A Cabernet Sauvignon served at 75°F/24°C or higher will be hot, alcoholic, and unpleasant to taste. If serving a Cabernet Sauvignon in a room this warm, place the wine in the refrigerator for 20 to 30 minutes until you achieve the desired temperature, or in a restaurant, place it in an ice bucket with a mixture of ice and water. Five minutes in the bucket should bring the temperature down to a desirable level of 65°F to 70°F/18°C to 21°C.

The best glass for serving Cabernet Sauvignon is a standard Bordeaux glass with a large enough bowl to swirl the wine to release its aromas.

Cabernet Sauvignon and Food Pairings

Cabernet Sauvignon is typically a dark, intense, and tannic wine that requires foods and cooking methods that stand up to its power. The classic food match with Cabernet Sauvignon is red meat. Almost any cut of beef or lamb cooked using any dry- or moist-heat cooking method is appropriate. Grilling, roasting, or sautéing higher-quality cuts of steaks, chops, and roasts work perfectly. The caramelization of the meat's sugars during cooking and the richness of well-marbled meat are perfect foils for the tannins in the wine. When less tender cuts of meat, like a lamb shank or pot roast, are braised or slow cooked, the liquid or sauce becomes the connection to the wine because the proteins and fats from the meat combine to offer a textural component. And if Cabernet Sauvignon is used as part of the cooking liquid, it is almost a guarantee that the marriage will work.

Grilled vegetables or potatoes basted in olive oil and herbs, particularly if cooked over charcoal, also go well with Cabernet Sauvignon. A great Cabernet Sauvignon served with barbecued steaks or chops and grilled vegetables is a perfect way to accommodate both vegetarians and meat eaters. Other options are bitter vegetables, like dark greens, Italian radicchio, eggplant, and peppers. Cabernet Sauvignon is also terrific matched with lardo, one of the great Italian antipasti, which consists of thinly sliced cured pork fat with salt, pepper, and herbs layered on toast or wrapped around a thin breadstick.

Chocolate and Cabernet Sauvignon can be a great match, but be very, very careful. The chocolate or chocolate dessert must be very dark, almost bitter, with little sugar. Semisweet chocolate is too sweet. Look for bittersweet chocolate that is at least 70 percent cocoa. To accompany chocolate, select the ripest style of Cabernet Sauvignon available. You need a voluptuous, jammy, high-alcohol wine that is sweeter than the chocolate. A Cabernet Sauvignon from California is a good place to start.

TOP 10 FOODS TO PAIR WITH CABERNET SAUVIGNON

1 Steak
2 Lamb chops
3 Charcoal-grilled meats and vegetables
4 Roasted prime rib
5 Mushrooms
6 Pot roast
7 Roasted peppers
8 Lardo
9 Bitter greens
10 Extra dark chocolate

Napa Valley is one of the premier Cabernet Sauvignon–producing regions in the world.
WINE INSTITUTE OF CALIFORNIA

SUMMARY

Cabernet Sauvignon is an intriguing red wine that serves as a model for other full-bodied red wines. Considered the premier red grape of France used in Bordeaux wines, it is now grown in almost every major wine-producing country. Most Cabernet Sauvignon is blended, even if labeled as a varietal wine. The more costly wines are made in two distinctive styles: Old World, demonstrating traditional winemaking characteristics all expressed in moderation, and New World, showcasing a modern interpretation of winemaking with ripe fruitiness and power.

Merlot

Originating in Bordeaux, Merlot was long considered the stepchild of Cabernet Sauvignon because it was thought of as a secondary blending grape. In reality, it is the most widely planted grape variety in France and is second only to Cabernet Sauvignon in wine production worldwide. Merlot produces a round, succulent, enjoyable varietal wine and takes the hard edge off other tannic varieties when blended.

Merlot is often combined with Cabernet Sauvignon because the grapes contain complementary characteristics. The grape varieties are dramatically different, but the sum of the two creates a distinctive wine that is thought by many to be better than the varietal wines made from either grape by itself. What Cabernet Sauvignon is missing, Merlot adds, and vice versa. For this reason, Merlot and Cabernet Sauvignon are almost always blended. Even when a label states that the wine is Merlot, in all likelihood it has been blended with some amount of Cabernet Sauvignon.

CHANGING CONSUMER TASTE

Merlot has long been subject to consumer whims. For decades, it was planted for use as a blending grape, and its popularity grew steadily. Then in the 1990s consumers in the United States, Chile, and Argentina jumped on the Merlot bandwagon following a broadcast on *60 Minutes* about the health benefits of red wine. Growers worldwide planted or grafted enormous quantities of Merlot to meet the mushrooming demand, which was fueled in large part by the U.S. market. California, Languedoc-Roussillon, and Chile became the leaders in this growth spurt. Why Merlot and not some other red variety? Merlot displays a soft fruitiness with a sensation of sweetness that new red wine drinkers find friendly and easy. It does not have the high acidity and challenging tannins of other red wines.

It was only a matter of time until supply outstripped demand and Merlot consumers switched to new, hot wine varietals, like Pinot Noir and Syrah. Especially in the United States, there was a glut of Merlot. Prices for the grapes dropped, and some growers began grafting Merlot over to these up-and-coming varieties. In spite of changing tastes, however, Merlot maintains its desirability with consumers because of its aromas, flavors, and ease of drinking. Now with a better balance of supply to demand, it still ranks among the world's most desired wines.

REGIONS AND
APPELLATIONS

Merlot is widely planted throughout the world. Because of its compatibility with Cabernet Sauvignon, it is commonly grown in close proximity to Cabernet Sauvignon–producing regions. (See Appendix 2 and Appendix 3 for maps of Merlot-producing countries and appellations, pages 380–447.)

France

France produces more Merlot than any other country. It is the dominant grape variety in Bordeaux, is the primary grape in some of the most luscious wines of Pomerol and St. Émilion, and plays a major role in virtually all Bordeaux appellation red wines. Why is Merlot dominant here? Because Merlot ripens early, it is almost guaranteed to fully ripen in Bordeaux's cool, variable climate.

A Merlot vineyard at Cheval Blanc in Bordeaux, France

The premier Merlot regions are the right bank appellations of St. Émilion and Pomerol. Smaller amounts of Merlot are produced in the Médoc and Graves, where Merlot is considered secondary to Cabernet Sauvignon. Historically, the left bank wines of the Médoc were considered superior to those of Pomerol and St. Émilion. However, the right bank wines are now considered on equal footing, and some of the wines sell at prices higher than those of the famous First Growths of the Médoc.

The bulk of Bordeaux Merlot is produced outside of these four appellations. Merlot-dominant Bordeaux and Bordeaux Supérieur wines are made from grapes grown in the Entre-Deux-Mers. The various Côtes appellations surrounding St. Émilion are also heavily planted with Merlot.

Merlot is also grown in a number of regions outside of Bordeaux. The other major area is Languedoc-Roussillon, where Merlot is made as Vins de Pays d'Oc varietal wine. In the warmer Southern Languedoc, growers are able to produce riper grapes at higher yields than their counterparts in Bordeaux. The wines are less complex and are sold at moderate prices internationally. Much of this Merlot is sold in bulk and bottled in the United States under an American label.

In the South-West, Bergerac is the largest Merlot-producing appellation, making Bordeaux-style blends. In Cahors Merlot is used as a blending grape for Malbec. In addition, a bit of Merlot is planted in Ardèche, Provence, and even in the Loire, where most of it is consumed locally.

SUPER TUSCANS

Innovative Tuscan producers, who found the Italian government's wine regulations overly restrictive, found ways around the laws. The producers wanted to make better-quality Sangiovese wines by reducing or eliminating the use of government-required poor-quality indigenous varieties and including international varieties, like Merlot or Cabernet Sauvignon, in their Sangiovese-based wines. To accomplish this, they decided to declassify certain wines and create proprietary names, which allowed the wines to be sold as Vino da Tavola (table wine), using the high-quality grape varieties they preferred, but unlike most ordinary table wine, at a very high price point. The press named these new wines Super Tuscans. Soon they were showing up all over Italy as well as the international marketplace.

Italy

Merlot is the third most planted red grape variety in Italy, behind indigenous Sangiovese and Barbera, but ahead of such notable native wine grapes as Nebbiolo and Negroamaro. In all, Merlot is planted in 14 of the 20 Italian regions. It is one of the largest producing varieties in the northeastern regions of Friuli, Trentino-Alto Adige, and the Veneto, where it has been planted for well over 100 years. Merlot is blended with other Bordeaux varieties or indigenous varieties, or is made into a varietal wine. Many are simple Old World–style wines. In some of the better vineyard sites, grapes produce wines comparable to those of Bordeaux. Gemola, a wine produced by Vignalta in the Colli Euganei appellation of the Southern Veneto, is a blend of Merlot and Cabernet Franc that echoes the quality of the premier wines of the right bank of Bordeaux.

Merlot is now also popular in Tuscany, particularly in the Maremma where it is used extensively in Super Tuscan wines (see Super Tuscans, at left), and it is also allowed in small amounts (up to 20 percent) in classic Chianti. The producer of Ornellaia in the Maremma makes a 100 percent Merlot called Masseto, which sells for more than $250 a bottle.

North America

Merlot is planted throughout California from the north to the south. The San Joaquin Valley (Central Valley) produces the most Merlot, which is used primarily for moderately priced wines sold in low-priced bottles, jugs, and bag-

in-box packages. Napa, Sonoma, Monterey, and San Luis Osbispo counties are the largest growers and producers of higher-quality bottled Merlot, but excellent Merlot can also be found in smaller appellations, such as Mendocino in the north and Temecula in the south.

Merlot is the most widely planted red variety in the Okanagan Valley in British Columbia, Canada.
BRIAN SPROUT/BLACK SAGE VINEYARD

The most prestigious U.S. appellation outside of Napa and Sonoma counties is Washington State where, many believe, the best Merlots in the New World are made. Merlot is the dominant red variety in Washington. The Columbia Valley, which is east of the Cascade Mountain range, is home to the second largest grape-growing area after California. The variety does well in a number of locations, but the subappellations of Rattlesnake Mountain, Red Mountain, and Wahluke Slope are highly regarded for their cultivation of Merlot.

A typical Merlot vineyard in the Red Mountain AVA, Yakima Valley, Washington
TYLER WILLIAMS/KIONA VINEYARDS

On the North Fork of Long Island, where *vinifera* vines were first planted in 1973, Merlot is the most widely planted grape variety. It was one of the first varieties used to make a Bordeaux style of wine on Long Island and is now the premier red varietal with characteristics more reflective of Bordeaux than California. The wines show fruity character with good acid and solid tannins.

Merlot is also the most widely planted red variety in the Okanagan Valley in British Columbia, Canada. Although farther north than Germany, the Okanagan Valley microclimate is able to produce red varieties, such as Merlot, because the mountain ranges on the east and the west of the valley protect the vines from excessive rain, and the valley receives long hours of summer sun, thereby allowing the grapes to ripen. Also, the lake effect moderates temperatures in both the winter and the summer.

Table 12.1 Prominent Merlot Appellations

COUNTRY	REGION/APPELLATION	SUBAPPELLATION	VILLAGE OR SUBAPPELLATION
France	Bordeaux	Libournais	St. Émilion Pomerol
		Fronsac	
		Côtes	Blaye Bourg Castillon Premières Côtes de Bordeaux
		Médoc	Margaux Pauillac St. Estèphe St. Julien Haut-Médoc
		Graves	Pessac-Léognan
		Entre-Deux-Mers	
	Languedoc-Roussillon South-West		
United States	California	Napa Valley	Stags Leap District Yountville Oakville Rutherford St. Helena Mt. Veeder Diamond Creek Calistoga
		Sonoma County	Alexander Valley Knights Valley Dry Creek Valley
		Mendocino County	Inland Valleys
		Central Coast	Santa Cruz Mountains Monterey County Paso Robles

	Washington State	Columbia Valley	Yakima Valley
			    Horse Heaven Hills
			    Red Mountain
			Wahluke Slope
			Walla Walla Valley

Italy	Tuscany	Maremma	
	Veneto	Colli Euganei	
	Friuli Sicily Trentino-Alto Adige Umbria		

Prominent Merlot Appellations, continued

COUNTRY	REGION/APPELLATION	SUBAPPELLATION	VILLAGE OR SUBAPPELLATION
Switzerland	Ticino		
Israel	Galilee	Golan Heights	
Australia	South Australia	Barossa Valley McLaren Vale	
	Victoria		
	Western Australia	Margaret River	
Chile	Central Valley	Curicó Valley Maipo Valley Maule Valley	
		Rapel Valley	Colchagua
Argentina	Mendoza	Uco Valley	
South Africa		Paarl Stellenbosch	
Spain	Castile and Leon	Ribera del Duero	
	Catalonia	Penedès	
New Zealand	North Island	Hawkes Bay	Gimblett Gravels

Other Countries

Although Spain's Merlot is not indigenous, the variety is well suited to the climate in many of the regions and is being tested as a varietal wine throughout the country. Northeast Catalonia is the most aggressive Merlot-planting region. It may have the best climate in Spain for Merlot because the grapes do not become overripe in its cool microclimates.

Merlot is also one of the most popular varieties planted in a number of Central and Eastern European countries, including Bulgaria, Moldova, and Romania. It is often blended with Cabernet Sauvignon or local indigenous varieties. Premium wine production has been held back in this region because of lack of technology and know-how, and the local population's acceptance and preference for sweet wines. A number of international winemakers have provided guidance and the wines are improving; however, they are not yet widely available in the international marketplace. It is possible to find them in ethnic markets or specialty food-and-wine discount stores.

In Chile Merlot is one of the dominant grape varieties, even though many of the vines that were originally thought to be Merlot turned out to be Carmenère (see Chapter 15, Carmenère, page 263). Many vineyards are still interspersed with both varieties, and much of the wine labeled Merlot is in fact a blend of Merlot and Carmenère.

Lucio Gomiero, owner/winemaker of Vignalta in the Veneto region of Italy, pours Merlot in his tasting room.

Argentina has fewer Merlot vineyards than Chile, as the grape variety takes a backseat to Malbec and Cabernet Sauvignon. It is also planted in Uruguay where it is blended with that country's prominent variety, Tannat.

Merlot is the dominant grape in the Hawkes Bay region on the North Island of New Zealand. Although the region is little known, it makes outstanding wines in a Bordeaux style by blending Merlot with Cabernet Sauvignon, Cabernet Franc, and Petit Verdot. Merlot is commonly the dominant grape variety because of its ripening ability in this cool climate. Although not spoken of as much as the more popular Shiraz and Cabernet Sauvignon, Merlot is also gaining a following in Australia. Cultivated in the cooler climate of Victoria, it is blended into Bordeaux-style wines, blended with Shiraz, and also made into varietal wines.

In South Africa, Merlot is the third most widely planted red variety after Cabernet Sauvignon and Shiraz. It is made into varietal wines and also blended into Bordeaux-style wines. It is considered an easy-drinking wine in South Africa.

Israel has begun producing excellent wine primarily from the major international varieties, including Merlot. Most of the harvest is made into a varietal wine but Bordeaux-style blends are also found here. The climate conditions are quite warm, which forces growers to plant vineyards at cooler, higher elevations.

Table 12.1 includes an abbreviated list of countries that produce highly regarded Merlot along with their most prominent appellations and subappellations.

MERLOT: THE GRAPE

Merlot, like Cabernet Sauvignon, is an offspring of Cabernet Franc. It is a grape that buds, flowers, and ripens at least a week, if not two, before Cabernet Sauvignon. Because the grapes ripen early, the risk of the grapes not achieving full ripeness is low, virtually assuring a good harvest. However, because it buds early, it is sensitive to late spring frost. Poor fruit set, also known as coulure, is a significant problem with Merlot. It is caused by cool, wet weather at flowering and drastically reduces yield.

Compared with Cabernet Sauvignon, Merlot is thinner-skinned, plumper, and fruitier, and has a much higher yield ratio (more juice to solids). Because of its thinner skin, it has less color and a lower tannin level, which makes a softer wine that is easier on the palate than Cabernet Sauvignon. When it reaches full ripeness, Merlot is riper and higher in sugar, but has lower acidity than Cabernet Sauvignon. If it becomes overripe, it becomes jammy, loses its acidity, and has no balance. It is less vigorous than Cabernet Sauvignon. Merlot is susceptible to downy mildew and prone to rot.

IN THE VINEYARD

Growers consider Merlot one of the easiest grapes to cultivate because it is dependable from year to year. It ripens early and can be picked while the weather is still warm, so it is not susceptible to problems associated with late-fall weather. It is also an adaptable grape variety, producing different characteristics depending on how it is handled in the vineyard.

Climate

Merlot is considered a variety that performs well in both cool and warm climates. However, cool, dry climates are preferred. In cool, wet conditions, it easily rots and develops unwanted mold.

Merlot ripens fully in cool conditions, in contrast to Cabernet Sauvignon, which does not always fully ripen. Although it ripens completely, it usually maintains its acidity and fresh fruit character while other grape varieties lose these traits.

Merlot's style changes in warmer climates. The warmer the climate, the riper Merlot grapes are likely to become. The wine develops a riper, sweeter, darker fruit character with softer acids. In hot climates, Merlot can easily become overripe, losing much of its acidity and producing less desirable overly ripe fruit flavors.

In Bordeaux, considered a cool climate, Merlot is planted primarily on the right bank, where it ripens fully but maintains its cool-climate character. In the Loire, to the north, Merlot is less successful because it struggles to ripen fully. In the warmer Languedoc to the south, the grapes grow in a climate similar to that of California, resulting in a riper, fruitier, softer style that has less acidity. In cooler Northern Italy, Merlot is typically less ripe, maintaining its acidity but with less of a ripe-fruit character.

Soil

Soils where Merlot is cultivated interact with the climate, resulting in different types of wine. Often lower-lying vineyards in warmer climates have more clay or even sand that absorbs heat and retains moisture, producing riper grapes with higher yields and less concentration. In the cooler climates of coastal, hilly, or mountainous areas, gravel or volcanic soils are more common, providing excellent drainage and contributing to a cooler-climate style of Merlot.

On the right bank of Bordeaux, the classic soil for Merlot is clay, sand, and limestone that remains cool and retains some moisture but drains well. In Northeast Italy, soils are more variable. The hotter valley floors and plains of Friuli, Trentino-Alto Adige, and the Northern Veneto combine sand, clay, and gravel. On the hillsides, soils are less fertile with better drainage and produce better grapes. In the Southern Veneto many of the best vineyards are on volcanic soil.

In the New World, higher, cooler hillsides with better drainage are more desirable than the warmer and more fertile valley floors. In most California Merlot vineyards soils are alluvial, ranging from sand to clay to gravel. In Hawkes Bay, New Zealand, Merlot is planted on both fertile soils and gravel. The best Merlot vineyards are located on the Gimblett Gravels, where as the name indicates, the soil is gravelly. In Chile, Merlot is cultivated in vineyards with sand, limestone, and loam (a combination of clay, silt, and sand).

Water Access

Growers must strike a delicate balance when providing water to Merlot. It requires adequate moisture but not too much. Water, either from rainfall or irrigation, is needed early in the growing season for fruit development and again close to the ripening phase. A total lack of moisture during fruiting prevents grapes from developing. On the other hand, too much water increases yield and the likelihood of overcropping, resulting in thin, vegetal wines. As harvest approaches, Merlot requires little or no water to achieve full ripeness and maintain acidity.

In Europe, where late-summer rains can cause serious rot and mildew, the grape's early ripening and harvest qualities help them avoid the potential wet weather. In the New World, where irrigation is commonplace, growers must pay close attention to the amount of water used as the grapes near harvest. Too much water plumps the grapes, increasing the tonnage, but lowering the quality of the crop.

Vineyard Management

A variety of rootstocks, vine clones, and training and trellising systems are available for Merlot. There is no definitive rootstock for Merlot, so its selection must be based on its fit with specific terroir, pest and disease resistance, vigor, and desired flavor profile. As a relatively young grape variety, Merlot has fewer clonal choices than varieties that have been in existence for a longer time, so matching these limited clones to rootstock and soil is important.

Climate, vineyard site, the grower's philosophy, and yield requirements impact vine-training selection. In France and other parts of Europe, the standard cane-pruned Guyot is used, whereas in the New World vertical shoot positioning (VSP) or other systems, such as Smart-Dysan or Scott Henry, are preferred. These New World systems grow higher and give the grapes greater access to sunlight.

Merlot yields are fairly broad, from under two tons per acre/35 hectoliters per hectare in Bordeaux to as high as six to eight tons per acre/105 to 140 hectoliters per hectare in New World countries. Shoot thinning, leaf pulling, and green harvest are typically used to provide access to sun exposure and control crop yield. Merlot can produce thin, watery, and overly vegetal juice if overcropped. Reducing yields intensifies the richness and flavor characteristics of the grapes, making a commercially desirable style of wine.

Harvest

The harvest date is particularly important to Merlot because of the delicate balance between maintaining acidity, achieving ripeness, and preventing overripeness. In Europe, the grapes are generally picked early to maintain their acidity, producing a lighter, more elegant style of wine with lower potential alcohol. The grapes show less fruitiness and more herbal character than grapes that ripen longer. Early harvest also helps to prevent moldy or rotten grapes that could develop if there is too much rain in the fall.

New World, and some Old World, producers pick Merlot much later. Longer hang time allows the grapes to achieve a riper fruit character and greater flavor concentration. However, acidity and balance are initially lost. In New World countries, this can be rectified during winemaking by legally adding acid back into the wine.

To prevent damage to the thin-skinned grapes, the best Merlot is hand-harvested. Nevertheless, many large producers, particularly in California, made the switch to mechanical harvesting long ago.

IN THE WINERY

The style of Merlot is influenced by grape characteristics when harvested, the differences between Old World and New World winemaking, and the specific winemaking techniques used. The grapes arriving in the winery initially define the style, but the winemaker also has a strong influence on the finished wine.

Cold Soak

Depending on the style desired by the winemaker, Merlot grapes may be allowed to cold soak for a period of time to allow the juice to macerate with the skins. Because Merlot skins are thinner than and not as dark as Cabernet Sauvignon, Merlot normally cannot achieve the color depth or the tannin level of Cabernet Sauvignon. Some winemakers bypass the cold soak and allow the grapes to immediately start fermentation. It is also possible to macerate Merlot after fermentation is complete.

Fermentation

Fermentation techniques are based on basic decisions made by the winemaker. As with other wines, the type of vessel, accessibility to air, yeast selection, and length of maceration all play a role in the style of Merlot produced.

To obtain a more elegant, lighter style of Merlot, skin contact is reduced during fermentation. For a richer, more Cabernet Sauvignon–like style, there may be more pumping over or punching down to extract as much color and tannins as possible. If the wine is aged in barrels, malolactic fermentation (MLF) takes place in the barrel naturally, or lactic acid bacteria are added to the wine if it remains in stainless steel. Because of the potential for low acidity with Merlot, MLF needs to be monitored closely.

Most Merlot is barrel aged.

Barrel Aging

Barrel aging is commonly used with Merlot in both the Old World and the New World. Some producers within the Bordeaux Appellation d'Origine Contrôlée (AOC) or Bordeaux Supérieur AOC are making an unoaked style, as are some New World producers. These wines are less expensive, and more food-friendly, and should be consumed fairly young. In Europe, the oak for the French barrels often used for barrel aging comes from a variety of different forests, and the oak source can add different characteristics to a particular wine. In the New World there is even more variety, with

American oak barrels added to the mix. Barrel aging for Merlot ranges from several months to as much as two years. Once barrel aging is complete, how the different Merlot lots are blended is the next major decision.

Blending

Merlot is historically a blending grape. Even when Merlot is sold as a varietal wine, rarely is it 100 percent Merlot. Most countries allow a certain percentage of other grape varieties in the blend.

The classic red Bordeaux combines Merlot and Cabernet Sauvignon. On the right bank Merlot is the dominant variety, with Cabernet Franc the secondary grape and Cabernet Sauvignon the third grape in the blend. Some Petit Verdot or Malbec may be added. Conversely, on the left bank, Cabernet Sauvignon is the dominant variety and Merlot is the secondary grape.

The United States allows up to 25 percent of other varieties to be blended with a varietally labeled wine without identification. In California Merlot is usually made into a varietal wine but is also a key ingredient in Meritage wines (see Chapter 11, Meritage Wines, page 187). Either way, Cabernet Sauvignon is likely to be blended with the Merlot. Merlot in other New World countries is produced in a similar fashion. In Italy and Spain Merlot is commonly blended with those countries' indigenous varieties.

MERLOT: THE WINE

Merlot is rounder, softer, and easier to drink than other red varietals. The softness on the palate matched with the sweet fruit character of berries and cherries makes it an easy wine to like. It lacks the high acid crispness found in Sauvignon Blanc and Riesling and the astringency and bitterness that can come from Cabernet Sauvignon. It is a wine that has stood in Cabernet Sauvignon's shadow, but it is Merlot that makes Cabernet Sauvignon blends so approachable. Jancis Robinson, noted wine writer, identifies Merlot as "Cabernet Sauvignon without the pain."

Merlot is used entirely as a table wine. In France, Italy, California, and other places where it is grown, it is also made into rosé-style wines, which are becoming quite popular. In California it is made into both dry rosé and an off-dry style, White Merlot, in the same manner as a White Zinfandel.

Color

The color of Merlot ranges from a light shade of red to a much darker ruby color when allowed to macerate extensively. If the grapes are harvested early and not cold soaked, the wine is a paler shade of red, similar to a dark Pinot Noir or Cabernet Franc. Because Merlot grapes are blending grapes, the color of Merlot is almost always influenced by its blending partner. It can turn to dark purple/blue when combined with Cabernet Sauvignon.

Table 12.2 Merlot Aromas and Mouthfeel

AROMAS			MOUTHFEEL
FRUIT/FLORAL	**HERBAL/VEGETAL**	**OTHER**	
Blackberries	Bell peppers	Cedar	Acidic
Black cherries	Black olives	Chocolate	Bright
Black currants	Black tea	Cola	Dilute
Blueberries	Bramble	Earth	Glossy
Boysenberries	Green olives	Floral	Intense
Plums	Green tea	Leather	Lean
Raspberries	Herbs	Meat	Light
Red cherries	Laurel	Minerals	Luscious
Red currants	Leaves	Smoke	Mouth-filling
	Mushrooms	Sweet	Opulent
	Oregano	Tobacco	Smooth
	Thyme	Vanilla	Supple
	Weeds	Violets	Tannic
			Velvety
			Watery

Aromas, Flavors, and Mouthfeel

Merlot's aromas start with fruit, including raspberries, cherries, blackberries, and boysenberries. The fruit aromas can range from fresh to ripe depending on grape ripeness at harvest. Other aromas that can be found in Merlot include a variety of herbs, vegetables, flowers, minerals, and earthy components. Riper grapes producing riper wines tend to allow fruit aromas to dominate the aroma palate. In the extreme some Merlots are called fruit bombs, with an intense fruitiness that masks all other aromas. Flavor intensity in Merlot is a combination of tastes, primarily acidity and sweetness, and the aromas that make the wine so consumer-friendly.

Merlot's mouthfeel is generally described using terms like *smooth*, *glossy*, or *supple*, but it can also be thick and rich. Merlot typically does not have the tannic structure that can introduce astringency. That is one reason it makes a good partner with the more tannic Cabernet Sauvignon. But ripeness at harvest and winemaking techniques, like maceration, can influence the mouthfeel of the wine. Table 12.2 lists aromas and mouthfeel that might be found in Merlot.

Wine Styles

Merlot is made into three styles: traditional, modern, and generic (see Table 12.3). Traditional and modern styles are primarily influenced by the climate and soil, but grape growing and winemaking techniques also impact the wine. Longer grape hang time, shorter or longer maceration, age of the barrels, and length, or even use, of barrel aging can affect style. Each of these techniques can be used worldwide to determine the wine style, so it is not so easy to say that all Old World Merlot is made in a traditional style and all New World Merlot is made in a modern style. Generic-style Merlot is made in both Old World and New World countries but is usually produced as a moderately priced quaffing wine.

Aromas and flavors in traditional styles include bright, fresh fruits, such as red cherries, red currants, and raspberries, as well as herbal notes like olives, tea, herbs, and bramble. The mouthfeel of traditional Merlots can range from lean and bright to glossy with supple tannins. The wine does not come across as abrasive as Cabernet Sauvignon. The dominant factor in developing this style of Merlot is a cool to moderate climate that does not encourage overly ripe grapes. Older barrels are typically used so that the grape aromas are not overpowered by the intense aromas that come from new oak. Acidity remains distinctive and the alcohol level remains below 14 percent. Elegance commonly describes traditionally made Merlot. A traditional-style Merlot is best represented by the Merlot-dominant right bank Bordeaux wines of St. Émilion. Other cool climates that can make a similar style are Northern Italy, New Zealand, and high-elevation vineyards in Napa Valley and Sonoma County.

Modern-style Merlots have a more pronounced fruit character with a richer mouthfeel. Darker and sweeter fruits, such as black raspberries, blackberries, and black cherries, are common. Even riper fruit has jammy or plummy flavors, and if the grapes are overripe, the wine develops aromas of raisins or prunes. Nonfruit aromas may include cedar, earth, chocolate, floral scents, and the aromas that come from barrel aging, like vanilla and smokiness. Climate for modern styles is warm to moderately hot and is conducive to achieving fully ripened grapes. Barrel aging for a modern-style Merlot usually calls for newer barrels because the riper fruit can stand up to the intense flavor characteristics that are absorbed from the new oak. The mouthfeel of a fully ripened modern style is opulent, intense, velvety, luscious, and thick. The texture of the wine becomes richer, softer, and silkier. Higher alcohol, more than 14 percent, is noticed by heat on the palate. Modern style wines come from California and Washington, Australia, and South America. However, finding similar styles elsewhere is common. Some of the most famous and expensive Pomerol wines of Bordeaux are made from almost pure Merlot and achieve a modern ripe, silky, rich style.

Table 12.3 Merlot Wine Styles

	TRADITIONAL (OLD WORLD)	MODERN (NEW WORLD)	GENERIC
VARIETAL/BLEND	Blended with Cabernet Franc and/or Cabernet Sauvignon	Mostly 100% varietal, some Cabernet Sauvignon	100% varietal or blended with other red varieties
CLIMATE	Cool to moderate	Warm to hot	Warm to hot
SOIL	Clay, limestone, sand	Alluvial, clay, gravel, other soils	Alluvial, varied
ALCOHOL	12%–13.5%	14%–15.5% or more	12.5%–14% or more
ACID	Moderate plus	Low to moderate	Moderate
AROMAS/FLAVORS	Black cherries, herbs	Blackberry jam, plums, oak	Red fruits
COLOR	Medium red to ruby red	Dark red	Medium to dark red
BODY	Medium to medium-heavy	Medium-heavy to heavy	Soft to medium

Examples of regional Merlot, each
produced in a different style, from left:
Sonoma County, California; Veneto, Italy;
Bordeaux, France; Washington State
TOM ZASADZINSKI/CAL POLY POMONA

Generic Merlots have fruit flavors, but when made from less ripe or over-cropped grapes they can be leafy, herbaceous, or weedy. Riper fruit reduces these herbal-vegetal characteristics. On the palate, the mouthfeel of these wines is soft but less developed, somewhat dilute and watery. This large category consists of wines produced in large volume, which are made from grapes that come from high-yield, mechanically harvested vineyards in California, Languedoc-Roussillon, and Chile. The winemaking is straightforward. There is no barrel contact. Oak alternatives, like chips or staves, may be added during fermentation. The wine is not aged but is processed and bottled as quickly as possible. The aromas are of sweet fruits and the texture is soft and moderately light. Alcohol is less than 14 percent.

ENJOYING MERLOT

Merlot is easy to say and easy to drink, and that is what has made it popular with so many wine consumers. California Merlot's ripe fruit and soft tannins make it approachable for the American palate. In Bordeaux and Northern Italy, its normal lighter, high-acid, lower-alcohol style goes well with food, which is of fundamental importance to the European consumer.

Buying Merlot

When you buy Merlot a key point is that most Merlot wines are blends. Labels on Australian and New Zealand Merlots, as well as Meritage wines (see Chapter 11, Meritage Wines, page 187), commonly identify the grapes used in the blend. Most Bordeaux wines and most varietal Merlots in the United States do not. Even if it says Merlot on the bottle, it is probably not 100 percent Merlot. There are exceptions. Some California, Washington, and other New World Merlots are pure, and Vin de Pays (VDP) wines from France, mostly from Languedoc, require 100 percent varietal purity and label their wines as varietal.

When purchasing Merlot, use style and price as guidelines. The best way to determine style is by looking at the bottle and noting where the wine is from: Europe, North America, or elsewhere. Wines from Bordeaux or Northern Italy are made in, or at least lean toward, a traditional style. Australian wines and Super Tuscans (see page 194) made from Merlot lean toward a modern style. Many better-quality Chilean wines are available in both modern and traditional styles. California and Washington wines are typically made in a modern style.

Looking at the alcohol level displayed on the label can also help determine the wine's style. Wines with less than 14 percent alcohol are more likely to demonstrate traditional character and less than 13 percent indicates an herbal, traditional style. Wine with an alcohol content of more than 14 percent is likely to reflect a modern character. Higher alcohol levels (14.5 to 16 percent) are an indication that the wine is probably a very ripe modern style.

Price influences the buying decision because within each price category a more limited number of styles and selections are available. Merlots in the lowest price range (under $10) are likely to be a generic style. Wines in this category are led by California appellation wines, Vin de Pays d'Oc wines, and wines from the general Central Valley appellation of Chile. Low-priced Merlots from Central Europe may also be found. Do not expect to purchase a full-fledged traditional- or modern-style Merlot at this price point.

Mid-priced Merlots ($10 to $30) are produced in more specific appellations (for example, Sonoma County rather than California) from lower-yielding vines. California or Washington wines at this selling price are likely to lean toward a modern style. Most Bordeaux and Bordeaux Supérieur wines are sold in this price range and are made in a traditional style. Excellent value wines from Italy, Chile, New Zealand, Australia, and South Africa are also available at this price point.

Merlots priced over $30, with some as high as $100 or phenomenally higher still for the best, provide a breadth of options in both traditional and modern styles from countries and appellations around the world. They receive more attentive winemaking, are aged in better and younger barrels, and have more distinct subappellation or vineyard designations. Most of these wines are estate-bottled. Classified wines from Bordeaux, including St. Émilion and Pomerol, and California wines from outstanding appellations and vineyards dominate at this price point. Washington State wines, Super Tuscans (see page 194) from Italy, and specialty wines from other parts of the world are also found in this price category.

Storing and Serving Merlot

Merlot is stored and served in the same way as Cabernet Sauvignon. Ideal storage temperatures are 55°F to 58°F/13°C to 15°C, or in the coolest, darkest closet or cupboard in your home. Merlot should be served at a temperature that refreshes, not much higher than perfect storage temperature and no higher than 65°F/18°C for a modern-style wine. Any higher and the wine loses its refreshing quality. If you purchase a bottle at a retail store the same day you plan to serve it, the ambient temperature of the store or your home is probably too warm. To bring the temperature down, put the bottle of wine in the refrigerator for 30 minutes to reduce it by eight to 10 degrees. If the wine seems too warm when ordered in a restaurant, request an ice bucket with cold water and a few ice cubes to accomplish the same task. Merlot should be served in a Bordeaux glass.

Merlot and Food Pairings

Merlot is a varietal that can pair with a broad range of foods, but be cautious. Not all Merlots go with every dish. Style is an important consideration when pairing Merlot with food.

The higher-acid, lower-alcohol style typical of lighter, traditional Merlots is food-friendly and goes with light and medium dishes. It can accompany a wide variety of foods, but is particularly good with moderately rich fish and almost any kind of poultry. Dark-meat chicken or duck is a good choice in almost any Western recipe, from braised thighs to fried legs. Pastas and risottos are ideal with Italian Merlot. Grilled vegetables and mushrooms by themselves or incorporated into other dishes are also excellent companions.

High-alcohol Merlots tend to have less acid and match better with richer, meatier dishes. Richer, riper, modern Merlot is best with the same foods that pair with Cabernet Sauvignon, such as braised chuck roast in a sauce of beef stock. Merlot and blackberries plays off the richness of the wine and the fruit of the grapes. Grilled meats, from pork to lamb to beef, all work effectively with modern-style Merlot. Pâtés and terrines are also good matches with richer Merlots.

TOP 10 FOODS TO PAIR WITH MERLOT

1 Braised beef dishes
2 Pasta dishes with a tomato-based sauce or drizzled with olive oil and fresh herbs
3 Roast pork
4 Grilled tuna or salmon
5 Mushroom risotto
6 Barbecued spring lamb
7 Fatty poultry like duck or goose
8 Chicken thighs braised in Merlot with winter vegetables and potatoes
9 Truffles (from the ground, not the chocolate kind)
10 Dark chocolate (with a rich, unctuous style of Merlot)

SUMMARY

Merlot is a luscious grape with texture, fruit, and richness that can be made into a varietal wine but is commonly blended with other grape varieties, like Cabernet Sauvignon, Cabernet Franc, and Sangiovese. It is perhaps best known as one of the two prominent red Bordeaux varieties.

Merlot is a chameleon grape that can adapt from cool to hot climates and change its style along the way. Winemakers can also influence the style of wines made from Merlot. The result, when done well, can be an expression of a particular place that showcases the fruitiness of the wine, whether made in a traditional or modern style. Although some critics deride Merlot because of a few badly made, innocuous wines, when well made, its enjoyable flavors make it one of the most widely loved, and consumed, varietals.

Pinot Noir

Whereas Cabernet Sauvignon has always reflected the power found in wine, Pinot Noir demonstrates wine's delicate but evocative, ethereal side. While Cabernet Sauvignon is sometimes called the king of wines to demonstrate its masculinity, Pinot Noir is referred to as the queen of wines because of its subtle, nuanced, feminine characteristics that continually change in the glass. Pinot Noir traditionally is a lighter, softer wine with mellow tannins that displays an amazing set of aromas and flavors. A newer style delivers richer, ripe-fruit character with rich tannins and higher alcohol.

Pinot Noir got a huge boost with the release of the movie *Sideways* in 2004. The movie accomplished for Pinot Noir what *60 Minutes* did for Merlot, and for red wine in general in the early 1990s: It created an explosion in demand that has only recently let up and has driven increased cultivation of Pinot Noir, and higher prices, worldwide.

REGIONS AND APPELLATIONS

Pinot Noir is grown worldwide but is dominated in quantity and quality by France, particularly Burgundy, California, Oregon, and New Zealand. Switzerland and Chile play a much smaller but still important role. (See Appendix 2 and Appendix 3 for maps of Pinot Noir–producing countries and appellations, pages 380–447.)

France

Burgundy is the home of Pinot Noir. There is no other grape variety that is so closely allied with a particular place. Every Pinot Noir produced elsewhere is measured against the greatest Burgundy wines, with winemakers everywhere trying to mimic their characteristics.

Burgundy is a small appellation by international standards and is exclusive because little expansion can occur. Red Burgundies come primarily from the Côte d'Or (slopes of gold), where Pinot Noir shares acreage with its other famous sister variety, Chardonnay. It is an area that is no more than 1.2 miles/ 2 kilometers wide and 34 miles/55 kilometers in length. The Côte d'Or is divided into two parts: the Côte de Nuits in the north and the Côte de Beaune in the south.

The Côte de Nuits produces Pinot Noir almost exclusively and a very small amount of Chardonnay. Villages in the Côte de Nuits from north to south are a who's who of great Burgundy appellations, including Gevrey-Chambertin, Morey-St-Denis, Chambolle-Musigny, Vougeot, and Vosne-Romanée.

The Côte de Beaune is better known for Chardonnay but actually produces considerably more Pinot Noir. Important Pinot Noir appellations in the Côte de Beaune are Aloxe-Corton, Pommard, and Volnay.

Louis Latour produces highly regarded Pinot Noirs at his estate in Côte de Nuits, France.

Above the Côte d'Or on a high plateau sit the Hautes Côtes de Nuits and the Hautes Côtes de Beaune, Appellations d'Origine Contrôlées (AOCs) that were created in 1961. Pinot Noir grown in the Hautes Côtes are different clones, the vineyards are cooler and more exposed to the elements, and the soils are less desirable than those of the Côte d'Or, but they do produce Pinot Noir of excellent value.

There are four AOC classifications and more than 100 AOCs for Pinot Noir alone. Burgundy, the smallest of the major appellations, has more AOCs than any other appellation. Why? Because each AOC has its own unique terroir—the foundation of Burgundy AOC classifications.

Pinot Noir is also produced in both the Côte Chalonnaise and the Mâconnais located south of the Côte d'Or. Although not in the same class as the wines of the Côte d'Or, in most cases, good grapes are cultivated that make good wines. In the Côte Chalonnaise, Rully, Mercurey, and Givry are noted communal AOCs.

Champagne is the largest Pinot Noir–producing region in France. All the Pinot Noir grown there is used to make sparkling wine. It is also planted in the eastern portion of the Loire, primarily in Sancerre, Menetou-Salon, and Reuilly, occupying 20 percent of the vineyard space. Pinot Noir is cultivated in Alsace, where it is the only red grape permitted, and it produces dry, light-style wines as well as Crémant d'Alsace blanc de noir and rosé. The Jura and Savoy regions bordering Switzerland also make a light-style Pinot Noir.

North America

After France, North America is the largest producer of Pinot Noir in the world. California and Oregon are by far the most noted states. Excellent Pinot Noir is produced in the Finger Lakes region of New York, the Leelanau Peninsula in Michigan and the northern part of Ohio along Lake Erie. In all cases, the wines are made in very small quantities. In Canada Pinot Noir is planted near Lake Erie in Ontario and in the Okanagan Valley in British Columbia.

California

Pinot Noir started to take off in California in the 1980s when great attention was paid to matching climate to the grape. Almost all quality California Pinot Noir comes from the following cool-climate appellations:
- Anderson Valley, Mendocino County
- Russian River Valley, Sonoma County
- Sonoma Coast, Sonoma County
- Carneros, Napa Valley/Sonoma County
- Santa Cruz Mountains
- Santa Lucia Highlands/Monterey County
- Santa Maria/ Santa Barbara County
- Santa Rita Hills/ Santa Barbara County

Oregon

In 1965 David Lett of the Eyrie Vineyard set out to find a location that would closely match the characteristics found in Burgundy, and his search led him to the Willamette Valley in Northwest Oregon. It has since developed the reputation as possibly the best appellation for Pinot Noir in the United States.

In this valley, the climate and soil characteristics are reflections of Burgundy, but it takes years of practice to develop the full capability of a new region. Finding the right microclimates, selecting the best clones, learning to manage the vineyard, and tweaking winemaking techniques have only been in practice for 40 years in Oregon, whereas Burgundy has been producing Pinot Noir for centuries.

New Zealand

The newest champion of the Pinot Noir grape is New Zealand. In a relatively short time, New Zealand, which first established quality Sauvignon Blanc, has now taken on Pinot Noir as its second great grape. On the North Island, Pinot Noir is produced around Martinborough. On the South Island, Marlborough, Nelson, Canterbury, and Central Otago are the prominent regions. Marlborough, a bit further along in its development, produces the most Pinot Noir in the country, but Central Otago has quickly become the most talked about appellation. The most southerly appellation in the Southern Hemisphere, it struggles to produce ripe grapes. But when they do ripen, the wines are some of the most intriguing Pinot Noirs on the market. Central Otago is following in similar footsteps to Oregon. Extraordinary Pinot Noir may occasionally be found, but it may be decades before Central Otago reaches its full potential.

Felton Road produces Pinot Noir in Central Otago, New Zealand, a region that sits on the edge of the southern wine-growing latitude.

Other Countries

In Europe, Switzerland, Germany, Austria, and Northern Italy are also noted producers of Pinot Noir. All generally produce a lighter, more traditional style because of the cooler temperatures in these countries and regions. They certainly struggle to achieve full ripeness in many vintages.

Pinot Noir is cultivated throughout Switzerland, primarily in the French-speaking region. Neufchâtel and Vaud along the north side of Lake Geneva produce a light style of wine. In the Valais, the largest wine-producing region, Pinot Noir is blended with Gamay to make a wine called Dôle.

In Germany Pinot Noir, known as Spätburgunder, is the most widely produced red wine, and its production is increasing at a rapid rate. Until recently, the quality was questioned primarily because the climate was too cold and it was difficult to get fully ripened grapes. With improved grape-growing and winemaking techniques, as well as the recent string of very warm vintages, quality has improved.

Table 13.1 Prominent Pinot Noir Appellations

COUNTRY	REGION/APPELLATION	SUBAPPELLATION	VILLAGE OR SUBAPPELLATION
France	Burgundy	Côte de Nuits	Gevrey-Chambertin Morey-St-Denis Chambolle-Musigny Vougeot Vosne-Romanée
		Côte de Beaune	Aloxe-Corton Pommard Volnay
		Hautes Côtes de Nuits Hautes Côtes de Beaune	
		Côte Chalonnaise	Rully Mercurey Givry
	Loire	Upper Loire	Sancerre
	Jura Savoy Alsace Champagne		
United States	California	Mendocino County	Anderson Valley
		Sonoma County	Russian River Valley Sonoma Coast Carneros
		Napa Valley	Carneros
		Santa Cruz County	Santa Cruz Mountains
		Monterey County	Santa Lucia Highlands
		Santa Barbara County	Santa Rita Hills Santa Maria
	Oregon	Willamette Valley	
	New York	Finger Lakes	
	Michigan	Leelanau Peninsula	
	Ohio	Lake Erie	
Canada	Ontario	Lake Erie	
	British Columbia	Okanagan Valley	
Switzerland	Valais Neufchâtel Vaud		
Germany	Baden		
Austria	Burgenland		
Italy	Trentino-Alto Adige		

Prominent Pinot Noir Appellations, continued

COUNTRY	REGION/APPELLATION	SUBAPPELLATION	VILLAGE OR SUBAPPELLATION
New Zealand	North Island	Auckland Martinborough	
	South Island	Marlborough Central Otago	
Australia	Tasmania		
	Victoria	Mornington Peninsula Macedon Ranges	
	South Australia	Adelaide Hills	
Chile	Southern Regions	Bió Bió	
	Aconcagua	Casablanca	
Argentina	Mendoza	Uco Valley	
Spain	Catalonia	Penedès	

In Austria, Pinot Noir, called Blauburgunder, is grown in very small quantities in the regions of Niederösterreich and Burgenland and made in a style reminiscent of Burgundy. In Italy most Pinot Noir (Pinot Nero) is grown in the Trentino-Alto Adige region in the far north of the country. It is also cultivated in Lombardy for the making of **MÉTHODE CHAMPENOISE** sparkling wine. Spain has approved the use of Pinot Noir in the making of rosado (rosé) Cavas. Pinot Noir does best in the cooler northeastern climate of the Penedès where most Cava is produced. Pinot Noir is also cultivated in Central European countries, such as Moldova, Macedonia, Romania, and Bulgaria.

In South America, up-and-coming Chile, followed by Argentina, produce small amounts of Pinot Noir. Excellent examples are starting to arrive in the marketplace from the cool coastal appellation of Casablanca and the southern Bió Bió. In Australia the island of the Tasmania, the better-known Adelaide Hills in South Australia, and the newer Victoria regions of Mornington Peninsula and Macedon Ranges provide the cool climate needed for Pinot Noir to be successful.

Table 13.1 includes an abbreviated list of countries that produce highly regarded Pinot Noirs along with their most prominent appellations and subappellations.

PINOT NOIR: THE GRAPE

Pinot Noir is one of the oldest grapes cultivated, dating back to before the Romans. The vine produces small, elongated, compact bunches that are dark purple in color. They are plump and juicy with thin skins and are prone to rot in humid and wet weather. The grapes bud early, making frost a potential problem. They are typically early-ripening, but if grown in a cool climate, ripening may not occur until September or even October. Pinot Noir vines are susceptible to overproduction and high yields, which can cause a quick decline in quality. The grapes can also struggle without enough sun but become overripe quickly if temperatures get too hot.

IN THE VINEYARD

There are no growers who pay more attention to their vineyards than Pinot Noir producers, and there is no place else in the world that prizes and promotes terroir more than Burgundy. To growers there, terroir not only includes the natural characteristics of climate, soil, and geography, but also the tools and methods used to cultivate their wine, with tradition playing a crucial role.

In the Old World, it is believed that the unique combination of characteristics of a particular place, which cannot be duplicated, is responsible for a wine's color, aromas, and flavors. Organic and biodynamic farming are new methods that are also becoming more prevalent. Some of the best Burgundy producers are incorporating these techniques to improve the quality of their vineyards and, concurrently, the quality of their grapes.

New World Pinot Noir producers have become equally fervent about some of these issues and concerns. What New World producers do not have, however, are generations and centuries of experience. Everything is new, and with only one vintage a year to learn from, accumulating experience is a slow process.

Selecting the right terroir is only the start of the grape-growing process for Pinot Noir. The next step to producing the best grapes is deciding which clone or group of clones to plant. Because of its age, there are more clones of Pinot Noir than any other grape. Pinot Noir has more than 200 different clones, whereas Cabernet Sauvignon has only a dozen at best. Each clone has different characteristics affecting color, aroma, flavor, body, yield, and disease resistance.

Climate

Historically, Pinot Noir has required a cool climate. The perfect climate conditions for growing Pinot Noir vary, but the underlying characteristic is cool temperatures, however they are achieved. The most desired characteristic for a Pinot Noir vintage is a long, cool growing season. Extremely cool climates only produce a lighter style of wine. With better growing practices, there are some warmer climates that are beginning to achieve success with Pinot Noir, but the end result is a richer style.

Burgundy

In Burgundy, the climate is generally cool because of its northern latitude, with the best vineyards elevated from the valley floor at 800 to 1,000 feet/250 to 300 meters on the western slope of the Morvan Hills. The region's long, warm summer days and cool nights are perfect for ripening. The east-southeast aspect, and in some cases a southern aspect, of the vineyards moderates direct sun so the grapes do not get too hot. The region is susceptible to late August–early

CRU VINEYARDS

Burgundy has added its own quality classifications to AOC requirements. The best vineyards in an appellation are identified as Grand Cru (great growth) or Premier Cru (first growth). Grand Cru wines, the highest level, come from vineyards that have exceptional soil and extraordinary reputations for producing consistently excellent wines. Premier Cru wines come from vineyards with excellent soil. These are the second-best sites.

The Côte d'Or contains 31 Grand Cru vineyards and hundreds of Premier Cru vineyards. Grand Cru vineyards account for only 2 percent of the total vineyards planted there; Premier Cru, for approximately 12 to 15 percent of the vineyards.

Bordeaux and Alsace also designate cru wines. Grand Cru or Premier Cru is printed on bottle labels if the wine came from cru-designated vineyards.

September rain that can be problematic for the harvest. Not every vintage in Burgundy is fabulous, as unfavorable weather conditions can cause occasional mediocre harvests.

The United States

In California and Oregon, climate plays a dominant role in appellations where Pinot Noir has been selected as a key variety. Pacific breezes keep temperatures cool, and the fog that develops most days is perfect for maintaining cool temperatures and extending the growing season. High humidity can be a problem because of the fog, but afternoon sun and ocean breezes allow the grapes to dry, limiting potential mold and rot.

In California the best Pinot Noir vineyards are located not more than 20 miles from the Pacific Ocean, where fog creeps into the valleys. All of these vineyards are shrouded in early-morning or late-afternoon fog with a burst of late-morning and early-afternoon sun for ripening. The risks to these vineyards include excessive rain damaging the delicate grapes and easterly driven heat waves, which ripen the grapes too quickly and produce overripe flavors. Elevated vineyard areas, such as the Sonoma Coast (1,100 to 1,500 feet/350 to 450 meters), sit above the fog, but temperatures are still moderated by the fog and its high perch.

In Oregon, the Coast Range on the west protects the Willamette Valley. The climate is mild with most rain coming during the winter. The summer is quite dry with long days, but remains cool because of the marine fog from the coast. Early in the region's development following several less than successful vintages, vintners wondered whether the valley's climate was too cool for Pinot Noir. In the last several vintages, however, summer temperatures have been quite warm, producing excellent wines. This could be another example of the impact of global warming.

A Pinot Noir vineyard at Navarro Vineyards in Mendocino, California

New Zealand

At first glance, the climate in New Zealand's Pinot Noir appellations does not appear to be similar to those of other Pinot Noir–growing countries. However, its southern latitude is equal to Burgundy's northern latitude, and they share the underlying cool-climate requirements of other Pinot Noir appellations.

The Martinborough region in the southern part of the North Island is cool but dry. Vines require irrigation to survive. On the South Island, the climate is cooler with increasing sun. In Marlborough and Nelson, grapes thrive in the long, cool summer with only limited risk of rain at the end of the ripening phase in the fall (April or May in the Southern Hemisphere). The Southern Alps mountain range that runs along the west coast protects these areas from adverse weather coming from the Tasman Sea.

The most southerly region for Pinot Noir is Central Otago. Surrounded by the snow-covered Alps, it is the coldest and driest part of New Zealand, with frigid winters and summers that barely get warm enough to provide adequate heat for growing grapes. Temperatures range only from 50° to 84°F/10° to 30°C. Vineyards all face north to maximize sun exposure. Daylight can last until 10 P.M., providing the needed ripening boost.

In Oregon's Willamette Valley Pinot Noir is planted in high-density vineyards.

Soil

There is probably no place where the soils are more famous than Burgundy. Limestone, clay, and marl (a combination of limestone and clay) soils provide the unique base for the cultivation of Burgundy wines and also provide excellent drainage. At the highest and coolest elevation, the Hautes Côtes sits on a high plain with hard limestone and little topsoil, which are not the best conditions for growing Pinot Noir. The midslope, considered the prime vineyard location, where all of the Grand **CRU** and Premier Cru vineyards (see Cru Vineyards, page 214) are located, has a blend of clay and limestone that provides perfect conditions for Pinot Noir. The flatter alluvial plain, at the base of the Côte d'Or, is mostly clay, receives too much moisture, and does not drain well. Much of the regional wine is made here.

In California the soils are as diverse as the state. They range from volcanic to limestone, sandy loam, and clay to granite. Growers are more concerned with the soil's ability to hold water and then allow it to drain in combination with irrigation than with the composition of the soil.

The soils of the Willamette Valley are old volcanic soils and marine sediment. Over the base is deep gravel and silt that provides good drainage. At 300 feet/100 meters elevation soils change. Above, gravel and silt are prominent; below, the soils contain more ocean sediment.

Martinborough, on New Zealand's North Island, contains a combination of stony loam over gravel. On the South Island soils vary. In Marlborough and Nelson, they are alluvial on the flat plains only a few feet above sea level. In Canterbury, soil contains the desired limestone as part of the chalk and loam soils. In Central Otago, the soils are composed of silt and loam.

Vineyard Management

Vineyard-management techniques in Burgundy and the New World differ somewhat, but Burgundy's success has led many New World producers to mimic as closely as possible the techniques used there.

In Burgundy's best vineyards, the vines are trained on single Guyot. The grapes are sometimes only a foot off the ground. Vertical shoot positioning (VSP) is used in many of the lower-quality, noncru vineyards. In the New World, VSP is most prevalent, but single Guyot or variations of both techniques are also used.

Burgundy's vineyards are generally more densely planted than New World vineyards. A typical Burgundy vineyard contains between 1,600 and 2,000 vines per acre/4,000 and 5,000 vines per hectare, but they can contain as many as 4,000 vines per acre/10,000 vines per hectare. A vine in Burgundy is planted about every square meter (approximately a square yard). In the New World, it is common to plant more widely spaced vines. Vines might be planted with up to four feet separation in rows that are as much as 12 feet or more apart to accommodate vineyard equipment, allowing less than 1,000 vines per acre/2,500 vines per hectare. Over time American growers are adapting their vineyards to more closely follow the Burgundy model with denser plantings.

The available canopy-management techniques are the same wherever Pinot Noir is grown. However, their application changes depending on the vineyard location and aspect as well as the grower's desire to reach a particular ripeness in the grapes. Because Pinot Noir is grown in cool, foggy areas, canopy-management techniques help to expose the grapes to the available sunlight. Leaf pulling is normally done to gain access to morning sun, as opposed to afternoon sun, which is too hot for these cool-loving grapes. Shoot-thinning opens up the grape bunches to the sun, as does training the vines upward.

Yield

Cool-climate vines, such as Pinot Noir, generally produce smaller yields than warmer climate vines. The typical yield for high-quality Pinot Noir in Burgundy is about 35 hectoliters per hectare, or the equivalent of two tons per acre. In California, Oregon, and New Zealand, it is about the same with a slightly wider range. This is a far cry from such varieties as Cabernet Sauvignon and Syrah, which produce an excellent crop at four to eight tons per acre/70 to 140 hectoliters per hectare. In South America, Pinot Noir has yields as high as four tons per acre/70 hectoliters per hectare. Pinot Noir in Champagne and other sparkling wine regions produces as much as six tons per acre/105 hectoliters per hectare.

Although Pinot Noir is one of the lowest-yielding varieties, growers reduce the yield even further to achieve more concentration in the remaining grapes. In Burgundy radical pruning, e.g. limiting the number of bunches per vine, is used to control crop size. Thinning the crop to achieve the desired yield can occur before or during veraison. In densely planted vineyards, the number of bunches allowed to remain on the vine is extremely low, minimizing total yield.

Harvest

Pinot Noir harvests are quite variable depending on weather and temperature. The optimal situation is a long, cool ripening phase, pushing harvest back to late September or early October. Hot weather or risk of rain can force harvest to take place in early September, while a late-summer rain may force a grower to risk waiting until well into October to reach the desired ripeness.

Pinot Noir is almost always harvested by hand because of the delicacy of the grapes. Small bins are used to prevent too much weight from crushing the thin-skinned grapes. Because the grapes are harvested by hand, diseased or poor-quality grapes are discarded in the vineyard and only the best grapes are transported to the winery.

Grape sugar level varies for Pinot Noir harvest. Historically, it was common to harvest Pinot Noir at 22° to 23° Brix, achieving a potential alcohol of well under 14 percent, and many producers still follow that model. But today many Old World and New World producers are allowing their grapes to hang longer on the vine to achieve riper flavors. Some Pinot Noir producers want to reach 25° Brix with a potential alcohol approaching 15 percent, or occasionally even higher.

IN THE WINERY

Many winemakers consider Pinot Noir one of, if not the most, difficult grape varieties to make into wine because so many things can go wrong in the winemaking. The type of yeast, lack of perfect cleanliness, barrel residue, and too much handling can all cause its desirable aromas and flavors to be lost. Upon the grapes' arrival at the winery, many winemakers do an additional hand-sort, removing any remaining unripe or diseased berries, or material other than grapes to make sure the winemaking starts with the best fruit possible. The variety is fragile, requiring constant monitoring; the wine can have beautiful aromas one day and disgusting off-aromas the next, or it can lose all its charm once bottled.

Cold Soak

Because Pinot Noir is a thin-skinned grape, it is difficult to obtain a lot of color and flavor from the skins. Cold soak is favored by many producers to maximize extraction of color, tannins, and flavor compounds. With Pinot Noir, maceration can last from a few days to a week or two. Fermentation starts when the vat is allowed to warm or cultured yeast is added to the must.

Fermentation

The winemaker can use different techniques to make Pinot Noir, resulting in differences in the wine. Fermenters can be oak vats or stainless steel tanks and can remain open or closed. Open fermenters, which increase oxygen contact, are common in Burgundy, but are also used elsewhere. They are typically smaller than those used for other red varieties. In Burgundy, the winemaker usually leaves stems on during the initial stages of fermentation to add complexity to the wine (whole-cluster fermentation). In the New World, this is far less common because many winemakers believe stems add undesirable vegetal character and tannins. Instead, they prefer carbonic maceration, which allows the wine to develop fruity flavors and brighter color.

Yeast

In Burgundy most producers, particularly smaller artisanal vignerons, allow indigenous yeast to ferment the wine, a riskier choice than using cultured yeast. Native yeast yields different characteristics from vintage to vintage and the length of fermentation can change dramatically, but the resulting complexity is highly desired. Native yeasts, if used, are likely to work slower than cultured yeast. It is possible to initiate fermentation with indigenous yeast but intervene with cultured yeast if any problems develop during fermentation.

Although some in the New World use native yeast, most rely on cultured yeast to ferment Pinot Noir. Cultured yeast strains provide a degree of certainty to the fermentation process and to the desired characteristics in the finished wine.

Fermentation Techniques

Fermentation temperatures for Pinot Noir are normally cooler for longer periods of time because of the delicacy of the grape. Punch downs are considered more delicate than pump overs and are used during both the maceration phase and fermentation. Length of fermentation for Pinot Noir is longer than for most varieties.

Pressing Pinot Noir can also vary depending on the techniques used. It can occur after cold soaking, during fermentation, after fermentation, and in some cases, during or after barrel aging. Stems, ripeness of the grapes, and desired style of wine all play a role in this decision. Malolactic fermentation (MLF) is used to soften the acid structure of the wine but typically takes place in the barrel during the aging period.

Barrel Aging

Pinot Noir is almost always barrel aged in French oak. Other kinds of oak, particularly American, is considered too aggressive for this delicate variety. Producers carefully select the forests for the oak used for the barrels because Pinot Noir responds differently to the distinctive characteristics of oak from each forest. Barrel aging lasts from eight to 24 months, with 12 months being the norm.

Rudi Bauer, winemaker at Quartz Reef Winery in Central Otago, New Zealand, checks the progress of Pinot Noir fermentation.

Different producers have different views on the most appropriate barrel age and toast. Most Burgundy wines are aged in older barrels, but some top Burgundy producers use a mix of old and new barrels. In California and Oregon a mix of barrel ages and toast levels is used to age Pinot Noir.

Pinot Noir must be handled gently while in the barrel. Pinot Noir may never be racked while in the barrel, and micro-oxidation is not used to reduce the amount of oxygen in contact with the wine. This is known as reductive winemaking. During the racking process, clarification is minimized because the only racking takes place at the end of barrel aging.

Blending

Different lots of Pinot Noir from different vineyards and different barrel-aging programs may be blended, but it is infrequently blended with other varieties. In Burgundy, it must be 100 percent Pinot Noir. In California and elsewhere, Pinot Noir is usually 100 percent varietal, but by law only 75 percent of the finished wine is required to be Pinot Noir. If a California Pinot Noir has more color, more body, a different array of fruit aromas and flavors, or higher alcohol than usual, it may have been blended with other, less costly varieties that add those components. The most common is Syrah, but others are also used. Low-priced Pinot Noir is also likely to have been blended with less costly varieties to lower the average production cost and selling price of a bottle of the wine.

Finishing

Finishing for Pinot Noir is more deliberate than for other varieties. Whereas other varieties go through stages of fining and filtering, many Pinot Noirs do not. Natural clarification is preferred because of the delicacy of Pinot Noir. The use of natural fining and filtration may be stated on the label with the terms *unfined* or *unfiltered*. Mechanical filtration is believed to rob Pinot Noir of some of its refined aromas and flavors. Pinot Noir is also likely to go through more bottle shock during bottling, so it must undergo a period of bottle aging before it is released for sale.

PINOT NOIR: THE WINE

Pinot Noir is an important contributor to Champagne and various styles of sparkling wines around the world (see Chapter 17, Sparkling Wines, page 276) and is also made into dry rosé wines, but Pinot Noir is best known as a table wine with the potential to dazzle the consumer with a broad array of aromas and flavors. It is a desirable wine because of its nuanced breadth of flavors, balanced acidity, and soft, approachable tannins, and it is the varietal most commonly spoken of in the context of terroir. Growing conditions and complex winemaking requirements make it one of the most interesting and provocative tasting experiences.

Color

Pinot Noir is noted for its light, elegant dark rosé to cherry red to garnet color. Because of its thin skins, it is difficult to achieve the dark colors that are found in varieties like Cabernet Sauvignon or Syrah. Old World Pinot Noirs usually demonstrate lighter shades of red, while in New World countries, the style is typically darker. Achieving deeper color requires extended maceration or even blending with a dark red variety such as Syrah.

Aromas, Flavors, and Mouthfeel

Pinot Noir has a broad array of aromas and flavors when compared with other international varieties. Fruity, floral, herbal, spicy, vegetal, and earthy aromas can all be found. The classic foundation aroma of Pinot Noir is some type of cherry. Oak can also play a dominant role. Depending on the toast of the barrel and length of aging, barrel influence on the wine can be quite profound, adding aromas of spices, especially cinnamon and nutmeg, caramel, toast, or smoke. Table 13.2 lists aromas and mouthfeel that might be found in Pinot Noir.

One characteristic that makes a traditional-style Pinot Noir so distinctive is the changing array of aromas that develop in the glass over time. One minute you may smell fruit and herbs, and a few minutes later a floral or earthy element may become dominant. Changes can continue as long as there is wine in the glass.

Table 13.2 Pinot Noir Aromas and Mouthfeel

	AROMAS			MOUTHFEEL
FRUIT	**FLORAL**	**HERBAL/VEGETAL**	**EARTHY (ANIMAL/WOODY)**	
Black cherries	Rose petals	Beets	Barnyard	Glossy
Cherries	Violets	Black olives	Caramel	Layered
Cranberries	Wild flowers	Black tea	Cedar	Light-bodied
Currants	Wilted roses	Caraway	Cigar box	Lingering
Raspberries		Cinnamon	Cola	Rich
Ripe tomatoes		Cloves	Forest floor	Round
Strawberries		Green tea	Gamey	Silky
		Green tomatoes	Leather	Smooth
		Mint	Mushrooms	Supple
		Nutmeg	Oak	Velvety
		Rosemary	Smoke	
		Spices	Toast	
		Weeds	Vanilla	

The mouthfeel of Pinot Noir is also different from that of other primary red varieties. Traditional Pinot Noir is typically lighter in body than other reds with softer tannins, pronounced acidity, and a sense of weight and texture. The wine lingers on the palate. Richer Pinot Noirs are heavier, more round, and more mouth-filling, and may be more tannic in some cases, with sweetness from the ripe fruit.

Wine Styles

Although Pinot Noir can be said to have different characteristics in every bottle that is opened, there are two foundation styles: traditional, commonly found in Europe, and modern, typically seen in the New World, particularly California. From that foundation, styles can twist, turn, and evolve based on regional characteristics and winemaking. Many believe that each region or appellation produces an individual style of Pinot Noir that can contain characteristics of each of the two core styles (see Table 13.3).

Traditional Pinot Noirs are subtle and evocative, demonstrating a restrained, but complex, fruit character. The fruits showcased are fresh red fruits, strawberries, cherries, raspberries, and cranberries. In addition there are floral, herbal, and earthy aromas, which are quite typical of Burgundian Pinot Noirs. The color is light to medium and the body is light but glossy. There is delicacy with an underlying strength to the wine. Acidity is integrated but distinguishable. Alcohol is less than 14 percent, commonly 13 to 13.5 percent.

Pinot Noir from the better vineyards in the Côte d'Or represent the benchmark for traditional Burgundian red wine. Colder regions in Alsace, Loire, Austria, Germany, and Northern Italy produce lighter, more delicate traditional wines. In some cases they may almost look like a dark rosé. Traditional wines can also be found in other appellations throughout the world, including New

Zealand, Chile, Canada, and New York, Oregon, the Midwest, and even California in the United States.

Modern-style Pinot Noirs are richer and darker in color. The wine might look more like Merlot or even Syrah. They have a more powerful but narrower ripe fruit profile, chewy tannins, and higher alcohol. The fruit character is richer and riper with black cherry and black raspberry aromas. The wine is medium- to full-bodied with concentrated flavors and softer acidity. The alcohol level is over 14 percent, sometimes approaching 15 percent, making the wine seem almost sweet.

Because of its warmer climate and American consumers' desire for riper, richer wines, California produces mostly modern-style Pinot Noirs. In general wines from warmer areas and warmer vintages and those subjected to more modern winemaking techniques take on more New World traits.

ENJOYING PINOT NOIR

Buying Pinot Noir

Knowledge of appellations and producers will help you to determine which Pinot Noir to buy. If you are looking for a modern, ripe style, purchase a California Pinot Noir. A general appellation, like Sonoma County or Santa Barbara County, might fit the style. Some Willamette Valley, New Zealand, and Côte d'Or Pinot Noirs can also be produced in a modern style.

If you want a more traditional style, then consider wines produced in less well-known regions or appellations. Burgundies from villages and specific vineyard designations in the Côte d'Or are the premium traditional wines. Burgundian Pinot Noirs from the Mâconnais, the Côte Chalonnaise, and the Hautes Côtes and Pinot Noirs from the Loire or Alsace are most assuredly traditional in style.

Alcohol content is a distinguishing factor between traditional and modern styles: less than 13.5 percent leans toward traditional, more than 14 percent leans toward modern.

Table 13.3 Pinot Noir Wine Styles

	TRADITIONAL	MODERN
VARIETAL/BLEND	100%	100%
CLIMATE	Cool	Cool to warm
SOIL	Limestone	Variable
ALCOHOL	12%–13.5%	14%–15% or more
ACID	Moderately high	Low to moderate
AROMAS/FLAVORS	Fresh fruits, earth, floral	Ripe fruits
COLOR	Pale to light red	Medium to dark red
BODY	Light	Medium-heavy

Examples of regional Pinot Noir, each produced in a different style, from left: Santa Barbara, California; Central Otago, New Zealand; Côte de Nuits, France; Willamette Valley, Oregon

Disparity in prices for Pinot Noir is quite dramatic. The best Côte d'Or village wines can reach $50 or more, the best Premiers Crus start at $100, and Grands Crus are several hundred dollars, with the most expensive shooting up well over $1,000. Hautes Côtes, Mâconnais, and Côte Chalonnaise Pinot Noirs may be the best-value red Burgundy wines. These general Burgundy AOC wines are moderately priced but may lack the distinctive character found in better-quality wines. Wines from the Loire, Alsace, or other European regions are also less expensive than Côte d'Or Burgundies.

Pinot Noirs from highly regarded California appellations generally have moderate to moderately high prices ($20 to $60). Larger county appellations, such as Santa Barbara County or Sonoma County, are less costly, and California appellation Pinot Noirs are the least expensive. Oregon Pinot Noirs have a similar price range to those from California, with Willamette Valley wines being the most expensive.

Some New Zealand Pinot Noirs are value-priced, but the best match the prices of California and Oregon wines. Pinot Noir producers from less widely known and distributed states and countries, like New York and Michigan in the United States, Australia, South Africa, and Chile, offer interesting Pinot Noir choices in a range of prices.

Storing and Serving Pinot Noir

Pinot Noir should be stored like any other wine at 55°F/13°C. In Burgundy, consumers prefer even cooler temperatures, from 3°F to 5°F/2°C to 3°C lower. Pinot Noirs of excellent quality with good acidity and alcohol below 14 percent can be aged for five to 10 years. Some red Burgundies are aged for much longer.

Modern-style Pinot Noirs that are rich and soft with high alcohol should be consumed fairly quickly, certainly in less than five years. Pinot Noir should be served at cooler temperatures (60°F to 65°F/16°C to 18°C) than other red wines in a classic Burgundy glass that has a large bowl with a closed rim to trap the aromas. Because many top-quality Pinot Noirs are not filtered, there is a good possibility that sediment will develop, so the wine requires decanting before serving.

Pinot Noir and Food Pairings

Pinot Noir is probably the most food-compatible red wine there is. It is versatile enough to go with an array of light meats, both fish and poultry, as well as with braised, roasted, and grilled red meats. It is exceptional when paired with creamy, hard, and aged cheeses. Mushrooms and other vegetables also work well.

The foods from each Pinot Noir–producing appellation make excellent companions to the wine. In Burgundy, Pinot Noir is served with braised beef or chicken. In New Zealand, grilled lamb is classic. In Santa Barbara, it is served with Central Coast barbecue of beef tri-tip and beans, and in Oregon, Pinot Noir is perfect with a slab of grilled salmon. And the wines and foods can be interchanged; a classic Burgundy dish like coq au vin can pair with a California, Oregon, or New Zealand Pinot Noir.

When selecting foods to go with Pinot Noir, keep in mind that because of its delicacy, traditional Pinot Noir cannot stand up to the rich and spicy foods that may pair well with heavier-style wines. Pinot Noirs can work well with pungent, but not spicy, international dishes. Soy sauce marinades, Korean barbecue, and savory dishes of the Pacific Rim can all be paired with Pinot Noir. Neither traditional-style nor modern-style Pinot Noirs are particularly good with shellfish and seafood like oysters or crab. However, richer Pinot Noirs can work well with richer foods like grilled or roasted lamb chops, steak, or braised meats.-

TOP 10 FOODS TO PAIR WITH PINOT NOIR

1 Beef bourguignon
2 Coq au vin
3 Vegetable gratin
4 Grilled salmon
5 Duck sausage
6 Roast pork
7 Venison
8 Brie
9 Sautéed wild mushrooms
10 Roast chicken

SUMMARY

Pinot Noir is a sensual wine that intrigues growers, winemakers, and consumers alike. Although it is produced in smaller quantities than Cabernet Sauvignon, Merlot, and Chardonnay, and generally costs considerably more, Pinot Noir has had the highest rate of increase in popularity of any major varietal over the last several years. To growers and winemakers, it is known as the heartbreak grape because it is difficult to grow and problematic to make into wine. To wine lovers, Pinot Noir can be the most fickle of wines. One wine elicits the most sensual experience with its subtle, elusive aromas and flavors. The next confronts you with flat, thin, dull aromas and flavors that provide a lifeless experience with no appeal. However, growers, winemakers, and lovers of the wine continue the search for a perfect Pinot Noir.

Syrah/Shiraz

Syrah is considered one of the noble red grapes of France and the world alongside Cabernet Sauvignon, Pinot Noir, and Merlot. Syrah is widely produced in both the Old World and New World and has found a middle ground for wine character between Cabernet Sauvignon and Pinot Noir. It is a wine that is richer and deeper in flavor than Pinot Noir, but with softer tannins than Cabernet Sauvignon. Syrah originated in the Rhône, France. In the New World it was first planted in the 1830s in Australia, where it is called Shiraz (shear-AZ, not shear-OZ). Cultivation in the United States followed sometime later.

In the last 20 years Syrah has shown a dramatic increase in popularity throughout the world. Syrah is a wine that meets the demand of the New World marketplace for ripe, soft wines with a punch.

REGIONS AND APPELLATIONS

The primary Syrah-growing region in the Old World is Southern France. In the New World almost every grape-growing region in Australia produces Shiraz. In the United States, California, followed by Washington, are important producers. It is also beginning to find acceptance in warm climates in Old World countries, such as Italy, Spain, and Portugal, and New World countries from South Africa to Chile, Argentina, and New Zealand. (See Appendix 2 and Appendix 3 for maps of Syrah/Shiraz-producing countries and appellations, pages 380–447.)

France

Many stories have been connected to Syrah's origin. Did it arrive in France from Persia, was it brought to France by the Romans, or did it have regional roots within France? DNA testing by Carole Meredith at the University of California, Davis, has determined that Syrah was derived from two obscure grapes, Dureze and Mondeuse Blanche, from Southeastern France.

The principle home of Syrah in France is the Rhône, followed by Languedoc-Roussillon, Provence, the South-West, and Ardèche. A vast increase in production over the past decade has come from the Southern Rhône and Languedoc-Roussillon.

The Rhône

Classic French Syrah comes from the Rhône, which most consider the bellwether of Syrah production. The valley runs north to south along the Rhône River below Beaujolais and is divided into two distinctly different parts: the Northern Rhône, where Syrah is the only red grape variety planted and is predominantly a varietal wine, and the Southern Rhône, where Syrah is just one of several red varieties grown and is used in red wine blends.

In the Northern Rhône five appellations produce Syrah. The best and most important are Côte-Rôtie and Hermitage. The others are Cornas, Crozes-Hermitage, and St. Joseph, which produce good to excellent Syrah as well. The Côte-Rôtie, adjacent to the village of Ampuis west of the Rhône River, is known for its distinct steep slopes and terraced vineyards. Hermitage, perhaps the most famous Rhône appellation, sits east of the river facing southeast overlooking the village of Tain, and is also sloped. The Northern Rhône is smaller than the Côte d'Or in Burgundy with even more limited production. Small volume coupled with high demand make some Northern Rhône Syrah wines almost as expensive as some Grand Cru Burgundies or Premier Cru Classé Bordeaux.

In the Southern Rhône, which produces 20 times more wine than the Northern Rhône, Syrah is used as a blending grape with Grenache. It is one of the important varieties in classic Châteauneuf-du-Pape and is almost always an ingredient in the more general Côtes du Rhône and Côtes du Rhône-Villages wines. Other Southern Rhône appellations making similar wines include Côtes du Ventoux, Côtes du Luberon, Gigondas, Lirac, Tavel, and Vacqueyras. Wines from all of these regions are blends of Grenache, normally the dominant grape, with Syrah, Cinsaut, Mourvèdre, and Carignan in varying combinations.

Other French Regions

Syrah is made into a varietal wine in both Languedoc-Roussillon and Provence, but it is also used in regional-variety blends. Languedoc-Roussillon produces both Appellation d'Origine Contrôlée (AOC) and Vin de Pays (VDP) wines (see Appendix 2, French Wine Classifications, page 380) with AOC wines made from regional blends that can include Syrah. Outstanding appellations within the Languedoc include Coteaux de Languedoc and Costières de Nîmes.

Vineyards in the Côte-Rôtie, one of the most notable appellations in the Rhône, France, are so steep they must be terraced.

The Shiraz vineyard at GroomKalimna is per-
fectly suited to the hot climate in Australia's
Barossa Valley.
RACHAEL EHRAT

VDP wines are primarily identified and labeled by variety, which include Syrah, and are largely exported to New World countries. There has been a surge in lesser-known VDP appellations, such as Ardèche west of the Rhône River and the Drôme Valley east of the Rhône, which both make quality wines from Syrah. A considerable amount is also grown on the island of Corsica in the Mediterranean.

Australia

Shiraz has achieved a high degree of popularity and success in Australia recently, but that has not always been the case. In the 1970s many Shiraz vineyards were ripped out because of lack of consumer interest in the variety. It was only at the end of the 1970s and the start of the 1980s that many growers and producers realized that old-vine Shiraz is a treasure. The country's hot climate is ideal for Shiraz, and it is now considered the nation's variety, as Zinfandel is to the United States. It is planted in all of the major regions of the country: South Australia, New South Wales, Victoria, and Western Australia.

The best Shiraz comes from appellations within South Australia: Barossa Valley, Adelaide Hills, McLaren Vale, and Clare Valley. Other regions also have appellations producing noteworthy Shiraz, including Heathcote in Victoria, Hunter Valley in New South Wales, and Margaret River in Western Australia. The producers in these appellations argue about who makes the best Shiraz. Although considered a warm, or even hot, country, Australia contains a range of microclimates that produce different styles of wines. With serious drought conditions throughout the country, water access and irrigation are increasingly important considerations.

The United States

Much of California is perfectly suited to Syrah because of its warm to hot climate in which the variety does particularly well. It is produced from Mendocino County in the north to San Diego County in the south. It thrives in Northern Napa Valley, the eastern part of Sonoma County, and the warmer inland valleys of Mendocino County. Central Coast producers have taken Syrah under their wings and planted it in those areas that are not suitable for Pinot Noir. Whereas Pinot Noir thrives in cooler, foggier climates closer to the Pacific Ocean, Syrah does best in the warmer eastern and flatter parts of Paso Robles in San Luis Obispo County, Monterey County, and Santa Barbara County. In the south, the appellations of Temecula and San Diego County are exceptional areas for cultivating Syrah. Perhaps the greatest number of cultivated vines are in the San Joaquin Valley (Central Valley) where the hot, dry weather enables the variety to produce large yields used for moderately priced wines.

Washington State is the newest region in the United States to have embraced Syrah. Most of the vineyards have been planted in the last dozen years, indicating that the vines have yet to reach their full potential. It is grown exclusively in the Southeast Columbia and Yakima valleys, particularly the Walla Walla AVA, noted for their hot days and very cool nights, which represent some of the greatest day-to-night temperature differences in the United States. This climate characteristic allows the grapes to achieve dark color and big fruit flavors from the excessive sunlight but also maintain bright acidity from the very cool nights.

Other Countries

In Europe, Syrah is becoming more popular in warmer regions, like Italy and Spain. It is also planted in Central and Eastern European countries, most importantly in Moldova where yields are high but the wines are not yet able to compete internationally.

In Northern Italy a few producers in Trentino-Alto Adige and Piedmont are cultivating Syrah in very small quantities. In Central Italy Tuscany is the most prominent growing region. Few Syrah varietal wines are produced there, but it is becoming more acceptable as a grape used to add color and flavor to Sangiovese, even to Chianti. More frequently it is included in Super Tuscan blends (see Chapter 12, Super Tuscans, page 194). In the south, Sicilian Syrah plantings are increasing and although it is still only cultivated in small amounts, there are some excellent successes with varietal wines in different parts of the island.

In Switzerland the Valais, located much farther north on the Rhône River, grows and produces very small quantities of Syrah that reflect the lighter style found in France's Northern Rhône.

In Spain Syrah is being widely tested because of its success in warm climates. However, relatively little Syrah is being planted. At this point Catalonia in the northeast is cultivating the most Syrah for use in blended wines. It is also increasingly being planted in several hot central regions (Castile-La Mancha, Valencia, Murcia) and Andalusia in the southern region. There are also plantings in the Alentejo in Portugal.

Table 14.1 Prominent Syrah/Shiraz Appellations

COUNTRY	REGION/APPELLATION	SUBAPPELLATION	VILLAGE OR SUBAPPELLATION
France	Rhône	Northern Rhône	Côte-Rôtie Hermitage Cornas Crozes-Hermitage St. Joseph
		Southern Rhône	Châteauneuf-du-Pape Côtes du Rhône Côtes du Rhône-Villages
	Languedoc-Roussillon		Coteaux de Languedoc Costières de Nîmes
	Provence Ardèche Corsica		
United States	California	Mendocino County	McDowell Valley Yorkville Highlands
		Napa Valley	Carneros Oakville St. Helena Mt. Veeder
		Sonoma County	Russian River Valley Knights Valley Carneros Sonoma Valley
		Monterey County	Santa Lucia Highlands
		San Luis Obispo County	Paso Robles
		Santa Barbara County	
	Washington State	Columbia Valley	Yakima Valley Horse Heaven Hills Red Mountain Wahluke Slope Walla Walla Valley

The popularity of Syrah has taken hold in several countries in the New World as well, including Chile, Argentina, South Africa, and New Zealand.

In South Africa, where the grape is known as Shiraz as it is in Australia, it is ideally suited to the country's warm climate and is on the path to overtake Cabernet Sauvignon and Merlot in popularity. The wines are modern in style and mostly big, ripe, and high in alcohol, but some producers are creating examples that have Old World finesse and elegance.

Chile's climate also suits Syrah. A range of styles are widely produced in the Curicó, Rapel, and Maipo Valleys of the Central Valley as well as in Aconcagua.

Prominent Syrah/Shiraz Appellations, continued

COUNTRY	REGION/APPELLATION	SUBAPPELLATION	VILLAGE OR SUBAPPELLATION
Italy	Tuscany	Maremma	
	Sicily		
Switzerland	Valais		
Australia	South Australia	Adelaide Hills Coonawarra Barossa Valley McLaren Vale Clare Valley Eden Valley	
	Victoria	Heathcote	
	Western Australia	Margaret River	
	New South Wales	Hunter Valley Mudgee	
Chile	Central Valley	Curicó Valley Maipo Valley	
		Rapel Valley	Colchagua
	Aconcagua		
Argentina	Mendoza		
South Africa		Paarl Stellenbosch	
New Zealand	North Island	Hawkes Bay	Gimblett Gravels

Argentina produces a lighter style than other New World countries, mostly in Mendoza, but it is unlikely to ever overtake Malbec in popularity. New Zealand Syrah does well in Hawkes Bay, mostly in the Gimblett Gravels microclimate, which has soils reminiscent of Châteauneuf-du-Pape.

Table 14.1 includes an abbreviated list of countries that produce highly regarded Syrah/Shiraz along with their most prominent appellations and subappellations.

SYRAH: THE GRAPE

Syrah is a warm, even a hot-climate grape, but it can also be cultivated in cool climates. The grape is small berried with dark skins that produce more **ANTHO-CYANINS** (color pigmentation) than most other grapes, resulting in wines that are equally dark in color, with rich, deep, black fruit flavors and firm tannins. Cool-climate Syrah has less color, but is still dark, with strong acidity, less-dominant fruit character, and more spiciness. The grapes are quite disease-resistant, although rot can be a problem in wet or humid conditions. Syrah buds late, which prevents frost damage, and is moderately early ripening, improving the grower's chances of missing late summer or autumn storms.

The Syrah grape that is planted in France and the United States is the same as the Shiraz grape planted in Australia. The Petite Sirah grape (proper name, Durif) found commonly in California is not Syrah. It is an offspring of Syrah with quite different characteristics (see Chapter 15, Petite Sirah, page 267).

IN THE VINEYARD

Syrah can thrive in a range of climates, but loves a Mediterranean climate best. It performs well in poor-quality soils and is loved by growers because it can produce large yields while maintaining good quality.

Climate

Sicily's hot Mediterranean climate favors warm-climate grapes, such as the Syrah planted at Spadafora in Sicily, Italy.

Syrah is a grape that thrives in the sun. Warm and hot is what it likes, but it can be successful in moderately cool climates with enough sun exposure. It is a grape variety that transplanted easily from the Rhône to Australia and eventually to most other New World countries, including the United States. Sunlight is a major advantage in most New World growing regions and Syrah can survive the dry conditions found in much of the New World. In the moderately cool Northern Rhône vineyards it grows successfully on south-and southeastern-facing slopes, which have excellent sun exposure. It does particularly well in the hotter flat and rolling plains of the Southern Rhône, Languedoc, Australia, and California. Flatter slopes, the valley floor, and open plains initiate warm-climate characteristics, but the vines also benefit from their deeper, and therefore cooler, soils, which may provide easier access to the water table. Steeper slopes offer better drainage, and higher elevations reduce temperatures, generating cool-climate characteristics. Proximity to a river may counterbalance the elevation and provide needed moisture.

Although Syrah can master almost any climate condition, weather is one of the factors that determine the crop quality. The Rhône is always at risk for spring mistral winds, summer rain, and hail; each can severely damage the crop.

In hot climates drought can diminish crop size. Even with irrigation during a drought, water access can be restricted, causing serious problems.

Soil

Syrah adapts to a wide range of soil types. Syrah is planted in sandy, loamy, clay, granite, schist, volcanic, and limestone soils, among others. It is cultivated in soil textures that range from silt to gravel to large stones to almost no soil at all. Thinner, granitic soils and sloped vineyards at higher elevations, like those of the Northern Rhône, maintain and reflect heat. In the New World deeper soils in the valleys and plains are more varied and heat summation may have more impact on flavor components. Poor-quality soils that make Syrah struggle for nutrients are generally preferred, but more fertile soils have demonstrated their ability to produce higher yields and an excellent flavorful crop as well.

A typical Syrah vineyard in the Red Mountain AVA, Yakima Valley, Washington
TYLER WILLIAMS/KIONA VINEYARDS

Water Access

Because Syrah is more drought-tolerant than some other varieties, it can be dry-farmed. In fact, dry-farmed vineyards produce most of the best Syrah wines worldwide. Syrah vineyards in the Old World are usually not irrigated although dispensation is granted for new vines in some appellations and in appellations experiencing drought conditions, which has occurred in Languedoc-Roussillon.

New World growers are able to manage water with irrigation to maximize both grape quality and higher yields. Australian vineyards are both dry-farmed and irrigated. Unfortunately an extended drought in Australia has left some irrigated vineyards without water, reducing the yield for many growers. In the United States almost all vineyards are watered.

Vineyard Management

Though far more limited in selection than for Pinot Noir, rootstock and clone selection for Syrah are increasingly important as growers are looking for specific characteristics from the grapes. In the case of Syrah they are seeking greater drought tolerance and yield control.

Historically, Syrah was head pruned just about everywhere, and some of the best vineyards still contain bush vines. In both the Northern and Southern Rhône most of the vineyards remain gobelet-trained (head-pruned), except where growers employ unusual trellising and training methods to deal with the extreme slopes of the Côte-Rôtie and mistral wind, perhaps one of the fiercest found in any of the world's wine regions.

In the Southern Rhône, and even more so in Languedoc, growers are employing New World training methods which are more cost effective to maintain, are easier to harvest, and even make mechanized harvesting possible. In many areas of the New World, variations of vertical shoot positioning (VSP) are replacing head pruning. Nevertheless, in Australia the old bush vines are some of the most cherished, and some high-quality wine producers are replanting vineyards with bush vines.

Yield

Syrah vines produce some of the largest crops of any variety, up to 20 tons per acre/350 hectoliters per hectare. However, the best normally yield two to three tons per acre/35 to 50 hectoliters per hectare. Some older vineyards in the Rhône and Australia yield even less, and only the lowest quality exceeds 10 tons per acre/175 hectoliters per hectare. Many believe that Syrah is one variety that can have high quality and high yields at the same time. However, the larger the crop's yield, the less distinctive the wine is likely to be. The smaller the yield, the more intensity of flavor develops in the grapes.

Because Syrah vines are capable of producing huge yields, deft canopy-management techniques are required. Shoot-thinning and leaf-pulling are used to increase sun exposure, and green harvest is used to control yield. Dr. Richard Smart, a prominent Australian viticulturist, advocates training methods that open the canopy to the sun. VSP allows for more sun exposure, increasing ripeness even for heavily cropped vines, and usually initiates larger yields of comparable quality to the grapes obtained from lower yields.

Harvest

Harvest time and method depend on the style and price point desired for wine produced from a specific crop. The best-quality grapes come from older low-yield vines that are hand-harvested, but there has been a significant increase and improvement in mechanical harvesting, which will certainly continue.

Early harvest from less-ripe grapes results in lighter style of wines. Late harvesting allows for greater ripeness, but runs the risk of overripe fruit, loss of acidity, and potential storm damage.

IN THE WINERY

Syrah follows standard red winemaking procedures. Color development, flavor profile, barrel selection, and blending are techniques that bring out Syrah's personality. Syrahs from cooler climates are more likely to be made into wines that have moderate color, higher acidity, and increased spicy character. Warmer-climate Syrahs are made into riper fruit and oak-driven styles.

Ripe grapes, long maceration, and barrel aging can intensify tannins. Acidity remains unless the grapes become overly ripe, but in the New World acid can be added back if the wine lacks balance.

Cold Soak

Dark-skinned grapes are distinctive to Syrah. The juice is typically cold-soaked with the skins before fermentation to extract color and tannins. Length of maceration varies from a day or two to as long as eight to 10 days depending on the amount of extraction the winemaker desires from both cold-soaking and fermentation. Some French producers leave partial stems on Syrah during this process to develop more complexity, but this practice is less common in today's market because it diminishes the fruit flavors. New World producers prefer using only Syrah berries to obtain a fruit-forward style of wine.

Fermentation

Most Syrah producers prefer fermenting the grape by itself. However, in some Northern Rhône appellations it is appropriate to ferment Syrah with a small amount (generally less than 5 percent) of the white variety Viognier (or a Marsanne-Roussanne blend in some appellations) to produce a wine with more floral character than a 100 percent varietal Syrah. This Rhône technique of blending small amounts of white Rhône varieties into Syrah before fermentation has been adopted by some producers in Australia, California, and elsewhere.

Carbonic Maceration

Carbonic maceration is used by some winemakers to help produce a fruit-forward style of wine. High fermentation temperatures are used to control tannins, with fermentation temperatures usually maxing out at 90°F/32°C. The fermentation process lasts from six to 12 days. In Syrah-growing regions of France, natural yeasts may still be used, but winemakers in the New World, as well as many in the Old World, inoculate with a wide range of yeasts that are able to handle the ripe, high-sugar Syrah grapes and convert them into high-alcohol wines. The practices of punching down and pumping over are used to reincorporate the skins with the juice for more extraction. Rotary fermenters are employed frequently with Syrah.

After fermentation, some New World producers use the reverse osmosis technique for ripe Syrah. This technique reduces the amount of alcohol in the wine but maintains the lush fruitiness. (The technique is also used for other high-alcohol wines.) Acidification, e.g. adding acid back into the wine, must be carried out to correct its balance.

Barrel Aging

Most quality Syrah is barrel aged. It accepts barrel aging easily because of its high level of color and flavor compounds. In France, French barrels are always used. In Australia and the United States, where varieties such as Chardonnay and Pinot Noir are almost always aged in French oak, American barrels are most commonly used for Syrah. Whether French, Australian, or American, more concentrated Syrahs are likely to be aged in new barrels (up to two years old).

Winemakers desiring less oak character select a barrel program that includes older barrels and age for shorter periods of time. Aging time for Syrah can vary from as little as six months to as long as 30 months. Price-sensitive Syrahs (less than $10 to $12 per bottle) are likely to use oak alternatives, such as chips, staves, and dust. Some producers maintain the natural fruitiness of the Syrah grapes by using only older barrels or no oak at all.

Blending

Syrah is often made as a varietal wine, but it is also a principal component of Rhône-style blends, the classic style of the Southern Rhône. In the Northern Rhône, where Syrah is the only red grape, Viognier is allowed in Côte-Rôtie (20 percent permitted by law), and the varieties are blended before fermentation. Hermitage and Crozes-Hermitage also allow the inclusion of small amounts of the white varieties Marsanne and Roussanne in their red wines.

In the Southern Rhône all wines are blends. The standard blend is dominated by Grenache, with Syrah, Mourvèdre, and Cinsaut making up the balance. Syrah is becoming a more dominant component in the blend, but it is still used in a smaller proportion than Grenache. In the Rhône several varieties are allowed, so it is possible to find some wines with other, more obscure grapes as part of the blend. In Châteauneuf-du-Pape, AOC regulations allow 13 different varieties in the blend (see The Châteauneuf-du-Pape Blend, at left). In Languedoc-Roussillon Syrah is also commonly blended with Grenache, Mourvèdre, and Cinsaut, as well as Carignan.

In Australia Shiraz is commonly blended with Cabernet Sauvignon as either the dominant or the minor variety. A style of wine called GSM is also produced and is a blend of Grenache, Shiraz, and Mourvèdre. Most recently, mimicking the style of the Northern Rhône, some producers are blending Shiraz with a small amount of Viognier.

In California and Washington State Syrah is usually made into a varietal wine, but Southern Rhône and Australian styles are also found. The law that requires only 75 percent of the primary grape in a varietal wine allows producers tremendous flexibility to blend in other varieties without informing the consumer. There are even a few producers making Northern Rhône–style wines with the addition of Viognier.

Stylized Wines

Syrah is used in making two different stylistic wines. In Australia sparkling Shiraz has become one of the most popular wines, and it is being copied in other New World countries, particularly the United States. Most sparkling Shiraz is made sweet but almost tastes dry because the sugar is balanced by the tannins, alcohol, and acid. It is made using either méthode champenoise or the Charmat process. The result is a dark red sparkling wine.

Syrah grapes can achieve great sweetness without rotting or breaking down, making them ideal for fortified port-style wine. The grapes have many of the same characteristics as classic Portuguese port varieties, and produce excellent

THE CHÂTEAUNEUF-DU-PAPE BLEND

Regulations permit 13 varieties in the blend.

8 RED VARIETIES

    Cinsaut

    Counoise

    Grenache

    Mourvèdre

    Muscardin

    Syrah

    Terret

    Vaccarèse

5 WHITE VARIETIES

    Bourboulenc

    Clairette

    Picardin

    Picpoul

    Roussanne

All Châteauneuf-du-Pape bottles are embossed with this seal.

wines that can replace a traditional port in any situation. These port-style wines are commonly made in Australia, California, and South Africa.

SYRAH: THE WINE

Until fairly recently, France and Australia produced the bulk of Syrah. Its increase in popularity has largely come from the New World because of the love affair so many consumers have with the rich, juicy style that is typically produced there.

Color

Syrah's dark skins, combined with extended maceration to extract the anthocyanins, make it one of the darkest wines. In the cooler Northern Rhône, some wines can almost pass for a dark Pinot Noir, but most varietal Syrah, whether from the Old World or New World, has a color range from dark red to blue and often almost black. It also shows tremendous concentration.

Aromas, Flavors, and Mouthfeel

One of the most desirable characteristics of Syrah is the range of aromas and flavors that come from the grape and its aging. Each climate—cool, warm, or

Examples of regional Syrah/Shiraz, each produced in a different style, from left: Barossa Valley, Australia; Paso Robles, California; Côte-Rôtie, France; Washington State

TOM ZASADZINSKI/CAL POLY POMONA

Table 14.2 Syrah/Shiraz Aromas and Mouthfeel

	AROMAS		MOUTHFEEL
FRUIT	FLORAL/HERBAL/VEGETAL	EARTHY/OTHER	
Blackberries	Black peppers	Barnyard	Fat
Black cherries	Cloves	Burnt rubber	Finesse
Black raspberries	Coffee	Chocolate	Grip
Cherries	Eucalyptus	Earth	Hot
Cranberries	Licorice	Game	Lean
Currants	Mint	Leather	Lush
Plums	Spices	Mushrooms	Mouth-filling
Prunes	Violets	Smoke	Rich
Raisins		Sweaty saddle	Tart
Ripe tomatoes		Tar	
Stewed plums			

hot—leaves its imprint on Syrah's aromas, flavors, and mouthfeel. As temperature in the vineyards warms and ripeness increases, the resulting wines can have aromas that change from black peppers, violets, and spices to black fruits, such as black cherries, blackberries, black raspberries, and black currants, to plums, licorice, and chocolate. The riper the fruit the more the actual fruit is emphasized, but very ripe fruit earns descriptors like chocolate, leather, meat, earth, game, licorice, and tar. Overripe character can highlight prune and raisin aromas. Table 14.2 lists aromas and mouthfeel that might be found in a Syrah/Shiraz.

The mouthfeel of a Syrah can range from lean and tart with sharp tannins in cool-climate wines to big, fat, and lush with tannins that grip in hot-climate wines. Riper grapes lead to wines that have higher alcohol and therefore are more mouth-filling.

Wine Styles

Syrah is classified into three basic styles: cool, warm, and hot climate (see Table 14.3). Terroir and irrigation impact style (see Impact of Irrigation, at left), but the characteristics of the Syrah grape and the resulting wine are most influenced by climate factors: heat days, sun exposure, aspect, and natural moisture or the lack thereof.

The Northern Rhône and other cool-climate zones, like Heathcote in Victoria, Australia, Carneros in Napa Valley, Santa Rita Hills in Santa Barbara County, and parts of Mendocino County achieve less ripe grapes with higher acidity, producing wines that have less fruit intensity, more pepper and spice, and lower alcohol. Although there are certainly fresh berry aromas, black peppers, cloves, and violets provide greater impact. These wines are likely to be more age worthy. Their color can be deep, brilliant red but rarely black. These lighter styles are often similar to a rich Pinot Noir.

IMPACT OF IRRIGATION

Very hot climates in the Australian appellations of New South Wales and Victoria, and the San Joaquin Valley (Central Valley) of California produce wines that do not fit the hot-climate style profile. The heavily irrigated vineyards in these regions result in very high yields. This in turn minimizes the fruitiness in the wine and dilutes the potential alcohol, producing a lighter, easier-drinking wine with lower alcohol and tannins than is usually found in hot-climate Syrah.

In the warm Southern Rhône and Mediterranean appellations the vines develop riper grapes, resulting in wines with ripe sweet fruit flavors. The color is quite dark. The VDP wines from Languedoc-Roussillon and those found in many New World appellations, like the Alexander Valley in Sonoma County or McLaren Vale in Australia, have more ripeness, more flavor intensity, more sugar, and lower acidity than cool-climate Syrahs. In Australia the wines have richer color and more powerful fruit flavors. Aromas of black cherries, blackberries, currants, and earthiness are characteristic. The mouthfeel is round and voluptuous with good, grippy tannins and the additional ripeness produces higher alcohol levels, up to 15 percent. These big, heady Syrahs are most popular today with the wine press and young consumers.

Syrah grapes from Australia, California, and Washington State are distinguished by the amount of potential sun they receive in most locations. That factor alone provides the dominant character of their wines. More sun and hotter days drive ripeness, creating more intense fruit flavors, more sugar, and higher potential alcohol in wines that have very dark, almost black color. Hot-climate Syrah is one of the ultimate ripe styles of wine. Along with California Zinfandels, they can achieve some of the highest alcohol levels of any wine. The wines can be unctuous and thick, similar in appearance and richness to port. Recognized hot-climate regions are Australia's Barossa Valley and Hunter Valley and Paso Robles and Lodi in California. The aromas and flavors show ripe fruit, particularly plums or even stewed prunes, with the essence of chocolate and licorice.

Table 14.3 Syrah/Shiraz Wine Styles

	COOL CLIMATE	WARM CLIMATE	HOT CLIMATE
REGIONS/APPELLATIONS	Northern Rhône, Coonawarra, Heathcote, Victoria, Western Australia, Santa Rita Hills, Santa Maria	Southern Rhône, Languedoc, McLaren Vale, Sonoma County	Barossa Valley, Hunter Valley, Paso Robles, Lodi
VARIETAL/BLEND	Varietal and blended	Varietal and blended with Grenache	Both varietal and blended
CLIMATE	Continental/mistral winds	Mediterranean/mistral winds	Cool to hot depending on appellation
SOIL	Granite	Variable—clay, stones, alluvial	Variable—clay, sand, loam, volcanic
ALCOHOL	12%–14%	13%–15%	14%–16% or more
ACID	Moderate to high	Moderate	Moderate to low
AROMA/FLAVORS	Black peppers, spices, violets	Ripe red fruits	Black fruits, plums, chocolate, licorice
COLOR	Medium to dark red	Dark red to black	Very dark blue to black
BODY	Light to medium	Medium to heavy	Heavy

The texture and body of the wine is dense, black, and concentrated with a soft, velvety, but hot mouthfeel. Alcohol in this style is typically above 15 percent and can easily exceed 16 percent.

ENJOYING SYRAH

Syrah is a wine to just flat-out enjoy. In most cases it is masculine in style but can be comforting and pleasant on all but the hottest summer nights. It is a wine that is satisfying with food but also can be enjoyed by itself, particularly for a cool late-afternoon picnic or on a wintry night sitting around a fire with good friends.

Buying Syrah

When you buy Syrah the important considerations are style and price. Evaluate the style characteristics that are important to you and try to select wines from a climate that is recognized for producing that style of wine. Based on the climate and appellation characteristics described above, you should be able to tell the difference between a light, cool-climate Côte-Rôtie or Crozes-Hermitage from the Northern Rhône and a dense, hot-climate Syrah from the Barossa Valley.

Another factor that plays a big role in the buying decision is price. While many $10 bottles of Syrah are available in the American market, it is unlikely that you will find rich, concentrated, almost syrupy wines with complex aromas and depth of flavor in a $10 bottle. Most wines at this price point come from high-yielding, irrigated vineyards from broad appellations. They are excellent values and great quaffing wines, but they will not be made from the same quality grapes or with the same winemaking procedures and barrel-aging regimens as finer wines.

Wines of high quality whose prices range from moderate to expensive are produced in most major appellations of the world and in every style. The best Côte-Rôtie, Hermitage, or Grange from Barossa Valley are all expensive. But there are many quality choices in the $20 to $60 range for every style of Syrah made.

Storing and Serving Syrah

Syrah should be stored and served just like other red wine. Storage temperature should be 55°F/13°C. Some wine professionals recommend that it be treated as you would a Cabernet Sauvignon; however, others recommend that it be served a little cooler, similar to a Pinot Noir.

Syrahs from warm and hot climates and from New World regions are normally consumed young, showcasing their ripe fruit flavors. These wines can be aged for two to five years without a problem. Cool-climate wines have the ability to age longer. Wines from the Northern Rhône have been known to age for a decade or more.

A Bordeaux-style glass works best with cool-climate Syrah. Some warm- and hot-climate producers prefer the larger bowl provided by a Burgundy glass to emphasize the powerful aromas of the wine.

TOP 10 FOODS TO PAIR WITH SYRAH/SHIRAZ

1 Butterflied and grilled marinated leg of lamb with fresh herbs
2 Parmesan or other hard cheeses
3 Smoked meats
4 Pappardelle pasta with a ragout of wild boar or braised beef
5 Rich, thick meat stews or chili
6 Regional American barbecue: ribs, brisket, sausages
7 Barbecued steak, pork, or sausages
8 Bean dishes like cassoulet with lamb or sausages
9 Braised dishes of beef or game cooked in Syrah
10 Roast prime rib

Syrah and Food Pairings

Unlike many other varietals, Syrah goes with a narrow range of foods because it needs strong or rich flavors to match the richness of most of the wines. Even cool-climate Syrah reaches out for richer foods. Red meat, especially beef, lamb, and game, is an excellent choice because it matches well with the wine's notable tannins. The meat can be grilled, sautéed, braised, or barbecued. Smoked meats are also good choices. The style of the Syrah should also be taken into account.

Lighter, more elegant cool-climate wine styles can pair with salmon and dark-meat poultry. However, Syrah should not be served with most fish and light-colored poultry. These choices are likely to be overwhelmed by the wine.

Heavier, warm-, and hot-climate styles are excellent matches for red meats, stews, and casseroles. Hearty starches, hard cheeses, and winter vegetables are also good accompaniments. Heavy Mediterranean dishes, such as cassoulet or Moroccan lamb, have a natural affinity with Syrah. For vegetarians a rich risotto with roasted winter vegetables and Parmesan cheese will pair well.

SUMMARY

Syrah has had an explosion of popularity over the last 20 years, particularly among New World consumers. Its popularity derives from the rich fruit flavors found in New World styles and its soft tannins, which make it an approachable wine. Syrah adapts to many different climates and soil types. It is generally considered a warm- to hot-climate variety that is planted around the world but is successful in cooler appellations. Whatever its style, Syrah/Shiraz is primarily a red-meat grape.

Other Red Vinifera Grape Varieties

We have just completed a review of four major international red varieties, all with a French history. Here we have selected 10 more red varieties that play a dominant role in some part of the world, are sold internationally, or hold a special place in local American markets. Three have French origin, four are Italian, two are Spanish, and one is Central European. Although French, Malbec is now dominant in South America, and Zinfandel, with roots in Croatia, is now considered a California variety. We also briefly describe 30 additional red *vinifera* varieties.

BARBERA

Although Nebbiolo is recognized as the glamour grape of the Piedmont region of Italy, Barbera is by far the most widely consumed wine in the region. The everyday wine of those living in Piedmont, it is affordable, blends well with the other Piedmont red varieties, Nebbiolo and Dolcetto, and adapts well to different vineyard settings not only in Piedmont, but around the world. The grape grows well in wide-ranging climates and can be made with or without oak aging, and its vibrancy works well with a variety of foods.

Countries That Produce Barbera

The most prominent Barbera wines come from Piedmont, but it is widely planted throughout Italy. Outside Italy, the largest plantings are in Argentina, where it has been mostly used in generic blended wines. More recently, excellent Barberas are being made in California, Australia, and even in Argentina. Very small, but successful, plantings are also being established in the Pacific Northwest.

Italy

Barbera is used in 50 percent of the red Denominazione di Origine Controllata (DOC) wine produced in the Piedmont region. Although indigenous to the area, it is an extremely adaptable grape that has been cultivated successfully throughout Italy, where it ranks second in terms of total acres/hectares of red grapes planted. There are extensive plantings in Lombardy, which is adjacent to Piedmont, as well as in Emilia-Romagna. Although there are some varietal wines from these regions, most of the wine gets blended into table wines. There are also significant plantings in Sardinia and Sicily, where, again, it most frequently ends up in blends. Because of the number of acres/hectares under cultivation, Barbera has achieved a much greater level of recognition than in the past.

The New World

Barbera is planted in several New World countries: the United States, Australia, Argentina, and South Africa. It has done especially well in California, particularly in the San Joaquin Valley (Central Valley) where it is used in generic blends. It is made into varietal wines in some of the better appellations, like Sonoma County, Napa Valley, Sierra Foothills, and the Central Coast. Most would consider Barbera to be more successful in California than the two other better-known Italian transplants, Sangiovese and Nebbiolo.

In the Pacific Northwest Barbera has been planted in Horse Heaven Hills near the Columbia River Gorge and in Oregon's Columbia Valley.

10 Common Red *Vinifera* Varietals

Barbera

Cabernet Franc

Gamay

Grenache/Garnacha

Lambrusco

Malbec/Auxerrois/Côt

Nebbiolo

Sangiovese

Tempranillo

Zinfandel

Red varietals commonly found in restaurants and retail stores, from left: Grenache blend, Rhône, France; Tempranillo blend, Rioja, Spain; Zinfandel, Napa Valley, California; Sangiovese, Chianti, Tuscany, Italy
TOM ZASADZINSKI/CAL POLY POMONA

Additional red varietals commonly found in restaurants and retail stores, from left: Nebbiolo, Piedmont, Italy; Gamay, Beaujolais, France; Cabernet Franc, Loire, France; Barbera, Piedmont, Italy
TOM ZASADZINSKI/CAL POLY POMONA

Argentina has been a prolific producer of Barbera because of Italian immigrants who brought it there. Until recently it was mostly used in making moderately priced blended wines, but its international popularity has caused local winemakers to start developing varietal Barberas.

Australia and South Africa are also producing varietal and blended wines from Barbera.

Characteristics

The Grape

Producers love Barbera because it is durable, vigorous, and pliable. It adjusts to diverse soils and climates and responds well to various winemaking techniques. It performs best in warm climates and grows well on calcareous, clay, or loam soil. This late-ripening variety maintains its acidity even when fully ripe. This characteristic is a major asset in hot climates where grapes usually gain sugar but lose acidity as they ripen. Barbera retains its quality even with high yields. Although it grows almost anywhere, vines on south-facing slopes result in more concentrated fruit.

The Wine

Barbera wine has deep ruby color, high natural acidity, very low tannins, and an intense, luscious fruitiness characterized by red cherries and red berries. In the winery it responds readily to both traditional and modern winemaking techniques.

The wine can be fermented in stainless steel, large casks, small barrels, or a combination of these. Fermenting and aging in oak changes its character. It may gain complexity at the expense of losing varietal traits.

Barbera has long been considered one of the great Italian value wines but can make a low-yield, concentrated wine that is more costly. It is often blended with other varieties, particularly Nebbiolo, or even international varieties like Cabernet Sauvignon, to produce a Super Piedmont style of wine. Barbera d'Asti produces what are believed to be the best examples of the wine. At a more modest level it is also used for table wine.

Enjoying Barbera

Barbera is a versatile wine for pairing with food. It is a perfect companion with poultry, richer fish dishes, pastas and risotto dishes, and cheese. Meat-filled pastas called *agnolotti* are a regional favorite. Barbera also demonstrates good acidity and astringency, which make it a perfect companion to grilled foods and roasts. Riper Super Piedmont or California styles are an excellent choice for grilled foods, roasts, and barbecue.

CABERNET FRANC

Cabernet Franc, considered the stepchild of Cabernet Sauvignon, is actually one of its parents, along with Sauvignon Blanc. It lacks the international cachet of Cabernet Sauvignon with growers because of the higher prices paid for Cabernet Sauvignon grapes, and is less popular with consumers who love the rich, ripe style found in so many Cabernet Sauvignon wines. Cabernet Franc can be found as part of a blend in Bordeaux or Meritage wines (see Chapter 11, Meritage Wines, page 187), or as a varietal wine in both the Old World and the New World.

Countries That Produce Cabernet Franc

Cabernet Franc is less well known than its Bordeaux siblings, Cabernet Sauvignon and Merlot, but it is grown widely in Europe and throughout the world. It is the most important red grape in the Loire, producing varietal wine. It is also used to make varietal wines by select producers in Italy, California, and Australia, but in other places it is used mostly as a blending grape for its traditional Bordeaux partners.

Europe

France is the dominant producer of Cabernet Franc, primarily in Bordeaux where it is used as a blending grape and in the central part of the Loire, where it is made primarily into varietal wines. In Bordeaux it is used primarily in right bank wines, which are dominated by Merlot. In the Loire it is produced primarily in the Touraine appellations of Chinon, Bourgueil, and St-Nicholas-de-Bourgueil and the Anjou appellation Saumur-Champigny. It makes excellent rosés throughout Anjou. It can also be found in France's South-West wine region, where it is blended with Cabernet Sauvignon and Merlot, in a style similar to a Bordeaux, or with the local indigenous variety Tannat.

In Italy, Cabernet Franc is cultivated in the northern regions of Trentino-Alto Adige and Friuli. It does well in these cooler climates but unfortunately it is losing ground to the better-known Cabernet Sauvignon. It is also found in the coastal Tuscan region of the Maremma where it is used in Bordeaux-style blends or is blended with Sangiovese into Super Tuscan wines (see Chapter 12, Super Tuscans, page 194).

Spain has recently started planting Cabernet Franc in Catalonia. Many believe that the Spanish grape Mencía, which is grown in Bierzo in the north-central part of the country, is a close genetic relative of Cabernet Franc. Because of the popularity of Bordeaux varieties, Cabernet Franc is also planted in several Central European countries along with Cabernet Sauvignon and Merlot. These are used in Meritage blends.

The New World

Cabernet Franc is planted widely throughout the New World. It is mostly cultivated where the other Bordeaux varieties, Cabernet Sauvignon and Merlot, are planted and Bordeaux-style or Meritage wines are made.

Because Cabernet Franc does well in cool climates, it is planted throughout the Northeast and the Great Lakes in the United States and in Eastern Canada, particularly in the Niagara Peninsula of Ontario. Some of the best varietal Cabernet Franc wines, reminiscent of the wines of the Central Loire, come from Long Island, New York, but they are available only in very small quantities. California is the largest U.S. grower of Cabernet Franc, sold primarily for use as a blending grape. However, a strong but small group of producers is making varietal Cabernet Franc in both a rich, ripe style and in a lighter Loire style.

Cabernet Franc is also grown in all of the major New World wine-producing countries, Argentina, Australia, New Zealand, and South Africa among them, where it is used in Meritage-style wines.

Characteristics

The Grape

Like Cabernet Sauvignon, Cabernet Franc has a small berry but it has less color in the skins and lower tannins. It ripens earlier than Cabernet Sauvignon, making it more desirable in cooler climates where late-season cool, wet weather can be problematic for grape quality. Soils change the style of the wine. Sand with gravel produces a light style. Clay, which holds moisture, moves the grapes to a richer style, and limestone creates the potential for even weightier wines with greater richness and body. Most French Cabernet Franc is planted on single Guyot close to the ground, making it difficult to overcrop. In the New World vertical shoot positioning (VSP) is more common.

The Wine

Cabernet Franc can produce a wine that is lean and light but it adds distinctive flavors when used as a blending component. It has less color than Cabernet Sauvignon and Merlot but offers some complex characteristics to a blend that are not found in either of the other major Bordeaux varieties. An herbal component is most noticeable. It is perhaps one of the most underrated wines because it is most commonly used in blends and is thought of as a relatively small part of Bordeaux blends. In reality, it is the second most-used variety in most right bank Bordeaux wines, particularly in St. Émilion, and the third most-used in most left bank wines from the Médoc and Graves. In Bordeaux Cabernet Franc adds its fruit character to blends even though it usually is the secondary grape to Cabernet Sauvignon or Merlot.

The Loire style demonstrates a more herbal character with higher acid and lower alcohol than in Bordeaux. In the New World where warmer climates generate riper grapes, Cabernet Franc showcases riper fruit, less herbal character, moderate acidity, and higher potential alcohol.

Cabernet Franc is invariably compared with Cabernet Sauvignon when made into a varietal wine. It lacks the power of Cabernet Sauvignon but has a beautifully complex nose of herbs, raspberries, currants, and violets. It lacks the typical ripe flavors and deep color of Cabernet Sauvignon, but it can be a perfectly balanced wine that works exceptionally well with a greater range of foods. Some detractors consider the wine's aromas to be weedy and bell pepper–like, which can be the case if the grapes are overcropped and underripe. But one person's weedy bouquet can be another's fresh, herbal character.

Enjoying Cabernet Franc

Varietal Cabernet Franc is a wine to be served with food, but it can be enjoyed by itself as well. It complements red meats, poultry, other white meats, and even fish. Pan-seared steaks, chops, and fillets of salmon or tuna work well. For a casual event Cabernet Franc is great with grilled hamburgers or barbecued chicken. Braised dishes and earthy vegetables, including mushrooms and root vegetables, pair with Cabernet Franc. Using fresh herbs in dishes brings out the herbal character in the wine. Anjou pears with blue cheese and walnuts is a classic Loire dish that is served with a Loire Cabernet Franc. With an Italian Cabernet Franc you might consider veal osso buco with soft polenta.

GAMAY

Centuries ago the duke of Burgundy banned Gamay Noir, known today as Gamay, from the prominent northern regions of Burgundy and relegated it to the southern regions of Beaujolais and Mâcon. Although it was assigned a secondary position to other varieties, today Gamay is one of the most popular grapes internationally.

Countries That Produce Gamay

Gamay is the dominant grape in only one location in the world, the Beaujolais subregion of Burgundy, where it yields more than 99 percent of the total grape crop and makes mostly young, juicy, fruity wines. Wine from the Beaujolais appellation has become the benchmark for Gamay-based wines around the world. Smaller amounts are planted in Mâcon, just north of Beaujolais, and in the Central Loire, primarily in Touraine, where it is made into varietal wine or blended with Cabernet Franc and Malbec (locally known as Côt). Gamay is also produced in varietal or blended wine in Savoy, and in the South-West small amounts are grown for blending.

Production of Gamay is limited outside France. It is most popular in Switzerland, where it is blended with Pinot Noir to make a jug style of wine called Dôle. Gamay is also found in Central Europe, and a bit is grown in Italy.

In California some wines have been sold as Napa Gamay and Beaujolais Gamay. However, Gamay grape vineyards are virtually nonexistent. Napa Gamay is now sold under the correct name of the grape used in the wine, Valdiguié (see page 269). Gamay Beaujolais was likely a poor-quality clone of Pinot Noir, and use of the name Gamay Beaujolais is no longer permitted.

Gamay is being planted in Ontario, Canada, on the Niagara Peninsula with some success.

Characteristics

The Grape

Soil plays an important role in Beaujolais. Most Gamay is cultivated on highly alkaline soils, resulting in high-acid grapes. An early-ripening variety, it is susceptible to spring frosts that can kill buds, but it can reproduce a second set of buds successfully. It is typically grown on gobelet to help control vigor, but still produces high yields. The grapes usually have low sugar levels, yielding low alcohol in the finished wine, but chaptalization increases alcohol content. There are many subvarieties of Gamay, but the classic is Gamay Noir à Jus Blanc.

The Wine

Producers in Beaujolais make the wine in three distinct styles: Beaujolais Nouveau, Beaujolais/Beaujolais-Villages, and Cru Beaujolais (see Cru Beaujolais, at left). Beaujolais Nouveau uses carbonic maceration without oak aging, which accentuates light, fresh-strawberry flavors. Maceration is short, producing early-drinking, low-alcohol wines. Beaujolais and Beaujolais-Villages go through a longer maceration, achieving greater depth of color, aroma, and flavor. This emphasizes the fruit character of the wine and softens the acidity. Beaujolais and Beaujolais-Villages are also not oak aged, so are still considered light and refreshing but with a bit higher alcohol than Beaujolais Nouveau.

CRU BEAUJOLAIS

Cru Beaujolais, considered the best of the region, comes from 10 selected villages with better granitic soils than most of the rest of Beaujolais. These wines are made in a style more representative of Pinot Noir than the typical Beaujolais. Each connoisseur of these wines is likely to have a favorite cru. Usually consumed from one to five years of age, some crus can age for up to 10 years.

Those marked with an asterisk tend to have the ability to age the longest.
Brouilly*
Chénas
Chiroubles
Côte de Brouilly
Fleurie
Juliénas*
Morgon*
Moulin-à-Vent*
Regnié
St. Amour

Cru Beaujolais is produced in the cooler north, which also has better soils and more restrictive yields. The grapes are allowed to ripen more fully, achieving higher sugar levels with higher alcohol potential. Longer maceration develops stronger tannins and some producers even barrel-age the wine. Cru wines have more concentrated flavors and body, producing wines that, although still moderately light, are quite a bit richer than Beaujolais or Beaujolais-Villages.

Enjoying Gamay

Beaujolais is a great starter wine if you are venturing into French red wines for the first time. Beaujolais and Beaujolais Nouveau can be as affordable as many inexpensive New World varietals. Cru Beaujolais is more serious with greater complexity, but the beauty is that even these "best of the best" wines are quite affordable.

Any young Gamay is a great match with the array of food found on the Thanksgiving table, including the turkey. It is also a perfect picnic wine with a baguette, some charcuterie (salamis, hams, and pâtés), soft cheeses, and fruit. Do not forget the small little pickles the French call *cornichons*.

Treat Cru Beaujolais like a Pinot Noir and serve it with either red or white meats. Grilled lamb or roast chicken are perfect. It also goes with salmon or tuna. Grilled vegetables, especially mushrooms, are quite compatible with these wines. Wines that are bigger and richer can even work with a grilled steak or mature cheeses.

GRENACHE/GARNACHA

A grape of Spanish origin, Grenache is the second most widely cultivated grape in its native land, where it is called Garnacha. It is the world's most widely planted red grape variety. Yet, it is a relatively unknown grape because it is commonly blended with other varieties, like Tempranillo in Spain and native Rhône varieties in Southern France.

Countries That Produce Grenache

The acres/hectares of Grenache under cultivation have increased tremendously in Spain and Southern France, which is responsible for a significant portion of the world's production. It is also planted in large quantities worldwide because it works so well when blended with other varieties.

Europe

Grenache is widely cultivated throughout warm southern appellations in Europe. It is a primary grape in France and Spain but is also planted extensively in Sardinia, Italy, where it is better known as Cannonau. It is also grown throughout Southern Italy, most notably Calabria and Sicily.

In Priorat, Spain, a mule is used to plow a Grenache vineyard because the hillsides are too steep for machines.

In Spain Garnacha is often blended with Tempranillo in red Rioja wine, and the Priorat produces some of the best examples of wines made from old-vine Garnacha. Other regions where Garnacha is planted include Navarre, Penedès, and Somontano in the north; Madrid, Tarragona, and Toledo in Central Spain; and Utiel-Requena in the south.

Grenache is the dominant variety in most of Southern France, primarily in the Southern Rhône and Languedoc-Roussillon, where it is normally blended.

The New World

Grenache has found a home in Australia but is also widely cultivated in California. In Australia it is produced as a varietal wine, but is better known as an important part of GSM (Grenache, Shiraz, and Mourvèdre) blended wines. Its most notable plantings are in South Australia in the Barossa and Clare Valleys. In California the Central Coast area around Paso Robles has become known for its Rhône-style blends, with Grenache playing an important role.

South Africa has had extensive plantings of Grenache for several decades. It is made into either a varietal wine or is blended.

Characteristics

The Grape

Grenache is a very productive, long-lived grape that performs well in warm and hot climates. It grows in dry and stony locations in Europe and sandy soils in California. It showcases fruitiness and ripeness, which makes it a perfect blending grape, similar to how Merlot is used in Bordeaux blends. Like Merlot its tannins are soft and it adds fruitiness when combined with less fruity varieties. Under dry conditions with restricted water it develops significant tannins. It is primarily grown on head-pruned vines in climates with little moisture.

The Wine

A 100 percent Grenache wine can be heavy, concentrated, and alcoholic, often attaining 16 percent alcohol as it does in the Priorat. These wines are strong, powerful, and fruity, with aromas of strawberries, blackberries, or raspberries and an intense cherry color. Most Grenache, however, is blended with other varieties, as in the well-known Rioja wines and Southern Rhône wines. It is the dominant grape variety in Southern Rhône blends, from simple Côtes du Rhône to the more elegant Châteauneuf-du-Pape.

Grenache is also the choice for rosé wines in both Spain and Southern France, including Tavel where it is used primarily as the base for the region's famous rosé wines.

Enjoying Grenache

A varietal Grenache or a Rhône-style blend goes with regional dishes like cassoulet. It also pairs with Mediterranean foods, herbs, and spices. A mixed grill of beef, pork, or even wild boar and vegetables, including various summer squashes, tomatoes, eggplant, and mushrooms, makes a perfect complement. In Spain the mixed grill substitutes lamb chops or butterflied leg of lamb for other meats and adds spring onions to the vegetables. A GSM from Australia is the perfect match for a thick char-grilled steak.

LAMBRUSCO

Lambrusco, a family of grape varieties, creates a wine that receives few accolades but may be the most popular exported Italian wine. This wine has introduced millions of consumers to Italian wines the world over. In the United States it had a large following during the 1970s and 1980s and is still widely drunk. Made in many styles, the best known is a light, fizzy, off-dry wine that is low in alcohol and most frequently comes in a screw-capped bottle or jug. It is a pleasant and refreshing, easy-drinking wine.

Countries That Produce Lambrusco

Lambrusco is produced predominantly in the north-central part of Italy in Emilia-Romagna. Most Lambrusco is made by cooperatives in large volume. The largest producer, Cantine Riunite in Reggio nell'Emilia, has more than 4,000 grower-participants and is one of the largest wine-exporting companies in the world.

Elsewhere in Italy, Lambrusco is planted in Lombardy, Piedmont, Trentino-Alto Adige, and Apulia. It has not been transplanted to other countries either in Europe or the New World, except for a variety known as Lambrusco Maesini, which is grown in Argentina.

Characteristics

The Grape

Lambrusco consists of as many as 60 clonal subvarieties, each presenting different characteristics. The best and most commonly used are Marani, Salamino, Maestri, Sorbara, and Montericco. Generally, the grape clusters are medium-sized, compact, and elongated. Grapes are medium-sized and spherical, with thin but tough dark blue skins and herbaceous taste. The grapes are generous producers that maintain high acidity. In the past, grapes were grown on vines that were trained high, in a manner similar to a pergola, to avoid mildew. Newer vines use the Geneva double-curtain trellising system.

The Wine

Lambrusco produces a wine that is ruby red with violet reflections and a perfume of violets, strawberries, and cherries. It is young, fresh, fizzy, and fruity. It is high in acid with very soft tannins. Ancellota, a local, indigenous grape that is genetically unrelated, is added to Lambrusco for its sweetness and flavor. It is the only non-Lambrusco variety allowed in the blend.

Lambrusco is made into primarily two styles: dry and sweet, or amabile. The most common domestic Italian style is dry, but amabile style is the dominant one for export. Most Lambrusco uses the Charmat process to create its bubbles, although there are some small producers who use the traditional method. Non-sparkling Lambrusco is also produced, although in fairly small quantities.

Enjoying Lambrusco

Lambrusco is a wonderful aperitif when served chilled before a meal. With a meal, it pairs with the style of foods that are served in Emilia-Romagna, including various cured meats, like salami and mortadella; egg pastas in cream-based sauces; and cheeses. It also works with salads like caprese (tomatoes, mozzarella, and basil) and other tomato-based dishes. Lambrusco, especially the amabile-style, is also a perfect partner for spicy foods, such as Tex-Mex, Thai, Indian curries, and Sichuan Chinese, as the fruitiness in the wine tones down the spice. It is also a good choice for a summer picnic or barbecue where a light, cool, and refreshing wine is desired.

MALBEC/AUXERROIS/CÔT

Malbec is a variety that has gone in and out of fashion. Historically, it comes from Southwestern France in the regions of Bordeaux and the South-West, but Malbec's new and preeminent home is Argentina, where it is the star variety.

Countries That Produce Malbec

Once the variety was transplanted to Argentina, Malbec was found to prefer the warm, dry climate that is characteristic of the Mendoza region. It took hold and consequently the country has become the dominant producer of Malbec in the world.

Malbec is also cultivated in other major New World wine-producing countries, including California, Chile, Australia, and South Africa. In these countries, it is used primarily as an addition to Meritage-style wines.

In Bordeaux, Malbec is one of the five primary varieties that are allowed in red wine blends. However, is has lost its luster and has largely been replaced with the other permitted varieties, especially Merlot. It is more successful and highly regarded in the South-West, where it dominates blends with local indigenous varieties playing a minor role. In Cahors, Malbec, commonly known as Auxerrois, is usually the dominant variety in blended wines (at least 70 percent).

Malbec is also produced in the Loire, primarily in Touraine, but also in parts of Anjou, where it is called Côt. It frequently is blended with Cabernet Franc and Gamay. It is rarely found elsewhere in Europe.

Characteristics

The Grape

Noted characteristics of the grapes are their large size, loose clusters, high tannins, and dark color. Yields can be high. Malbec is most susceptible to coulure (poor fruit set that causes reduced yields) and rot. Some believe it lacks aroma and finesse. Malbec changes with the weather. In warm temperatures, the grape offers deep color and a desirable ripeness. In cooler weather, it loses character.

The Wine

There are three styles of Malbec: two Old World styles, one found in Cahors in France's South-West region and the other in the Central Loire, and the New World style found in Argentina. Malbecs produced elsewhere generally mimic the Cahors or Argentinean styles.

In Cahors, Malbec has intense tannins, high acidity, and strong color and flavors with a notable tobacco character. It plays either a dominant or a minor role in blends and is also made into a 100 percent varietal wine. Some Malbecs are still made in a classic rustic style with high tannins that contain significant levels of antioxidants, which are believed to be heart-healthy.

The Malbec of the Loire is made in a much different style than that of the South-West because the Loire has much cooler temperatures, causing the wines to have less color and richness than the Cahors wines. It is a minor grape in the red wine blends of Touraine and Anjou, but is used to a greater extent in the rosés of Anjou.

Because the warmer climate produces riper grapes, Argentinean Malbec generally has softer tannins and lower acid levels than French Malbec. Its approachable New World style is characterized by riper fruitiness and higher potential alcohol.

Enjoying Malbec

The best foods to serve with richer Malbecs are like the wine, heavy and rich. Red meats work well whether the wine comes from Cahors or Argentina. Lighter preparations, or even duck or chicken, work better with a Loire style. A roast tenderloin of beef is an excellent companion. Classic foods of the South-West are cassoulet, lamb, and foie gras. Argentina is one of the major beef producers in the world. There the classic match is a grilled steak or spit-roasted haunch of beef served with chimichurri, or meat-filled empanadas.

NEBBIOLO

To Italians Nebbiolo is "the king of wines and the wine of kings." Many believe it is most deserving of being considered a noble grape. It is an incredible, complex, intoxicating grape. It results in a wine that is almost always very expensive and best when consumed after 10 to 15 years of aging.

Countries That Produce Nebbiolo

Nebbiolo has never earned worldwide recognition except by connoisseurs because it is planted in limited quantities. In Italy it is primarily grown in Piedmont where it has achieved great renown. The premier Nebbiolos come from the Barolo and Barbaresco subregions. Outside Piedmont it is produced in the adjacent regions of Lombardy and Aosta Valley, but those wines receive little acclaim.

Outside Italy attempts have been made to cultivate Nebbiolo in California, South America, and Australia, but with little success. An occasional wine might show promise, but consistency has been a problem.

Characteristics

The Grape

Nebbiolo is a thin-skinned, high-acid, tannic grape that produces some of the most aromatic and long-lived wines in the world. Piedmont sets the standard for Nebbiolo. Production is limited and restricted. It is one of the most difficult varieties to grow because it is sensitive to weather, soil, and vineyard-management practices. Vineyard location greatly influences its characteristics and results in subtle differences in the grapes and the wines. The grape prefers clay and limestone soil that has a high degree of calcium and hillsides with a south or southwest exposure. It buds early, making it susceptible to late-spring frost, and is harvested into late October or even early November, making it susceptible to autumn rains.

The Wine

Nebbiolo-based wine has a complex floral, herbal, or earthy bouquet. It is full-bodied, rich, and intense. The aromas associated with the variety are tar and roses, but violets, dried cherries, truffles, licorice, and spices, especially nutmeg and cinnamon, have been identified in Nebbiolo. The skins contain little color, so the wine is very light. As it ages, it becomes rosy orange but remains very pale. Darker color in a Nebbiolo is a good indication that it has been blended with some Barbera, Cabernet Sauvignon, or Merlot.

Traditional-style Nebbiolo requires maceration with skins and juice for as much as 20 to 30 days. The wine then matures in Slovenian oak casks for up to four years and in some cases longer.

Modernists have shortened maceration time to as little as three to four days, using punch downs, rotary fermenters, or in-tank paddles to obtain color and extract. Aging then occurs for as little as one year in French oak barrels. The result is a wine that is fruitier, with softer tannins and acids than the traditional style. These wines are more of a Super Piedmont style, not 100 percent Nebbiolo. They are blended with a small amount of Barbera, Dolcetto, or international varieties like Cabernet Sauvignon, which softens the tannins and results in a richer style of wine.

Traditional or modern, Nebbiolo must be aged in the bottle to achieve a balanced wine. Modern-style Nebbiolo is softer and easier on the palate than traditional styles. Therefore, it does not require lengthy aging. Traditional Nebbiolo should be aged much longer than modern style to mellow the wine. Some Nebbiolo can mature for as long as 30 or 40 years.

Enjoying Nebbiolo

A platter of local Piedmontese salami and lardo are a common match for a Nebbiolo wine. A pot-roasted piece of beef in wine (called *brasato*) with a dark wild mushroom (or truffle) risotto or soft polenta as a base are also a perfect complement to a Nebbiolo. The high acidity and high tannins in the wine are a good contrast to any roasted meat or vegetable dish.

SANGIOVESE

Sangiovese is Italy's signature grape. It can be found in almost every region of the country, but it earned its reputation in Tuscany where it is the base grape for Italy's best-known wine—Chianti. Although extremely successful in Italy, Sangiovese has struggled when planted in other parts of the world.

Countries That Produce Sangiovese

More acres/hectares are dedicated to Sangiovese than to any other variety in Italy. An indigenous grape, Sangiovese has its primary domain in Tuscany where it goes by many different names: Brunello, Prugnolo, and Morellino, which are all Sangiovese subvarieties. It is also common in surrounding regions, including Emilia-Romagna, Abruzzo, and Umbria, but is planted all the way to Sicily.

Outside Europe it is fairly broadly planted, with everyone trying to replicate its success in Tuscany. However, producers have had only modest results. It is planted in California, Australia, and Argentina in significant amounts. Piero Antinori of Tuscan fame had a joint venture with Atlas Peak Winery in the Atlas Peak American Viticultural Area (AVA) in Napa Valley with the largest planting outside Italy. The wines never met expectations and the vines were either grafted or replanted to other international varieties, like Cabernet Sauvignon. Plantings in Australia and Argentina have also been slow to develop. Its success in the New World may be hindered because it is planted in hotter vineyards than those in Tuscany.

Sangiovese, an indigenous Italian grape, thrives in the vineyard at Sauvignola Paolina in Tuscany, Italy.

Characteristics

The Grape

Sangiovese is a high-acid, tannic grape that is difficult to ripen. Even though it ripens late, it maintains its acidity. It is short on color in its skins, so it is often blended with other reds to give it deeper tone. The variety is planted on different soil types throughout Tuscany, but limestone gives it the best body and flavor. It is a vigorous producer so it has to be pruned carefully.

The climate in Northern Tuscany can be too cool and rainy to produce Sangiovese with character, so Southern Tuscany, with its warmer climate, is better suited to the variety. It is one of the most malleable grape species and has mutated into many subvarieties, each of which has adapted to its particular terroir and has radically different characteristics. Sangiovese is an aromatic grape with a primary bouquet of sour cherries, but it also has herbal, spicy, tea leaf, and tar characteristics, and an earthiness that reflects Tuscan forests.

The Wine

Sangiovese-based wines come in many versions, each from different clones. Chianti, considered the classic Tuscan wine, is the most familiar wine. Historically in Chianti, it was blended with the red varieties Canaiolo, Colorino, and Mammolo, and the white varieties Trebbiano and Malvasia. In an effort to improve the quality of Chianti, white varieties have been discarded and the international varieties Cabernet Sauvignon, Merlot, Syrah, and Cabernet Franc are now allowed to be blended with Sangiovese, and can compose up to 20 percent of the blend. The result has been an increase in the popularity of Chianti.

Other versions of Sangiovese-based wines are Brunello di Montalcino, Rosso di Montalcino, Vino Nobile di Montepulciano, and Morellino di Scansano. The most highly regarded is Brunello di Montalcino. Made from the Brunello clones, it is produced within a fairly small geographic zone surrounding the beautiful hill town of Montalcino—and nowhere else. The grapes come from older, better-sited vineyards than do grapes for Rosso di Montalcino. Brunello di Montalcino is made from 100 percent Sangiovese grapes, and is the epitome of Sangiovese—deep red, powerful, and elegant. Regulations stipulate that the vines yield no more than 80 quintals (approximately 8.8 tons) per hectare. The wine must age in a combination of barrel and bottle for four years, three months, plus an additional year for Riserva.

Rosso di Montalcino, also from Brunello clones, comes from the area around Montalcino, but the vines are younger, are not planted in the best vineyard settings, and are allowed to produce higher yields. The wine requires only one year of aging. Vino Nobile di Montepulciano, a wine made from Prugnolo Gentile, is blended with Canaiolo. It is not as powerful as Brunello di Montalcino but is less acidic than Chianti. Morellino di Scansano is a different clone, Morellino, which is produced in the Southern Tuscan region called Maremma.

Sangiovese is also bottled and labeled as Sangiovese with a Tuscany appellation. Less expensive than Chianti and very drinkable, this wine is guaranteed to come from Tuscany. Finally, Sangiovese is often used as the base grape variety for the famed Super Tuscan wines that are sold by proprietary name and are blends that include Bordeaux varieties. Sangiovese can be harsh if not handled correctly. It improves with maceration on the skins and malolactic fermentation (MLF). It ages well in new oak, which gives it a hint of vanilla, but most are aged in the classic large, old botti which provides subtler oak influence.

Enjoying Sangiovese

Sangiovese's distinguishable cherry aroma, high acidity, and moderate alcohol make it a red wine that pairs well with many foods. It is best known as the complement to foods of Central Italy, including tomatoes, beans, Parmigiano cheese, wild boar in any form, and the famous bistecca alla Fiorentina (beefsteak of Florence), a very thick, wood-fire–grilled steak, cooked rare and garnished with fresh herbs, salt, and fresh olive oil. Add to this a range of pastas, pizzas, polentas, and the options for pairing food with Sangiovese are practically limitless.

TEMPRANILLO

When one thinks of Spanish wine, Rioja, made from Spain's premier native grape Tempranillo, immediately comes to mind. The grape variety probably originated in Rioja, acquiring different names as it spread throughout the country and causing consumers some confusion. Tinto Fino, Tinto Madrid, Tinto del Pais, Tinto de Toro, Cencibel, Tinta Roriz, and Ull de Liebre are just some of the synonyms.

Countries That Produce Tempranillo

An indigenous grape, Tempranillo is cultivated throughout much of Spain, but historically it was not widely grown in other parts of the world. Recently, however, growers in other countries have been successfully cultivating Tempranillo because they have learned that it performs well in many different environments.

Europe

Tempranillo is produced in almost every region of Spain, but is most prolific in the cooler central and northern regions. It is best known in wines from Rioja and Ribera del Duero, but excellent wines from Tempranillo are produced in Aragon, Navarre, Catalonia, Castile and Leon, Castile-La Mancha, Valencia, Murcia, and Montilla-Moriles.

Spain's Iberian neighbor Portugal is the second-largest Tempranillo producer in Europe. In Portugal the grape is known by the name Tinta Roriz. It is used in both red table wines and fortified wines. Italian producers have also started planting the grape.

The New World

In the New World there is excitement about Tempranillo because of early success with the variety. Argentina is the most prolific producer of Tempranillo. Until recently it was used as a blending grape in modest-quality wines. It is now being grown to produce high-quality wines that will be marketable internationally. Tempranillo has gained several awards in California and increased planting of the variety continues in Oregon and Texas. Australia is the other country where Tempranillo is achieving a measure of success. It is planted in small quantities throughout the major wine regions of the country.

Characteristics

The Grape

Tempranillo is a thick-skinned, early-ripening black grape that is low in tannins and acidity. It prefers calcareous or sandy clay soil. In cooler conditions it becomes more acidic; in warmer climates it develops higher sugar levels and color. However, one of the reasons that Tempranillo is such a great wine grape is that it maintains its acidity as it ripens. The vines' density must be managed to maintain the grapes' quality. Historically, most vines were head-pruned, but more recently newer vineyards are trellised and spur-pruned. It is an extremely adaptable grape and has evolved several subvarieties, each with individual traits that are suited to growing conditions in different vineyards.

The Wine

Tempranillo wines can be powerful and flavorful with moderate alcohol levels,

---

RIOJA CLASSIFICATIONS

Rioja wines can be categorized under different age classifications. These classifications appear on the label and are helpful when selecting a wine.

Joven are young, unaged wines.

Crianza are wines that must be aged at least 24 months, with at least six months in oak.

Reserva are wines that must be aged at least 36 months, with at least 12 months in oak.

Gran Reserva are wines that must be aged at least 60 months, with at least 18 months in oak.

deep ruby color, and intense aromas. It is usually blended with other varieties that contribute acid, aromas, and flavors, resulting in very complex wines. The classic Rioja blend consists of Tempranillo, Garnacha, Mazuelo, and Graciano. In other regions and countries local varieties may be used, and international varieties are increasingly appearing in the blends.

The Spanish government established stringent regulations for aging Rioja (see Rioja Classifications, at left). The differences in the required aging periods have considerable impact on the characteristics of the wine. When young and unaged, Tempranillo results in a light style with aromas of strawberries, raspberries, blackberries, black cherries, and currants. Aging changes the wine to a heavier style containing notes of plums, coffee, spices, vanilla, raisins, cinnamon, and leather.

Tempranillo matures beautifully in the bottle, producing light- to moderate- to full-bodied wine. The oak-aged wine can be held for long periods, up to 30 years, without oxidizing. Tempranillo is good when consumed young but is at its best as old, oak-aged wines. It is also used for rosés.

Tempranillo grapes harvested at Viñas del Cenit in the Toro region of Spain.

Enjoying Tempranillo

When you are in Logroño, Rioja, it is difficult not to pass a tapas bar offering little plates of regional food specialties with the local wines of Rioja. Little sandwiches or kabobs of fresh sardines, anchovies, chicken, rabbit, chorizo, mushrooms, or any of the local cured meats and cheeses are a perfect complement to Tempranillo. The most classic dish served with this varietal is lamb—whether chops, steaks, or a leg—grilled over a grapevine or wood fire.

ZINFANDEL/PRIMITIVO

Though Zinfandel is not an indigenous California grape, it is believed to be the first wine grape of note that people drank and enjoyed in the state. The Mission grape variety, credited with being the earliest grape cultivated in California, had been around for hundreds of years, but Zinfandel was the first variety in the state to produce wines that people enjoyed for their aromas and flavors rather than merely for their alcoholic kick.

Zinfandel's popularity has experienced an ebb and flow since it was first planted in the United States. It was a favorite wine from the time of the California Gold Rush through the late 1800s. During Prohibition, it was the variety of choice of home winemakers who could legally make up to 200 gallons of wine per household (see The Wine Brick, page 260). In the 1970s and 1980s, it began losing ground to the onslaught of Chardonnay, Cabernet Sauvignon, and Merlot. Many old Zinfandel vineyards were uprooted and replanted with more popular international varieties.

THE WINE BRICK

During Prohibition each head of household in the United States could legally make up to 200 gallons of wine each year. Growers in California harvested grapes, dried them, and formed them into bricks of dried must, and shipped them east. On the label they stated, "Do not add water—it will cause fermentation, creating alcohol." Even someone who did not understand fermentation could figure out how to create alcohol by reading the label.

Wine bricks were a way for ethnic Europeans to continue to enjoy their own homemade wine during Prohibition.
MUSEUM OF HISTORY AND ART, ONTARIO, CALIFORNIA

What happened next is one of the great stories of the California wine industry. Bob Trinchero of Sutter Home Winery was drawing off juice from some Zinfandel that was just starting to ferment in order to intensify the color and tannins of the juice remaining on the skins. The light pink must that was withdrawn from the tank was unable to complete fermentation, leaving a light pink wine with residual sugar. Rather than dump the "unusable" pink wine, it was sold off in the winery tasting room. It was an instant hit with novice wine drinkers and those who liked sweeter wine. It was soon bottled at Sutter Home, labeled as White Zinfandel, and almost immediately it was copied by other wineries.

This led to a resurgence in the grape's popularity as a blush wine in the late 1980s and early 1990s, saving many old Zinfandel vineyards. In the late 1990s, Zinfandel experienced another revival as a trendy, affordable dry red table wine that matches with a broad range of foods.

Countries That Produce Zinfandel

Much research has been conducted in the past decade to determine the origin of Zinfandel. Many thought it came from Southern Italy where the variety Primitivo is planted. It was eventually discovered that Zinfandel and Primitivo are, in fact, the same grape. Over time, clones have evolved that changed some characteristics of the vines in each location, but they are synonymous. Further research uncovered a DNA match with Crljenak Kaštelanski, a variety indigenous to Croatia. Although we now know that it is Croatian, Zinfandel is perceived as America's grape.

The vast majority of Zinfandel is made in California from Mendocino County to San Diego County. It was the most widely planted red variety in the state into the 1990s until it was overtaken by Cabernet Sauvignon. Today it is the second most-planted red variety in California.

It has been planted in several other states, with no more than a few dozen acres cultivated in any of those states. Most states besides California do not have the number of heat days needed to ripen the Zinfandel grapes. However, a few select vineyards and wineries outside California produce excellent examples of the variety.

In Europe the original grape, Crljenak Kaštelanski, is still grown in Croatia, but most Primitivo, as it is called in Italy, is cultivated in the Southern Italian region of Apulia. Primitivo was historically not held in much regard, and most of it was blended into basic table wine. Partially because of the discovery that it is the same as Zinfandel, however, it has acquired higher status and today is both used in blends and bottled on its own as a varietal wine.

Because of the popularity of Zinfandel in California, many winemakers have taken cuttings to other New World countries. Today most have at least a small number of Zinfandel acres planted from cuttings transplanted from California. Mexico has demonstrated some success with the variety in the Baja Peninsula just south of California.

Characteristics

The Grape

Whether Primitivo in Apulia or Zinfandel in California, the grape can be grown in different climates and soils, using varied vineyard-management and winemaking techniques, to result in widely varying styles of wine. However, it is a vine that is classically grown on poor soils in hot climates. The best are bush vines that are head-pruned.

Zinfandel ripens unevenly and it is common to find unripe grapes on the same vine as overripe grapes, making harvest difficult. Most growers of high-quality Zinfandel allow the grapes to achieve full ripeness, causing some to be overripe with very high sugar and potential alcohol.

In California, Zinfandel is produced in hot-climate, moderate-climate, and even a few cool-climate vineyards, but they all have ample sun. There are several vineyards in California that are from 100 to 140 years old that are still yielding quality grapes from head-pruned vines. Grapes destined for White Zinfandel or inexpensive varietal table wines are taken from high-yield vines that are trellised and mechanically harvested.

The Wine

Different styles of Zinfandel are enjoyed for different reasons. White Zinfandel is a fun wine, while red Zinfandel is more serious. White Zinfandel is light and fresh, and always has a touch of sweetness, unlike rosés that are classically dry. It is popular with novice wine drinkers because of its spritzy, sweet, soda-pop character. It is easy to drink and low in alcohol, providing a transition for those new to wine. The American taste for sweet beverages makes it a popular low-alcohol option at adult events.

The common California Zinfandel style today is dry and full-bodied, with spicy, peppery aromas. High sugar levels in the grapes create high-alcohol wines with jammy, ripe-fruit aromas. High-alcohol Zinfandels tend to be soft, round, and smooth with heat on the palate. This is a style that appeals to many wine consumers today, particularly young consumers who love the ripe, fruity flavors and are not put off by the high alcohol levels.

Years ago Zinfandel was a starter wine for many young people primarily because of its cost. A full-bodied Zinfandel could be purchased for far less than a Cabernet Sauvignon, Pinot Noir, or Merlot. Today it is still considered a value wine, generally selling at lower prices than comparable international-varietal wines.

Primitivo ripens well in Southern Italy but typically does not achieve the ripeness and high alcohol found in California. The style tends to be more rustic with wines unaged or aged in botti.

Enjoying Zinfandel

In the past Zinfandel was rarely found on white-tablecloth-restaurant wine lists. In part because it was inexpensive, it became the wine served with burgers and pizza. Today Zinfandel is appreciated in fine-dining establishments and is served along with many kinds of food.

White Zinfandel is a good aperitif wine and its acidity makes it compatible with many dishes. Lighter foods or those with a little spice or heat work well.

Full-bodied Zinfandels are different. Think power with power. Foods should be rich enough to stand up to the richness of the wine. Red meat, e.g. steaks and chops, work extremely well. A favorite accompaniment is prime rib of beef. It is still a great wine to go with pizza and burgers. Because of its intense ripeness, it can also be nice for after-dinner sipping in place of port. Many producers in California make a port-style wine from Zinfandel. It is rich, luscious, and sweet and matches well with chocolate and blue-veined cheeses.

ADDITIONAL RED VINIFERA GRAPE VARIETIES

The following grapes include French, Italian, Spanish, Greek, German, Austrian, and American varieties, some of well-known and distinct origin, and others a bit obscure. Some may be difficult to find in retail stores. They are more likely to be sold in specialty wine shops or food stores that specialize in the products of a particular country.

Agiorgitiko A popular Greek grape variety grown predominantly in Peloponnesos, Agiorgitiko is a dark, fruity grape that produces an aromatic, fruity wine with low acidity. It is made in a dry, varietal style or blended with other indigenous or international varieties. One of the most desirable Greek rosé wines comes from this grape. The wine is a natural companion with lamb and classic Greek dishes, like moussaka.

Aglianico Some believe that Aglianico's source is Ancient Greece; others, that it is a native Italian grape. In either case, it thrives in the poor, volcanic soils of Mt. Vulture, an extinct volcano, in Basilicata and Campania, Italy. It is also becoming the dominant grape in other southern regions. Aglianico is a thick-skinned, tannic, almost black grape with complex, perfumed flavors. It requires full MLF and aging to soften its tannins, and produces a full-bodied, dense, rustic, and powerful wine with strong tannins and good acidity. The primary aromas are of black fruits, such as cherries, currants, and blackberries. It has a ruby color when young, but like Nebbiolo, it can get an orange tinge as it ages. Two of the most important wines are Taurasi in Campania and Aglianico del Vulture in Basilicata. Foods that go well with Aglianico include hearty country fare, such as tomato-based pastas, cheeses, pork salami, and sausages.

Carignan/Carignano/Cariñena/Mazuelo Although it originated in Spain, Carignan has established itself in Languedoc-Roussillon and the Southern Rhône.

It is widely planted on Sardinia in Italy, where it is called Carignano, prized in the Priorat in Spain, where it is known as Cariñena, and used as a blending variety with Tempranillo in Rioja under the name Mazuelo. In California it is produced in the Santa Cruz Mountains, where it has won several awards over the years. This black grape contributes high alcohol, tannins, acids, and color to wines when blended with other varieties, usually Grenache and Syrah. Carignan-based wines are dark, robust, and balanced and are terrific matched with dark-meat poultry, pork, and lamb.

Carmenère Although a native of Bordeaux, Carmenère is one of the most widely planted and successful red grapes in Chile. In fact, Chile is the only major producer of Carmenère in the New World. Carmenère has been correctly identified in Northern Italy where it was previously thought to be Cabernet Franc. A finicky grape, Carmenère requires good soil, water at the right time, and a long growing season to protect its docile acidity. In Chile it is commonly blended with other Bordeaux varieties. The wine is round and rich with a velvety texture on the palate. It showcases a wonderful array of fruit aromas from cherries to blackberries to dark plums as well as notes of earth, leather, tobacco, and chocolate. The ripe fruit flavors make it an exceptional choice with roasted meats or grilled steaks and chops.

Charbono Charbono is an obscure grape variety made famous in the 1950s and 1960s by Inglenook Winery in the Napa Valley. It is thought to be of Italian origin, similar to Dolcetto, Barbera, or Bonarda, but actually it is most likely the French Charbonneau from the Savoy region. Several producers have continued to grow and make wine from this very dark grape. It produces a dark, inky, tannic wine with good acid, and it can age for decades. It is made into a varietal wine in California, where only about 100 acres/40 hectares are planted. Charbono partners well with grilled foods.

Cinsaut (also sometimes spelled Cinsault) A high-yielding, drought-resistant grape, Cinsaut is best known in Languedoc-Roussillon, which is considered its historic home. It is produced in a diverse range of countries, including those in the Middle East and Africa. Its best wines come from low-yielding vines and the wines tend to be light, soft, and aromatic when young. It is used predominantly as a blending grape with Carignan or in Grenache-based Southern Rhône blends. It adds immediate fruit on the palate as well as perfume and suppleness to the blends. Cinsaut is also frequently used in making a very good dry rosé. Rosés of Cinsaut work well with picnic foods, and the reds pair with pissaladière, a classic French-style tart with onions and olives that is similar to a pizza.

Colorino Colorino is one of Tuscany's indigenous varieties that can be added to Sangiovese in a Chianti blend to boost color and tannins. When wine regulations were changed to permit international varieties in the Chianti blend, Colorino lost out to Cabernet Sauvignon and Merlot. More recently several producers have moved away from these international varieties because they are too powerful and have returned to Colorino, which is more complementary to Sangiovese. It is also an addition to another Sangiovese-based wine, Vino Nobile di Montepulciano.

Corvina Corvina is the principal grape in the much-maligned blended wine Valpolicella and also is the base for the more highly regarded dried-grape wine Amarone. An ancient indigenous variety, Corvina grows in a limited area in the Veneto. It is a dark red, thick-skinned grape that ripens late. The variety is aromatic and fruity with high acidity, low tannins, and light color. A vigorous grape, it is often overproduced, leading to a poor-quality crop. Corvina is always blended with other varieties, particularly Rondinella and Molinara, to give wine body and color. Its bouquet is primarily described as that of tart red cherries. Because Valpolicella is so common and is often made into mediocre wine, connoisseurs have frequently dismissed it as a serious wine, considering it a light red, good for quaffing. However, producers today are creating Valpolicella that deserves notice. The wine is best with typical Italian food, such as pasta, risotto, cheese, and cured meat.

Dolcetto Dolcetto is primarily produced in Piedmont, where it is the third red variety of prominence after Nebbiolo and Barbera. In Italy, it is generally considered a light quaffing wine, although some winemakers are attempting to make rich, ripe styles. Small amounts are produced in California as a light, easy-drinking wine. Dolcetto is moderate in acidity and sugar, but its skins generate an intense dark purple color. Most Dolcetto is youthful with fruity characteristics. However, long maceration gives higher tannins, which increases aging potential and results in a richer style. Although it is almost always made as a varietal wine, a few winemakers blend it with Barbera and even Nebbiolo. In Piedmont Dolcetto is served with salami, lardo, simple pastas, roast chicken, and a classic dish called *bagna cauda*, a slow-cooked sauce of anchovies, garlic, and olive oil that is served communally with raw and cooked vegetables. It also works as a picnic or barbecue wine.

Gaglioppo A grape variety that is only produced in Southern Italy, Gaglioppo thrives along the Adriatic coast and is the dominant red variety in Calabria. The grape ripens easily, has high acidity, and produces wines moderate in alcohol (less than 14 percent). The grape is dark purple but results in wines that are medium to dark red depending on the length of maceration during fermentation. It is made mostly into a varietal wine that is unoaked, with varying levels of aging in the tank and bottle. It is also made into a lighter rosé wine, and when blended with Cabernet Sauvignon, it may be oak aged. The wine can be moderate- to full-bodied with good acid and tannins. Richer styles are excellent accompaniments to roasted and braised meat dishes, and lighter styles go with cured meats and tomato-based pasta dishes.

Graciano Found primarily in Navarre and Rioja in Spain, Graciano, an indigenous grape, adds finesse, floral aromas, and structure to blended wines. It results in highly prized low-alcohol, high-acid wines with excellent character and extract. They are fruity and fresh, with intense color and strong aromas. The wines are tannic when young and become richly colored and perfumed when aged. These factors make it a desirable blending grape.

Lagrein Lagrein is an indigenous dark red variety from the northern part of Trentino-Alto Adige. The cool-climate grapes are high in acid and tannins. The wine is also high in acid but has soft tannins with bitterness in the finish that can be removed by aging in wood. Lagrein has deep color; is spicy, earthy, and herbaceous with bright red fruit flavors; and is moderate in alcohol. To soften Lagrein, it is blended with Merlot, which is widely planted in the same locale. It also makes excellent rosés. The wines go exceptionally well with rich meat soups, stews, and braised dishes.

Lemberger/Blaufränkisch This late-ripening grape variety comes from the Burgenland region of Austria, where it is called Blaufränkisch. It is planted widely in Central and Western Europe as well as the United States, most notably in Upstate New York and Washington State, where it is known as Lemberger. The grape has dark ruby color with high acidity and tannins, and the wines are medium to dark in color. Styles include a spicy, fruity wine that might be mistaken for Gamay or Cabernet Franc from the Loire. If allowed to macerate during fermentation, the wine achieves a darker, richer style that can be oak aged. Because of its high acidity, the wine goes well with a range of foods from dark-meat poultry to roasts of veal and beef. It also works with rich, creamy pastas and risottos.

Mencía Although its origin is uncertain, Mencía has been planted in Bierzo, Spain, for so long that it is considered a native grape. If the grapes ripen adequately, Mencía-based wines are complex, concentrated, and acidic with soft tannins and rich, inky color. They are fresh, fruity, and fragrant, and are meant to be drunk young. Mencía is often blended with Garnacha Tinta to add color and complexity and extend maturation. This wine goes with grilled pork and lighter red meat dishes.

Mission The first grape brought from Europe to the Americas in the 1500s, Mission was planted by the Jesuits. Franciscan monks, who developed missions (hence its common name) along the California coast, expanded its cultivation. Although the grape is red, it has little color and was not typically used for red wines. However, it was favored for its versatility. It made young, high-acid rosés, fortified wines, brandy, and raisins. Today it is used almost solely for fortified wines: angelica and sherry. When brought from Spain, it was called Criolla, and it can still be found by that name in South America, where it is also related to the Pais grape. Recent DNA testing has shown that it is exactly the same as Listan Prieto, an almost extinct Spanish grape.

Montepulciano The Montepulciano grape is often confused with the wine Vino Nobile di Montepulciano, which is produced from a Sangiovese clone in the Tuscan hill town of Montepulciano. The variety Montepulciano is cultivated widely in the Central Italian regions of Abruzzo and Marche. Much is made into table wine but some makes better-quality wines, like Montepulciano d'Abruzzo and Rosso Cònero from the Marche, both blends of Montepulciano and Sangiovese. The grape is ruby red and produces an easy-drinking wine with lush, inky color and soft tannins. The wine goes well with pizza, cured meats, or red meats.

Mencía vineyards in Bierzo, Spain, sometimes consist of only a couple of rows.

Mourvèdre/Monastrell/Mataro Known as Monastrell in Spain, this indigenous grape is widely planted there. However, it is more widespread in southern France, where it is called Mourvèdre. In Australia and California it is called Mataro, although the use of the name Mourvèdre is increasing in California. In France it is primarily blended with Grenache and Syrah, and is a key component in Châteauneuf-du-Pape and Côtes du Rhône blends. In California it is combined with Rhône varieties, usually Grenache and Syrah, and is made into a varietal wine as well. It is also blended with Grenache and Syrah in Australia to produce a wine known as GSM. Mourvèdre has thick skins and dark color and thrives in hot climates and sandy soil. Wines are dark, robust, and high in alcohol with rich tannins. Because alcohol levels can be as high as 18 percent, the wine is often blended with lighter varieties to reduce the alcohol. It has aromas of red and black fruits, including currants, cherries, plums, and blackberries. Its most notable descriptor is meaty, along with characteristics of leather, earth, and mushrooms. It should be served with roasts, dark-meat poultry, cassoulet, or mushrooms and winter vegetables.

Negroamaro An ancient grape variety, Negroamaro, meaning black and bitter, is grown prominently in the Apulia region of Italy, where the most well-known wine is Salice Salentino. The grape has thick, dark skins and produces a wine that is equally dark with earthy aromas and a tannic, bitter finish, making it an excellent wine with the foods of the region. It can be made as a varietal, including outstanding rosé, or blended with Malvasia Nera, which adds acidity and provides balance. The wine is an outstanding partner for tomato-based pastas and pizza. In Apulia it is served with roasted lamb or goat.

Nerello Mascalese Nerello Mascalese is an ancient grape produced on volcanic soils on the slopes of Italy's Mt. Etna, an active volcano on the island of Sicily. The grapes are dark in color, producing sweet cherry–like fruit with high tannins. It is commonly fermented in botti. Wines are acidic and tannic with spicy, earthy character. Nerello Mascalese is made into a varietal wine or more commonly blended with its sister variety, Nerello Cappuccio, or with Nero d'Avola. It pairs well with fish grilled over grapevines. It is also a perfect accompaniment to pistachio-sauced pasta made with pistachios grown on the hillsides of Mt. Etna.

Nerello Mascalese growing at Benanti on the slopes of Mt. Etna in Sicily, Italy

Nero d'Avola Sicily has one thing that Nero d'Avola craves—heat. An indigenous grape, it has adapted perfectly to the island's extremely hot, dry climate, where the rolling hills are often scoured by searing winds coming from the North African deserts. Nero d'Avola (also called Calabrese) is the highest-quality grape cultivated in Sicily and its most extensively planted red grape. A red grape so dark it looks black, Nero d'Avola is often compared to Syrah. Generally, it develops good sugars and good acidity. However, if left to ripen for too long, the sugar level gets too high, resulting in wines with excessive alcohol levels and low acidity. The wines are lush, exotic, and robust. They are inky-colored with aromas of violets and black fruits. They are full-bodied with moderate tannins and age well. Nero d'Avola from Cerasuolo di Vittoria, the historic area for this variety where it is blended with Frappato, is recognized for its outstanding quality. Sicilian cuisine includes seafood and spicy North African dishes, such as lamb with couscous, that pair well with this wine.

Petit Verdot Petit Verdot is one of the minor but important varieties used in producing a Bordeaux blend. Always added in small percentages, it is vital because it adds color to the finished wine. The grape is planted around the world wherever other Bordeaux varieties are grown. In some locations it is made into a dark, syrupy varietal wine that is almost black in color.

Petite Sirah (Durif) Petite Sirah is the offspring of Syrah and a little-known grape called Peloursin. Its principal name is Durif, but it is called Petite Sirah in California, which produces the lion's share of this variety worldwide. Australia, though the second-largest producer, cultivates a much smaller volume of the grape. Petite Sirah is small and thin-skinned with compact clusters that are quite susceptible to rot. It is successful in California because of the generally dry climate and the benefit of lots of sun. Historically, Petite Sirah was used as a blending wine, to add body and strength in weak vintages or to give wine a kick of color or enhance aroma. Although still used for blending, it is more frequently made into a powerful varietal wine. It produces a dark, inky, aromatic wine with big tannins and big alcohol. It contains ripe fruit flavors of blackberries and currants with full-bodied lushness. Red meats naturally match with Petite Sirah. Grilled steaks and chops, roasts, and braised meats are perfect as are simpler dishes like pizza, barbecue, and burgers. Sausages, mushrooms, and grilled vegetables also pair well.

Pinotage Pinotage was created by crossing Cinsaut and Pinot Noir. This unusual cross became one of the most widely planted varieties in South Africa, but it has not gained a following elsewhere. Producers made many mistakes with the young variety and wines were initially of poor quality, but low yields from older vines combined with good vineyard-management practices later began to produce good grapes. Furthermore, tinkering with winemaking methods has improved the quality of the wine. This is a wine to serve with beef.

Refosco/Refošk This ancient variety is probably native to the Friuli region of Italy but is also found just across the border in Slovenia and Croatia. In a region of predominantly white wines, Refosco is the dominant red indigenous variety in Friuli. It is a late ripener with characteristics that range from wild berries to currants and plums depending on ripeness. It is an acidic varietal with noticeable tannins that is normally drunk young. In Friuli it is served with rabbit, braised lamb shanks, mushrooms, or polenta dishes.

Rubired A cross of Alicante Ganzin and Tinta Cão, Rubired was created at the University of California, Davis, to provide color to port-style wines. It is now the fourth most-cultivated red variety in California. Because it is one of the few varieties that produces red instead of clear juice, it is used for its color in jug-wine blends, and more recently to produce a product called Mega Purple (see Chapter 5, Color Enhancement, page 93).

Sagrantino An intense, dark red grape primarily grown in Umbria, Italy, Sagrantino has only recently received significant attention from some producers

Tannat grapes from Irouléguy in South-West France produce dark, tannic wines.

in the region. The grape has thick skins with small berries, producing must that has very rich and dark color, good acidity and flavor extract, high sugar content, and distinct tannins. The wines have a cherry and smoke aroma and can be blended with the lighter Sangiovese to create a less-intense wine. It is also used to make a dried-grape passito wine. Sagrantino should be served with stronger dishes including game.

Tannat A very dark, black-skinned grape from the South-West region of France, Tannat produces one of the darkest and most tannic wines made (called Madiran) and has similarities to Petite Sirah. It is commonly blended with Cabernet Sauvignon and Merlot, but may be combined with other indigenous varieties from the South-West. Wines from Madiran and Irouléguy are considered the best. The historic, rustic wines made from Tannat are thought to be the highest in antioxidants and are the benchmark wine of the French Paradox (see Chapter 3, The French Paradox, page 31). The variety is now found in California and is also cultivated broadly in Uruguay and Argentina. Tannat is perfect with duck in every form; piperade, a red pepper sauce; and beans.

Teroldego Teroldego is grown almost exclusively in the Trentino region of Italy, where the best are sold as Teroldego Rotaliano. Restricted yields produce the highest-quality grapes and wines. Teroldego grapes are dark purple and produce wines with dark fruit, tar, and coffee aromas. It is a high-acid wine with strong tannins on the finish. Barrel aging is used to soften the tannins. A very small amount of the grape is grown in the Central Coast appellation of California. Teroldego complements red meats, but in Trentino it is served with rich braised dishes, game, or goulash.

Valdiguié A light red grape, Valdiguié originated in Southern France. In the South-West it is known as Gros Auxerrois (not to be confused with Malbec, which is also called Auxerrois). In the United States Valdiguié produces large yields of fruity grapes. Carbonic maceration is commonly used to produce a light-bodied, juicy wine that has low tannins and purple-red color, with a nose of black cherries, similar to Gamay. Low in alcohol, Valdiquié makes a perfect quaffing wine.

Zweigelt Zweigelt is the most widely produced red grape in Austria, and was created by crossing the Blaufränkisch and St. Laurent varieties. It buds late and ripens early, which is perfect for the very cool climate of Austria. It is also produced in Northern Central European countries and has found a home both in Eastern and Western Canada, where the climate is very similar to that of Austria. Because Zweigelt is a vigorous producer, controlling crop size is required to produce a quality wine. Skins are dark purple, producing a wine dark in color but fairly light on the palate. Usually drunk young, Zweigelt works well with pork dishes.

SUMMARY

Consumers have a broad range of choices when selecting a red wine, from light and fruity wines to big, rich, tannic, and brooding options that come from all corners of the world. These alternative varieties, in many cases indigenous to the countries that produce them, are quite different from the popular international varieties. These choices frequently lead to less well-known countries and regions, which provide an interesting opportunity to learn about different parts of the world through their wines. And as with the white wines of the world, there are a number of food styles that are a perfect complement to each of these red varietals.

16

Hybrid and Native American Varieties

Hundreds of wines are made from *Vitis vinifera*, and many of these receive great acclaim from wine connoisseurs and consumers, but wines from hybrid and native American grapes should not be overlooked. In many cases, they provide interesting alternatives to the well-established varietals.

Many of these varieties and wines are indigenous to parts of North America, and some people consider them equal in quality to *vinifera*-based wines. A number of these wines fit perfectly with many consumers' preference for sweet wines.

FOXY AROMAS

Foxy is a term that is often associated with wines made from native American varieties, particularly *labrusca*-based wines. Critics say these wines have an animal or musky characteristic as opposed to fruity or floral aromas. Fans of these wines say the critics are used to drinking *vinifera*-based wines and are unable to appreciate the different, enjoyable, and pleasant characteristics of American wines. Crossing indigenous varieties with *vinifera* often eliminates or reduces the foxy aromas and flavors.

HYBRID VARIETIES

A grape hybrid is created when two or more different Vitis species are crossed. Hybrids can occur naturally in the vineyard, but more often scientists propagate new varieties to obtain specific characteristics in the new grapes. Hybridizing different varieties results in a wine that exhibits the best qualities of the species used.

Most frequently French *vinifera* varieties are crossed with native American varieties. The presence of *vinifera* in the cross reduces the foxy smell of some native American grapes (see Foxy Aromas, at left) and increases *vinifera* aromas and flavors, while the presence of native American grapes develops hardy vines that can withstand cold winters or hot, humid summers and resist disease. Hybrids can also consist of two or more native American species; for example, a *labrusca* variety can be crossed with a *riparia* variety to obtain specific traits in the grapes.

Those hybrids that were created in the United States are often referred to as *American hybrids*, even if they contain *vinifera* grapes. Those that were developed in France became known as French-American hybrids. In the United States native and hybrid grapes are planted widely in the East, South, and Midwest as well as the eastern half of Canada.

White Hybrids

The following are the most common white hybrids cultivated in the United States and Canada.

Cayuga White A cross of the French hybrid Seyval Blanc
and Schuyler, an American grape, this hybrid was developed at Cornell University in Upstate New York and is
now grown throughout the Northeast. The vine is moderately tolerant of cold temperatures and is disease-
resistant. It is vigorous and produces high yields. Cayuga
White is found in a range of table wines from off-dry to
sweet as well as sparkling wine and is a common base
in blended wines from Upstate New York. The wine has
medium body, delicate aromas, and good balance.

Chardonel This productive hybrid was created at Cornell
University in 1990 by crossing Chardonnay and Seyval
Blanc. It is a hardy vine that survives cold temperatures.
Chardonel produces superior-quality wines that are
pleasant and delicate, with the best having Chardonnay-like characteristics.
The grape has good sparkling-wine potential. It performs particularly well in
Southern Michigan, Pennsylvania, and Arkansas.

*New York oenologists were pioneers in
creating and using hybrid and native
American grapes.*

www.taggphotography.com

Edelweiss Edelweiss, an early-ripening hybrid, was developed in Minnesota by
crossing the native American varieties Minnesota 78 and Ontario. The vine
is disease-resistant and is able to produce good grape crops even when
planted in frigid climates. It is normally made in styles ranging from off-dry
wines to ice wines. Its sweet white wines have mild, fruity flavors.

Seyval Blanc Seyval Blanc, a French-American hybrid of Seibel varieties, is a
cool-climate grape that does well in North America, particularly Canada.
The disease-resistant vines provide moderate to high yields, but production varies depending on the local soil and climate. Seyval Blanc is the most
popular white variety in England, which has a cool, rainy climate, similar to
that of the Northeastern United States. It makes an excellent dry white wine
that demonstrates good acidity and is well balanced with citrus, melon, and
green-apple aromas. Malolactic fermentation (MLF) and barrel fermentation
give the wine Chardonnay characteristics.

Traminette This hardy hybrid, a cross of Gewürztraminer and Joannes Seyve
23.416, maintains the character of Gewürztraminer when cultivated in cold
climates. Wines can be dry or sweet, but either way they maintain high acidity, balanced fruit, and the aromas and flavors of Gewürztraminer. They have
high acidity, contain some residual sugar, and are very aromatic. The spicy,
fragrant wines have recently won top awards at well-regarded wine competitions. Hybridized in New York, Traminette was only released for propagation
in 1996 and is being grown primarily in New York and Michigan.

Vidal Blanc Vidal Blanc is a hybrid, which, although developed in France, has made
its greatest strides in North America, both in the United States and Canada,
because it grows well in cold climates. It is a cross of the French *vinifera* variety Ugni Blanc and the hybrid variety Rayon d'Or (also known as Seibel 4986).

Vidal Blanc grapes
www.taggphotography.com

The vines are hardy, vigorous, and disease-resistant. In northern regions it is recognized for producing excellent ice wine. In more southerly areas, like Virginia or Missouri, it is made in a range of styles from dry to sweet.

Vignoles Also known as Ravat 51, Vignoles is a French-American cross of Seibel 8665 and Pinot de Corton (Pinot Noir). It is a hardy, productive, late-ripening hybrid. It can achieve high levels of ripeness while maintaining good acidity and is also prone to developing *Botrytis cinerea*. It makes intriguing dry wines, sweet wines, and ice wines. Most Vignoles is grown in Upstate New York and in Missouri.

Red Hybrids

The following red hybrid varieties are planted widely in the Eastern, Southern, and Midwestern United States.

Baco Noir Baco Noir, a French-American hybrid of Folle Blanche and an unidentified native American grape, is planted in both France and the eastern part of the United States and Canada. It is a hardy and very vigorous grape, but is susceptible to rot. Baco Noir has fruity rather than foxy aromas, which indicates that it was probably not hybridized with *labrusca*. Skin contact is commonly used to heighten its color, and it requires MLF to soften its acids. Baco Noir is commonly aged in American oak barrels. The wines are simple and fruity, and if ripe grapes are used, these dark wines potentially yield high alcohol levels.

Chambourcin Although planted in many parts of the world, Chambourcin is a French-American hybrid of unknown parentage. It performs well in damp climates that have moderate winters, is a vigorous producer, and is considered one of the better red hybrids. The highly rated wine is dry, shows dark-fruit character, and is best when consumed young. It is spicy with cherry notes. Chambourcin has found a home in the Eastern United States, especially Virginia.

Chardonel and Norton are planted at the OakGlen Winery in Hermann, Missouri.
CAROL WARNEBOLD

Chancellor Chancellor, a French-American hybrid of Seibel parents, is cold-resistant and produces high yields but is susceptible to mildew. It is grown throughout the Eastern and Midwestern United States and in Canada where it is primarily used in generic red blends with other red hybrid varieties. The wines are light and pleasant.

DeChaunac Another French-American hybrid of Seibel parents, DeChaunac is very winter-hardy and was quite popular in the early 20th century in Northeastern Canada and the United States. The vine is disease-resistant and produces high yields. Generally, the wine has neutral aromas and flavors, but oak aging adds complexity. Although it is sometimes made into a varietal wine, it is more often used as an ingredient in red blends.

Marechal Foch Marechal Foch is a winter-hardy and disease-resistant, early-ripening French-American hybrid. Although its origins are not certain, it is probably a cross of Gold Riesling and a *riparia/rupestris* hybrid. It lacks the typical foxy character associated with other indigenous grapes. Carbonic maceration is commonly used to obtain fruity wines, and it can be oak aged, which helps it achieve a good degree of complexity. Its berry aromas make it a good grape for blending with herbaceous and acidic varieties. Marechal Foch is often compared favorably with Beaujolais and is a desirable replacement for Pinot Noir in regions with harsher climates. However, because non-*vinifera* parents are included in the cross, it will likely never attain the level of acceptance that these *vinifera* varieties enjoy. Marechal Foch grows in a range of climates from Minnesota to Virginia.

NATIVE AMERICAN VARIETIES

Native American grapes are those that are indigenous to the United States. They are the same genus, *Vitis,* as European varieties but are of different species than *vinifera*. Many indigenous species are cultivated, but they are often considered better for nonwine products, like juice and jelly, or for table grapes. However, some have been accepted for use in wines where *vinifera* varieties are unable to grow successfully, and a few are showing results with exceptional wines.

The common native species associated with winemaking are *aestivalis, labrusca, riparia, rotundifolia, rupestris,* and *berlandieri. Riparia, rupestris,* and *berlandieri* were exported to Europe because their rootstock is phylloxera-resistant (see Chapter 4, Phylloxera, page 53) and are the core of all rootstock now used worldwide. However, their grapes are not used to produce wine.

Many native American varieties are actually hybrids that combine a native species, typically *labrusca* or *aestivalis,* with a *vinifera* variety. These were the earliest hybrids developed in the United States and were used as the basis for other hybrids in both the United States and France. They are typically planted only in the United States and are considered to be native grapes.

Descriptions of some of the most common native American varieties follow. Several of these produce red grapes but they are used exclusively for white or rosé wine.

Niagra grapes
www.taggphotography.com

White Varieties

Catawba Catawba is a pinkish-red grape that is a hybrid of *labrusca* and *vinifera*. The vines tolerate cold winters, are hardy and vigorous, and produce late-ripening, low-sugar, and high-acid grapes. Similar in nature to Concord, much Catawba is made into juice or jelly or used as a table grape, but it is also made into delightfully sweet wines and sparkling wines. It also makes clean, crisp white and rosé wines. It has the foxy aromas characteristic of *labrusca*. Catawba is a perennial favorite at wine competitions.

Delaware Delaware is a dark-skinned, early-ripening red grape that is a hybrid of *labrusca, bourquiniana*, and *vinifera*. Because it was crossed with *vinifera*, it is not as foxy as some other native American varieties. A vigorous vine, it tolerates cold weather but is susceptible to phylloxera and fungal disease. It is planted in both the East and Midwest of the United States and makes white wines that range from dry to sweet but is preferred for sparkling wine. The wines are pleasant and fruity and generally have some residual sugar.

Niagara Niagara is a *labrusca* cross of two native grapes, Concord and Cassady, plus two other indigenous grapes. It is a vigorous and productive white grape that holds up to cold temperatures. Niagara is a low-acid, low-sugar grape that makes floral, strongly flavored wines. The grape produces white wines with the noted foxy character and is used mostly for off-dry and sweet wines in New York State.

Scuppernong (Muscadine) Originating in the Southeastern United States, Scuppernong is a sweet white grape that performs well in warm, humid environments. It is vigorous and disease-resistant. It results in well-structured wines with distinct flavors. Scuppernong is known for dessert wines that are similar to Muscat.

Red Varieties

Concord Think Welch's grape jelly or grape juice. More than any other wine, Concord wine smells like the grapes. It is the only grape considered 100 percent *labrusca* and has pronounced foxiness. The vines produce blue-black grapes, and can withstand cold temperatures, are disease-resistant, and productive. It is a late-ripening, low-sugar, high-acid grape that performs in more diverse soils and climates than any of the other native grapes. It is almost entirely made into sweet red wines and some sparkling wines. It is best known as one of the major varieties used in making kosher wines on the East Coast.

Isabella Although its origins are unknown, Isabella is credited by most as a cross of a native American species and *vinifera*. This black grape was first planted in South Carolina in the early 1800s where it adapted to the heat and high humidity. Similar to Concord, it presents the same foxy character of *labrusca* varieties. It is used for white and rosé wines.

Norton Considered by many to be the finest native American red grape variety, Norton can produce wines equal to those from *vinifera* vines. The grapes are deep, dark blue to purple in color and acidic. Norton grows well in humid climates because of its resistance to fungal diseases, making it a dominant grape in mid-Southern and mid-Atlantic states. A variety from the species *aestivalis*, the grapes lack the foxy character of *labrusca* varieties. The red wines are full-flavored and can have characteristics similar to those of Zinfandels. The grape has competed internationally with the best of *vinifera* in blind tastings, but unfortunately it still retains the stigma attached to any non-*vinifera* variety. Norton performs particularly well in Missouri, Arkansas, and Virginia and is considered by many to be the best wine with barbecue.

SUMMARY

In the United States, native American and hybrid grapes are grown in locations where the better-known *vinifera* are unable to survive. Many of these varieties make excellent wines, even equal in quality to *vinifera* varietals. They are available in the Eastern, Southern, and Midwestern states but are rarely sold in the Western United States. When traveling in the United States or Canada, sampling these unique wines can open you up to new experiences and increase your understanding of wine.

Concord grapes
www.taggphotography.com

Sparkling Wines

What is a festive occasion without including the bubbles and brightness of sparkling wine? Sparkling wine can put anyone in a mood to celebrate, but enjoying it does not have to be limited to events such as weddings, graduations, and the start of a new year. Sparkling wine can just as easily be a refreshing drink shared with friends on a Tuesday night.

According to legend, hundreds of years ago the monk Dom Pérignon discovered the process for making sparkling wine. Whether or not he actually created Champagne, which is perhaps not entirely true, he did advance the use of cork closures and was influential in making the wine of the Champagne region produced from underripe grapes taste good. In any case, someone found that adding sugar and yeast to the wine made from these grapes could restart fermentation. This secondary fermentation created carbon dioxide (CO_2), producing bubbles, and alcohol. Sparkling wine was born.

To this day, vintners trap CO_2 in a sealed bottle during the secondary fermentation, and the CO_2 naturally dissolves into the wine. The result is a high-acid sparkling wine. This style of wine was given the name of the region where it was produced, Champagne, and the process, which is still used for high-quality sparkling wines today, came to be known as *méthode champenoise*.

Although many use the term *Champagne* generically to mean sparkling wine, by law true Champagne can only be made in France's Champagne appellation. Outside Champagne, you can find many excellent-quality sparkling wines, but they are called by different names, such as *crémant, mousseux, Cava,* or the general term, *sparkling wine.* Sometimes producers outside Champagne label their sparkling wines "Champagne." However, many of these wines do not use the true Champagne varieties and are frequently made using bulk processes instead of the méthode champenoise. Consequently, the difference in quality and price between these sparkling wines and Champagne is significant.

The use of the term *méthode champenoise* is also restricted to the Champagne region. In other European Union (EU) countries and wine regions producers who are creating sparkling wines using this technique, as well as anyone selling sparkling wine in the EU, are required to use the term **TRADITIONAL METHOD** instead. However, some producers outside of Europe chose to ignore this requirement and state that their wines are made by méthode champenoise.

GROWING GRAPES FOR SPARKLING WINES

Sparkling-wine production worldwide is closely tied to cool climates where grapes struggle to ripen. Cool temperatures are necessary to maintain acidity because low temperatures prevent grapes from ripening too quickly. This allows the wine to maintain its crispness, which marries well with its sparkle and is critical to the pleasure in tasting and drinking Champagne and other sparkling wines.

Champagne is not only known for its cool climate but also for its chalky limestone soil, which adds a distinct character to the wines. Other appellations where sparkling wine is produced have different soils that help create a typical style for each area.

Variety Selection

Although you can make sparkling wine from any red or white grape variety, the best tend to be made from those that naturally maintain high acid levels. The benchmarks of Champagne are Chardonnay, Pinot Noir, and Pinot Meunier, but Chenin Blanc of the Loire, Pinot Blanc of Alsace, and Riesling of Germany are highly acidic as well. In the appellation of Penedès in Catalonia, Spain, the classic Cava is made from Spanish indigenous varieties, Parellada, Macabeo, and Xarel-lo. Macabeo and Xarel-lo are particularly good examples of varieties that maintain high acidity. In California, the classic Champagne grapes, Chardonnay and Pinot Noir, are typically used as well as Pinot Blanc, Pinot Gris, and a few others. Nonaromatic varieties like Chardonnay work best for drier-style wines, whereas aromatic grapes like Riesling work better for sweeter wine styles. In Australia producers use the dark, black grape, Shiraz, to make sparkling wines, which is somewhat unique in that sparkling wines are not typically made from dark red grape varieties.

Vineyard Management

Vineyard-management techniques are of less consequence for sparkling wines than for still table wines because the grapes are harvested before full ripeness develops. The key is to pick the grapes when acids are high and sugar is low (generally around 18° to 20° Brix) but the grapes have ripened enough to have lost their herbaceous character. If grapes ripen fully, they lose too much acidity, and the first fermentation achieves too much alcohol. To achieve the ripeness levels needed for sparkling wine, grape varieties should be carefully selected and planted specifically for this style of wine. Some vintners decide to make sparking wine because the grapes for their table wines did not ripen adequately.

Grapes have historically been hand-harvested, as is done in Champagne, because whole clusters with grapes that have unbroken skins are more desirable. But with the advent of more delicate mechanical harvesters, machine picking of grapes destined for sparkling wines is increasing outside France.

MAKING SPARKLING WINES

Sparkling wines can be made using a number of methods: The méthode champenoise (traditional method), the bulk Charmat process, the transfer method, the Russian continuous process, and carbonation. The two most common methods used to make sparkling wines commercially are the méthode champenoise and the Charmat process.

Méthode Champenoise (Traditional Method)

Initial Fermentation

Grapes are harvested at sugar levels of 18° to 20° Brix then brought to the winery and pressed. Both red and white grapes are pressed prior to fermentation for sparkling wines without color like NV brut and blanc de blanc (see Types of Sparkling Wines, page 282). This minimizes both color, particularly from red grapes, and possible astringency. In Champagne, much of the pressing actually occurs in the vineyard to ensure this outcome. In fact, even with all of the new technological improvements in pressing equipment, many of the finest producers use an old-fashioned basket press to gently press the grapes, maximizing the free-run juice. For rosé and red sparkling wines, the grapes are allowed to macerate in contact with their skins prior to pressing to achieve the desired color.

The juice is then fermented just as it is with any table wine until it is totally dry. Different varieties are fermented separately, and individual vineyards or blocks within vineyards are also fermented separately if they showcase different terroir and grape-variety characteristics. The resulting base wine component has only 10 to 11 percent alcohol and is barely palatable given that it is so high in acid and its fruit flavors so meager. But at this point, it is not a wine to be consumed. Rather, wines from various lots are combined to create a desired blend (CUVÉE).

Blending (Assemblage)

Blending the different lots of grape varieties, terroirs, and vintages to arrive at the desired base wine is a key step in making sparkling wine. The Champenoise refer to this step as *assemblage* (to assemble). Combining components from each lot allows the vintner to develop complexity and create a house style.

Incorporating some lots of barrel-fermented wine adds dimension and character to the cuvée. Most sparking wines, even Champagne, are nonvintage. This is because during the blending process multiple vintages are used to create the final mix. This is done for the purpose of fashioning a unique style, but it also allows for creating consistency from year to year.

Bottling and Secondary Fermentation

The base blend is put in the bottle from which the finished wine will be consumed. After the wine is in the bottle, a blend of sugar (a form of chaptalization), yeast, and yeast nutrients, called liquor de triage, is added to initiate the secondary fermentation, creating CO_2 in the bottle. The fermentation also adds an additional 1 to 1.5 percent alcohol to the wine, boosting the alcohol content to about 12 percent. The bottle is sealed with a crown cap, just like the ones used on bottles of beer, and is stored on its side. The secondary fermentation takes from eight to 12 days, depending on temperature—the cooler the temperature, the longer the fermentation. During this time, the death and decay of the yeast cells (a process known as autolysis) gives the wine greater complexity. The wine develops approximately five atmospheres of pressure in the bottle, so the process requires a thicker, heavier bottle than those used for regular table wines.

Perhaps the most important step in making sparkling wine is when the winemaker blends the various lots into a final cuvée.
MOËT & CHANDON

Aging

Once the secondary fermentation is complete, the dead yeast cells fall out of solution and settle to the bottom side of the bottle, where they remain while the wine ages. Most wines typically age for an additional 12 to 36 months after the secondary fermentation is complete. However, some wines labeled "late disgorged" are aged for an extended period, sometimes several years. This period of aging or *en triage* is critical to the development of the toasty, breadlike, caramel aromas and flavors that méthode champenoise wines demonstrate.

During aging sediment settles to bottom of bottle.
TOM ZASADZINSKI/CAL POLY POMONA

Riddling (Remuage)

When the wine has aged long enough, the next process is removing the sediment from the bottle. The bottles are shaken and turned, and the base of the bottle is slowly elevated over several days until it is standing upside down on its crown cap. All of the sediment eventually falls into the neck of the bottle, forming a plug and leaving perfectly clear wine. In Champagne, historically, individuals called remueurs spent 30- to 40-year careers turning 25,000 to 30,000 bottles a day, and riddling (remuage) was a six to eight week process. Most riddling today is done mechanically and is completed in several days. Bottles are placed in large wire cages, called gyropalettes, and a computer is programmed to turn the pallet automatically.

Riddling Rack: Sparkling wine bottles are turned by hand at Champalou in Vouvray, France (left); Gyropalette: More typically, sparkling wine bottles are turned mechanically (right).

During riddling sediment settles to neck of bottle.

TOM ZASADZINSKI/CAL POLY POMONA

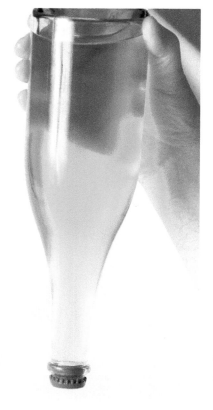

Disgorgement (Dégorgement)

The sediment must be removed from the neck of the bottle using a process called disgorgement. Historically done by hand, today it is all mechanized. The neck of the bottle is inserted into a brine solution cooled below the freezing point, which freezes the sediment plug. Then, with the neck of the bottle pointing downward, the bottle is opened in a continuous motion and the crown cap is removed at the last moment. When the crown cap is popped off, the pressure from within the bottle causes the sediment plug to explode out of the bottle. At the same time that the cap is opened, the bottle is turned right-side up so as to lose as little wine as possible after the sediment plug pops out. Typically a small amount of wine is lost during disgorgement.

Dosage

In order to replace the wine lost during disgorgement, the winemaker adds a dosage. The dosage can be made up of wine, sugar, and, in some cases, a touch of brandy or even Cognac. The final dosage recipe is a trade secret of each producer. Not only does this step fill the bottle, but it also gives the wine its desired level of sweetness. The particular wine style is determined by the dosage. Sparkling wine is the only wine where it is appropriate and allowed by U.S. law to add sugar to make the wine sweeter.

Corking

Once the bottle is filled with the dosage, it is sealed with a cork that is much fatter than a regular table-wine cork. When compressed and put in the bottle, it provides a much tighter fit. Even with the tight fit, the cork must then be secured with a metal cap and a wire cage to ensure that it remains in the bottle. The neck of the bottle is then foil-wrapped. Without the cage and left at room temperature, the cork would quickly explode out of the bottle from the pressure of the trapped CO_2.

Storing and Aging

At this point the wine is aged from as little as a couple of weeks to as much as three months before being released. This time allows the added dosage to fully incorporate into the wine. Wines are shipped as soon as possible after this resting period. Once finished, most sparkling wines improve with slight aging to allow for the dosage to marry with the wine and should be consumed within a few years. However, in some cases, vintage sparkling wines can be aged for several years. Some top vintages of the best Champagnes have been aged for several decades.

Charmat Process

Charmat process (named after the inventor of the process) wines start much the same way as wines made in the traditional method, with the creation of a base wine. But rather than bottling the wine before secondary fermentation, Charmat-process wines are placed into a stainless steel tank capable of withstanding six to eight atmospheres of pressure. The same mixture of yeast and sugar is added to start a secondary fermentation, which creates the same level of CO_2 and pressure produced by the traditional method. After fermentation the yeast sediment naturally settles to the bottom of the tank, leaving a totally clear wine. The wine is then racked from one tank to another under pressure. The sediment is left behind and the clear wine is then ready for dosage. A dosage of sugar and wine, which will determine the final style, is added to the tank. The wine is then bottled using a counter-pressure filler to maintain the same level of CO_2 in the bottle as in the tank.

In the Charmat process sparkling wine is placed in a pressurized tank for secondary fermentation.

The benefit of the Charmat process is that it is much faster and less expensive, requiring less labor to make and process the wine, than the traditional method. However, sparkling wines made using this process are early-drinking wines that often do not hold their bubbles in the glass as long as sparkling wine made using the traditional method. Some producers are capable of achieving a fine bead and exceptional mousse (bubbles or fizz) formation through more attentive handling of the cuvées and longer periods of aging in the tank. The most highly recognized wine made with the Charmat process is Prosecco primarily from the Veneto region of Northern Italy. Many Charmat-produced wines are available at a moderate price point in the United States and around the world.

Other Methods

The transfer method is another technique that is sometimes used to create sparkling wines. After the wine goes through a secondary fermentation in the bottle, it is transferred to a large tank where it is filtered, dosage is added, and then it is placed in a different bottle. This process is typically done when producing half bottles and splits, the small bottles found on airlines and in some restaurants. It is also used with the traditional method when preparing large-format bottles that are at risk of breakage.

The Russian continuous process moves the wine through a series of five tanks. This process is used extensively for wines of modest quality.

Carbonation is a simpler method in which the wine is fermented once and CO_2 is injected into the tank, just as in the process of making a carbonated soft drink. This method is used in lower-quality sparkling wines.

Types of Sparkling Wines

Sparkling wine can be made using different grape varieties and at different quality levels. Each of these types can also be produced in different sweetness styles.

Nonvintage (NV) The classic sparkling wine, led by Champagne, is a blend of Chardonnay, Pinot Noir, and Pinot Meunier from different vineyards and multiple years. Each producer has its own secret recipe. Outside Champagne, different varieties may be exchanged for one or all of the classic varieties. This type will not improve with aging and should be consumed relatively soon after purchase.

Vintage Vintage sparkling wine is made only in the best years, and the vintage is designated on the label. In Champagne, 100 percent of the wine used must come from the stated vintage. Outside Champagne, in the rest of France, only 85 percent of the wine must be of the particular vintage. In California, 95 percent must be from the named vintage, and in Australia, 85 percent must be from the named vintage. Of all of the types of sparkling wines, vintage is the most likely candidate for aging.

Prestige Cuvée These are the most coveted and most expensive sparkling wines. Prestige cuvées are made from even more strictly selected grapes than vintage wines. The most well-known prestige cuvées are Dom Pérignon and Roederer Cristal.

Blanc de Blanc This is a white sparkling wine made only from white grapes. The classic Champagne blanc de blanc is made entirely from Chardonnay. In the New World, other white varieties may be blended with Chardonnay.

Blanc de Noir This is a white sparkling wine made using primarily, if not all, red grapes. In Champagne it is made with Pinot Noir or Pinot Meunier, sometimes both, and sometimes with the addition of a small portion of Chardonnay. Outside Champagne, Pinot Meunier is less widely used and Pinot Noir is more common, in addition to a small percentage of Chardonnay or other white varieties. The color of a blanc de noir ranges from golden to faint pink. A soft pink is acceptable in most regions, but in Champagne any pink color in a blanc de noir is considered a flaw.

Rosé This is a soft-pink-colored sparkling wine that is made from red varieties, or at least is dominated by red varieties, most commonly Pinot Noir. The color can range from light pink to a more intense rosé color. Although usually darker than blanc de noir, there are some rosés that are lighter in color than some blanc de noirs.

Sparkling Red Sparkling red wine, often made with Shiraz or other dark red grapes, has the same bubbly character as other sparkling wines, but its color is deep red. In Australia it is one of the most popular wines produced. This

Table 17.1 Sparkling Wine Styles

STYLE NAME	RESIDUAL SUGAR (G/L)	DESCRIPTION
Extra Brut, Brut Nature or Natural	0–6	The driest style. Not typically seen in the United States. More commonly found in Europe.
Brut	0–15	The driest style found in the United States. The wine can range from dry to off-dry. The most common style internationally.
Extra Dry	12–20	Always contains a noticeable level of residual sugar. Quite common in the U.S. market.
Sec	17–35	Sec is noticeably sweet.
Demi-Sec	35–50	Sweet.
Doux	Above 50	Very sweet and rarely seen.

wine type is also made in California using Syrah or other dark red varieties, but it is illegal in Champagne.

Styles of Sparkling Wines

Style is defined by the amount of residual sugar in each wine. The terms used internationally to describe style indicate sweetness levels ranging from very dry to very sweet. However, within each style there is some variability, so different wines within a particular style may not be exactly the same.

Each type of wine, including nonvintage, blanc de blanc, and so on, can be made in any sweetness level. Sweetness is determined by the grams per liter of residual sugar in the wine. Most sparkling wine types are made in brut or extra-dry styles. The other styles, extra brut, sec, demi-sec, and doux, are more frequently seen in Europe; less so in the United States.

When you purchase sparkling wine, it is important to understand what each term means because the descriptive name does not always match the wine's sweetness level. For example, *extra dry* is off-dry (semi-dry), and *sec* (the French word for "dry") is medium-sweet. Learn the style terms and their sweetness levels listed in Table 17.1 so you will not be disappointed in your purchase.

SPARKLING WINE REGIONS AND APPELLATIONS

Champagne represents the best of sparkling wines because of its heritage and outstanding quality. Champagne produces more sparkling wine than any other region in the world. Sparkling wines are also made throughout France, most notably in the Loire and Alsace, as well as the cooler regions of Italy, Germany, Austria, Spain, Australia, South America, South Africa, and the United States.

France

In France the region or appellation is used as the name of the wine produced there. The term *sparkling wine* is never used. Virtually every type and style of sparkling wine available is produced in France.

Champagne

Some of the world's most famous vineyards rest in the Champagne region.
WILFRED WONG

The most famous of all sparkling wines are those from Champagne in the North of France. In this cold region, grapes do not ripen consistently enough to obtain adequate sugar to make drinkable table wines. They are too high in acid and too low in alcohol. However, they are ideal grapes for sparkling wine.

Champagne has just one appellation; there are no sub-appellations or village appellations. It consists of five sub-regional departments, each producing wine from all three allowed varieties but usually dominant in one. Côte des Blancs and Côte de Sézanne produce Chardonnay-dominant wines, Montagne de Reims and the Aube produce Pinot Noir–dominant wines, and in the Marne Valley Pinot Meunier is dominant. All the wine from these five regions is labeled Champagne.

There are currently over 300 villages and 84,000 acres where Champagne grapes can be grown. There are over 260 houses, 44 cooperatives, and more than 5,100 growers. A Champagne house is a large producer/**NEGOTIANT** who is allowed to purchase grapes in volume. These houses produce more than 70 percent of all Champagne and close to 90 percent of all exports. In 2008 the Champagne region was allowed to expand to include an additional 40 villages and 2,500 acres of land. Overnight the land newly declared within the appellation zoomed to a value of $750,000 per acre. Including Champagne houses, cooperatives, growers, and proprietary store brands over 12,000 different wines are made each year in Champagne. Seventeen Grand Cru villages and 41 Premier Cru villages produce about 30 percent of the Champagne made.

Other French Appellations

The most well-known sparkling wines outside Champagne come from Alsace (Crémant d'Alsace, mostly made of Pinot Blanc), the Loire (Crémant de Loire, mostly made of Chenin Blanc or a blend of Chenin Blanc and Chardonnay), and Burgundy (Crémant de Bourgogne, made from Chardonnay and Pinot Noir). Producers must use Crémant, not Champagne, to identify their sparkling wines.

Italy

Italy is known for a fairly wide array of sparkling wines made by both the traditional method and the Charmat process. Italian sparkling wines are made in styles from dry to sweet. All of the sparkling wines mentioned below come from the cooler northern regions.

Lombardy

The same styles are produced in Lombardy as in Champagne, from brut to demi-sec. They also make a bone-dry style of sparkling wine called Pas Dose, which means no dosage. Franciacorta, a Denominazione di Origine Controllata e Guarantita (DOCG) zone within Lombardy, is the premier producer of traditional-method sparkling wines in Italy. The sparkling wines of Franciacorta are made following the rules of Champagne using Chardonnay, Pinot Noir, and Pinot Bianco (Pinot Blanc), which replaces Pinot Meunier.

Piedmont

Piedmont produces more sparkling wine, primarily from Moscato d'Asti and Asti (formerly known as Asti Spumante), than any other region in Italy. Although they are both made from Moscato Bianco (Muscat), Moscato d'Asti and Asti are two entirely different wines. Moscato d'Asti stops its fermentation early, typically making a sweeter, less carbonated, and less alcoholic wine (4 to 6 percent). Asti ferments longer, increasing its CO_2 level and alcohol level (up to 9 percent), resulting in a wine with less sugar. Both styles almost always have some degree of sweetness and some are quite sweet (demi-sec) dessert-style wines. Although modestly regarded from a professional viewpoint, these are downright pleasant to drink.

Piedmont is also known locally for producing traditional method wines made of Chardonnay and Pinot Noir. A most delicious wine is a sparkler, Brachetto d'Acqui, that is made from the red grape Brachetto in a sweet dessert-wine style with freshly scented rose aromas. It pairs extremely well with dark chocolate.

The Veneto and Emilia-Romagna

An Italian sparkling wine that is becoming wildly popular is Prosecco (the grape name as well as the name of the wine), produced in the Veneto. Prosecco is made using the Charmat process and comes from the communes of Conegliano and Valdobbiadene. It is typically made in a brut or extra-dry style and is normally served as a light aperitif or with lighter Northern Italian dishes. Sold at an affordable price in the United States.

In the region of Emilia-Romagna, Lambrusco (see Chapter 15, Lambrusco, page 251), a dark red indigenous variety, is used to make a modestly regarded but incredibly popular style of sparkling wine in an off-dry (amabile) style. It was so successful that it became the most widely sold Italian imported wine in the United States in the late 1970s and 1980s. It also sells well in Germany and other Northern European countries. Typically made using the Charmat process, it is mostly produced by large cooperatives, but high-quality small producers are making Lambrusco using the traditional method.

Spain

Spain is the land of Cava, which many say is the greatest value in traditionally made sparkling wines. Well over 90 percent of the Cava produced comes from the Penedès appellation of Catalonia. The classic indigenous grapes used in making Cava are the white varieties Macabeo, Parellada, and Xarel-lo. Indigenous red varieties, Trepat, Monastrell, and Garnacha, are used to make rosé-style sparkling wine.

Recently Spanish wine law has permitted the use of the classic Champagne varieties, Chardonnay and Pinot Noir, in the Cava blend. Cava is dominated by two producers, Freixenet and Cordoníu, both from the same town, Sant Sadurni d'Anoia, in the Penedès. In the last several years they have purchased many smaller Cava producers and brands, and now completely dominate Cava in both the domestic and the international markets. Natural and brut styles are available within Spain, but extra dry Cava sells exceptionally well in the United States.

The Penedès is the leading producer of Cava in Spain. The three Cava grape varieties are grown in vineyards around Sant Sadurni d'Anoia, home to numerous Cava bodegas.

Germany

German vintners produce a sparkling wine called Sekt. Most Sekt is made using the Charmat process and is heavily consumed in the domestic market. Much of the Sekt is produced from modest-quality grapes imported from other countries. However, under a relatively new law, if all of the grapes come from Germany the bottle can be labeled "Deutscher Sekt." The grapes most commonly used to make Sekt are Riesling, Müller-Thurgau, Pinot Blanc, and Pinot Gris (Ruländer). Pinot Noir has more recently been used to make rosé Sekt. Much of German Sekt is labeled as *trocken* (dry), but can have a range of residual sugar, resulting in different sweetness levels. There has been a surge of small, new producers who are making high-quality Sekt using the traditional method and selecting single varieties of good-quality grapes, such as Riesling.

The United States

The United States has become a prolific producer of sparkling wine, even though it has been difficult to raise its level of consumption in this country. Most Americans still consider sparkling wine a special-occasion beverage. Virtually all sparkling wines made in the United States are produced in either the brut or extra-dry style.

Sparkling wines have been produced in America since the mid-1800s. After Prohibition, with the advent of the Charmat method, many sparkling-wine producers switched to the more efficient and cheaper bulk process even though it produced a lower quality of wine. Today, many of the lowest-quality sparkling wines made in the United States are made using the Charmat process but are still labeled as Champagne, to the consternation of the Champenoise.

In 1965 Jack and Jamie Davies purchased the old Schramsberg Winery in the Napa Valley with the goal of restarting the production of high-quality sparkling wine in the United States using the traditional method and the classic varieties, Chardonnay and Pinot Noir with Pinot Blanc replacing Pinot Meunier. The winery continues to produce outstanding traditional method wines today.

Following the Davies's initial success, many of the most famous Champagne and Cava producers started to establish branches throughout Northern California, including Napa Valley, Sonoma County, and Mendocino County. Moët & Chandon entered the market first with Domaine Chandon, followed by Piper-Heidsieck's Piper Sonoma, Louis Roederer's Roederer Estate, Mumm's Mumm Napa Valley, and Taittinger's Domaine Carneros, as well as Cordoníu in Napa Valley and Freixenet's Gloria Ferrer in Sonoma County. All label their wines as "sparkling wine," not Champagne or Cava.

John Culbertson was the first to establish a traditional method facility in the small Southern California appellation of Temecula in 1991. Now called Thornton Winery after its current owner, the winery still produces traditional method sparkling wine. South Coast Winery, also located in Temecula, is making high-quality Charmat process wine under the guidance of Jon McPherson.

Outside California, Washington's Chateau Ste. Michelle, under the name Domaine Ste. Michelle, is one of the largest sparkling-wine producers in the United States, making wine using the traditional method. Argyle in Oregon, known for outstanding Pinot Noir, is also noted for high-quality traditional method sparkling wine.

Sparkling wine bottling line at South Coast Winery in Temecula, California

Beyond the Northwest, several outstanding producers are using the traditional method. Of particular note is Gruet, located in Southern New Mexico, which is affiliated with the Gruet family vineyards in Champagne, and Mawby Winery, which has been making sparkling wine in the Leelanau Peninsula north of Traverse City, Michigan, since 1984. There are a number of producers in other states that make sparkling wine as well. Small producers typically use the traditional method, resulting in wine with higher price points. Large producers tend to use the Charmat process or transfer method and can sell their wines at low prices, usually well under $10 per bottle. Constellation Brands, the largest wine company in the world, produces huge volumes of American "Champagne" using the Charmat process under several brands, including Cook's, J. Roget, and Great Western.

Australia

Australia has also made sparkling wines since the mid-19th century, but has only become a participant in the international market in the last 30 years. Australia's sparkling wines are made using both the transfer method and the traditional method. The most desired wines are made of the classic Champagne varieties, Chardonnay, Pinot Noir, and Pinot Meunier, in a dry style. Moët & Chandon played an instrumental role in developing a quality wine trade in Australia, arriving there in 1985, 12 years after the company came to Napa Valley. There are now a number of good producers of sparkling wine in Australia that are both domestically owned or that have international affiliations.

One type of wine that is unique to Australia is sparkling Shiraz, a dark red sparkler made in fruity and oak-aged styles. Although much of it is made with Shiraz, a number of producers are making sparkling red wines with Cabernet Sauvignon, Merlot, and Grenache. The wines are full-bodied and can be somewhat tannic, with typical sparkling wine fizz and a touch of sweetness on the finish. Copies of the sparkling Shiraz style can be found elsewhere in sparkling-wine regions of the world, particularly in California.

South Africa

The Charmat process and even carbonation were historically widely used to produce sparkling wine, but only recently have South Africans begun to make strides toward better quality in their sparkling wine. Currently it is made in a number of different appellations using both the Charmat process and more recently the traditional method under the local name Cap Classique. Chenin Blanc, Sauvignon Blanc, and even the indigenous Pinotage were the primary sparkling-wine grapes, but Chardonnay and Pinot Noir are increasingly finding their way into South African sparkling wines. Most wines are made in a dry style.

Other Countries

Virtually all European countries, including England, Portugal, and Switzerland as well as Central and Eastern European countries, make sparkling wine. Most use indigenous and classic Champagne grapes, both white and red. The grapes may be domestic or imported, and in some cases even grape concentrate is used. Quality and method of production vary widely: Many countries use the Charmat process, fewer use the traditional method. Depending on the producer, the wines range from dry to medium-sweet to sweet styles. Most of these wines are sold within their individual countries or to their close neighbors, so trying to find them in U.S. markets is difficult. But when you travel to any country, it is a good experience to seek out top producers and try the various styles of wine.

Argentina is the leader in sparkling-wine production in South America. It provided the first outpost for Moët & Chandon with Bodegas Chandon in 1959. Sparkling wine is made using both the traditional method and the Charmat process, primarily in a dry style, and most of it is sold within South America.

ENJOYING SPARKLING WINES

Buying Sparkling Wines

When buying sparkling wine, you want to consider type, style, and cost. Different types are made with different grapes. A blanc de blanc made from Chardonnay will be light and ultracrisp. A blanc de noir or a rosé from the Pinot Noir grape will be richer. Nonvintage is likely to fall somewhere in between.

Style has an even more important effect, especially if pairing the wine with food. Drier wines, like extra brut and brut, are more compatible with a broader range of nonsweet foods. As with white and red table wines, sweetness is most effective with desserts or spicy seasoned foods.

Cost may be the greatest concern because of the price disparity. Champagne provides the most diversity in style but expect to pay the most for true Champagne. The least expensive NV brut or extra dry typically starts at $30 and some of the top vintage wines or prestige cuvées may cost $200 or more. Outside Champagne, French Crémants of the Loire, Alsace, or Burgundy, Spanish Cava, Italian Prosecco, California sparkling wines, and sparklers from any other Old World or New World country are more affordable. Prices for both well-made traditional method and Charmat process wines typically range from $15 to $25 but can be as inexpensive as $10. If you are looking for a sparkling wine that will be used in a mixed drink, like a Mimosa, a bulk-processed sparkling wine at an even lower price will work.

Storing and Serving Sparkling Wines

Most sparkling wine is meant to be consumed as soon as possible after purchase. The wine will not improve with aging and will begin to lose its freshness and its bubbles. The one exception to this rule may be vintage Champagne in especially good years. Some vintage Champagne can be kept for decades, although its style will change and it will lose some of its effervescence.

Sparkling wine should be stored as you would any other table wine, at a temperature of 55°F/13°C to 58°F/15°C. Then before it is served, it should be prechilled, preferably to below 50°F/10°C. Sparkling wine should be poured at temperatures of 45°F/7°C to 50°F/10°C.

Wherever it comes from, sparkling wine should always be served in the same type of glass. One of the primary components of this style of wine is the bead (bubbles), so a glass that showcases them is desirable. A flute, a tall cylindrical glass, is the common choice. Some prefer a tulip-style glass that adds shape to the flute. A regular wineglass can work as well and actually provides more aromatics, but loses the visual impact of the bubbles. Sparkling wine is also different from table wines in that you typically do not swirl the glass because it causes the effervescence in the glass to fade quickly.

Sparkling Wine and Food Pairings

Although best known for toasting at special occasions, sparkling wines are the perfect beverage to serve with breakfast, lunch, or dinner as well as any time in between. They are arguably one of the most food-compatible wines available.

Salty foods are a perfect complement to sparkling wine's high acidity. It is a wonderful wine to serve with briny fresh oysters straight from the ocean. Many consumers think that sparkling wine is an elegant wine that should only be served with "fancy" foods. This is not the case. Brut or extra brut is great with salty snack foods, like potato chips, popcorn, or salted nuts.

Rich, high-fat foods balance the acids in the wine. A simple grilled sausage goes well with sparkling wine, but it also works with pâtés, creamy cheeses, duck, and even roast beef. Richer wines with more body, such as blanc de noir or rosé made from Pinot Noir grapes, stand up to the richness of these foods.

Bubbles also complement lighter dishes like vegetable or meat salads with an acidic dressing. Marinated asparagus, a difficult vegetable to pair with wine, is good with a brut sparkling wine. Sweeter styles of sparkling wine work with some Asian dishes. Extra dry or even demi-sec accompany Thai noodle dishes that are not too heavily spiced, Chinese fried shrimp fritters, or dim sum.

The classic Mimosa made from fresh orange juice and a moderately priced extra-dry sparkler is ideal for brunch. Served with fresh fruit, pastries, quiches, and other types of egg dishes, it makes a casual event seem festive.

The sweetest styles of sparkling wine, like Italian Brachetto d'Acqui or sparkling Shiraz, go well with chocolate, and Asti is a match for fresh fruit desserts.

TOP 10 FOODS TO PAIR WITH SPARKLING WINE

1 Potato chips with NV brut
2 Fresh oysters with blanc de blanc
3 Eggs Benedict with mimosa
4 Grilled sausages with mustard with NV extra dry
5 Asparagus risotto with Prosecco
6 Butter-braised lobster or crab with blanc de blanc
7 Roast duck with rosé Champagne
8 Jamón Ibérico with Cava
9 Fried chicken with blanc de noir
10 Dark chocolate with Brachetto d'Acqui or Asti

SUMMARY

Sparkling wines are a beautiful expression of a unique winemaking style. With the number of producers from many countries making high-quality sparkling wines, they are becoming a substitute for still wines as well as a special-occasion beverage. The benchmark, Champagne, still holds a special place for overall quality, but by no means is it the only region producing wonderful sparkling wines. The Champenoise have applied their skills to make high-quality sparkling wines the world over. Meanwhile, producers of sparkling wine from other European regions and the New World are making wines similar to Champagne, but also producing wines from different grapes and using other methods, notably the Charmat process, to make products that meet consumers' demand for quality at an affordable price.

Sparkling wine styles, from left: Crémant, Prosecco, Traditional Method (California), Champagne, Cava

TOM ZASADZINSKI/CAL POLY POMONA

Dessert and Fortified Wines

This chapter focuses on sweet wines that can accompany dessert or substitute for dessert, as well as on fortified wines. Dessert and fortified wines are both fairly small categories in the United States, but they are important in other parts of the world, particularly Europe.

A dessert wine is any wine that has at least 6 percent residual sugar. In addition to their high concentrated sugar levels, many dessert wines also have low to moderate alcohol (7 to 14 percent). There are two primary ways to arrive at the desired sugar concentration: using grapes so concentrated with sugar that there is enough to create the necessary alcohol and also leave the wine sweet, and arresting fermentation with sulfur dioxide (SO_2).

A fortified wine is any wine that has had grape-based spirits (unflavored brandy) added to it to stop fermentation. The amount of sugar in the wine when spirits are added is retained in the wine. The grapes for dessert-style fortified wines do not require extreme sugar levels because fortification stops fermentation, retaining the sugar.

DESSERT WINES

As general awareness of wines increases, there has been a corresponding interest in dessert wines. Many winemakers enjoy the challenge of making a high-quality dessert wine. Dessert wines can be made from any grape variety; however, most dessert wines are made from high-acid white varieties or include them in the blend. The acid balances the sweetness of the wine. Riesling, Sémillon, Sauvignon Blanc, Chenin Blanc, Muscat, and Furmint are *vinifera* varieties frequently selected for dessert wines. In North America hybrids and native American varieties like Vidal Blanc, Vignoles, and Concord are also used.

Late-Harvest Wines

With late-harvest wines, most of the work is done in the vineyard. Extended hang time allows the grapes to maximize their sugar content. Late-harvest wine throughout the world achieves sweetness levels from 6 to 12 percent sugar and higher.

Making Late-Harvest Wines

Late-harvest wine grapes are not harvested until a high-sugar concentration has been reached. Some grapes are left on the vine until they reach a sugar level of 35 percent (35° Brix) or more. When picked, the grapes have normally started to raisinate (begin to shrivel like raisins), causing moisture loss and further concentrating sugars.

Hand-harvesting is required because only select grapes that have achieved the right level of sweetness are picked. Sometimes three or four passes through the vineyard are required to harvest all the grapes. Once picked, grapes must be handled delicately prior to fermentation to prevent crushing the grape skins and losing the valuable nectar. A late fall rain diluting the juice, birds and critters eating the sweet fruit and reducing the yield, and possible infection from mold and disease can ruin the harvest. The risk is high and the yields are low, so late-harvest wines can be expensive.

At the winery, grapes are hand-selected to ensure that only the best ones are chosen. This is followed by long, slow pressing, which maximizes the amount of juice obtained from the grapes.

Fermentation starts normally but is stopped naturally or artificially to prevent all of the sugar from being converted to alcohol. Fermentation stops naturally when the alcohol level gets high enough to kill the yeast, leaving residual sugar in the wine. The winemaker can also add SO_2 to kill the yeast and stop fermentation artificially.

Late-harvest wine may or may not be barrel aged, depending on the desired style of the winemaker or producer. Nonbarrel-aged wines tend to have a younger, fresher character, whereas barrel-aged wines are rounder with more complexity.

Botrytized Wines

Botrytis cinerea, commonly known as noble rot, is a fungus, a form of bunch rot that attacks grapes in the vineyard under certain conditions. Many types of bunch rot are detrimental to grapes, but not *Botrytis cinerea*. It adds a quality to the grapes that gives the finished wine an exquisite aroma and flavor of honeyed fruit, creating one of the most desirable characteristics in a dessert wine.

Making Botrytized Wines

Botrytis cinerea–affected wines are created in the vineyard, not the winery. They are made exclusively from white grapes that are infected with *Botrytis cinerea*. The mold develops in late summer and fall under the unusual conditions of

Moldy grapes infected with Botrytis cinerea *result in exquisite sweet wines.*

heavy fog, which creates high humidity, followed by sunlight, which dries the grapes on the vines. The mold infects the grapes by piercing the skin, which has the beneficial effect of allowing moisture to escape from the grapes, intensifying the sugar level.

This type of dessert wine is produced most frequently from grapes in humid climates like those of Sauternes and Barsac in Southern Bordeaux, the Rheingau in Germany, the Neusiedlersee in Austria, and Tokaj in Hungary, but it can be made from grapes grown under similar conditions around the world.

At harvest, grapes must be handpicked to ensure that only those infected with the mold and ripe enough are selected. Several passes through the vineyard may be required to complete the harvest. Grapes are brought to the winery quickly to minimize any deterioration or loss of nectar from the delicate fruit. Because there is so little juice in the grapes, they are pressed multiple times using old-fashioned basket presses (see Chapter 5, Pressing, page 81).

Fermentation typically occurs at low temperatures in stainless steel tanks, in barrels, or in new French oak barriques. Because of the high sugar content, fermentation can last for an extended period of time. The wines are aged for up to 36 months in old or new oak barrels. Château d'Yquem in Sauternes, the most famous producer of this style of wine, only uses new oak barrels. Artificial introduction of *Botrytis cinerea* has been attempted with limited success.

Botrytized wines are some of the most expensive in the world. Factors affecting cost include the risk of losing a harvest to bad weather by extending the time the grapes are left on the vine, labor-intensive harvesting, extended production time, and the use of new French oak barrels.

Ice Wines

Ice wine is created when grapes are left on the vine well beyond the normal harvest date, until they are frozen. In a few locations in the world, the climate brings on quick freezes, making ice wine possible. This style originated in Germany but is made in a number of countries, including Austria and Canada, which is now the largest producer of ice wine in the world.

In Germany and Austria Riesling is the desired variety for making ice wines, but others can be used as well. In Canada they are made with Vidal Blanc, a French-American hybrid, followed by Riesling, Gewürztraminer, and Ehrenfelser.

Making Ice Wines

Vineyards, or blocks of vines within vineyards, are preselected based on the likelihood that they will reach low enough temperatures at harvest. In Germany and Austria a plastic shroud is placed over each vine or even each bunch to prevent birds and storms from damaging the grapes. In Canada the entire vineyard is netted on the top and sides to prevent birds, deer, and bears from eating the grapes. *Botrytis cinerea* is not a desirable characteristic in ice wines because it reduces the amount of juice and the natural taste of the grape.

In 2000 Germany, Austria, and Canada, the three major ice-wine-producing countries, agreed to standards for ice wines. The standards state the minimum level of sugar at harvest must achieve 32° Brix. Canadians have established an even higher standard of 35° Brix. Grapes may be harvested at different degrees of ripeness depending on the producer's style. Less ripe grapes have more acidity, resulting in greater longevity in the wine. Riper grapes provide more sweetness and greater body.

Ice wines are made from grapes that have frozen on the vine.

www.taggphotography.com

Grapes are not picked until the temperature drops to 18°F/–8°C (16°F/–9°C in Canada). Only perfect grapes without any indication of mold are selected. All picking is done by hand. To pick at the proper temperature, harvest may start under lights in the middle of the night.

When the grapes are harvested, they are immediately taken to the winery where they are pressed, sometimes outside to maintain proper cold temperatures. A basket press is commonly used because it presses the grapes more delicately. As the grapes are pressed, sugar content is measured. First-press juice is the sweetest because the grapes are completely frozen, and pressing only releases the nectar. As pressing continues, the juice becomes diluted from the melting ice in the grapes and is less sweet.

Pressed juice for ice wine must be no less than 32° Brix (35° Brix in Canada), but it can be as high as 40° or even 50° Brix. Diluted juice, caused by the melting ice, that comes out below 32° Brix (35° Brix in Canada) gets blended into late-harvest wine.

Fermentation lasts from a few weeks to a few months. In Europe wines may be fermented in stainless steel or oak barrels. In Canada stainless steel is more common. When finished, the alcohol level must be at least 8.5 percent, but 10 percent is normal. Too high an alcohol content robs the wine of fruit character. Final sugar must range between 200 grams per liter and 300 grams per liter, with 250 grams per liter typical.

Dried-Grape Wines (Passito)

Making wines from dried grapes is the oldest form of making dessert wines. Centuries before other methods were developed, winemakers used grapes that dried naturally on the vine. Even today, some producers leave ripe grapes on the vine until they lose their moisture and begin to raisinate prior to being picked and dried, bringing out the sweetness in the grapes.

Italy is by far the largest producer of this ancient style of wine, but it is also made elsewhere. As an example, Willi Opitz from the Neusiedlersee of Austria makes Opitz One, a wonderful dried-grape wine (*Schilfmandl*) from the variety Zweigelt.

Making Dried-Grape Wines

At Isole e Olena in Tuscany, Italy, grapes are dried on bamboo racks to make Vin Santo.

Grapes are always hand-harvested. They may be picked before, at, or after full ripeness has been achieved. Grapes harvested before full ripeness keep stronger skins and resist rot as the grapes dry. They also maintain high acidity and flavor freshness.

Harvested grapes are brought to drying rooms at the winery. They are laid out on bamboo mats, hung from long racks, or more commonly today, placed on wire-mesh trays, which are cleaner and help prevent rodent infestation. In the past grapes were often placed in attics, but today producers have drying rooms with good air circulation, temperature control, and supervision to detect any mold or rot. In some cases *Botrytis cinerea* takes hold. Drying takes from a few weeks to several months. During this time, the grapes dehydrate, which concentrates the sugar and causes a chemical change in the grapes' aromas and flavors.

The dried grapes are pressed tightly using a gravity press and are fermented in oak casks. Because of the sugar concentration, the yeast may act in an unusual fashion. The fermentation can stop and start, and it can last from several months to as long as two or three years. Dried-grape wines can be aged in old oak barrels and allowed to oxidize like a cream sherry, creating a traditional style. Winemakers prevent oxidation in newer styles, resulting in a fresher style of wine.

Dessert Wine Regions and Appellations

Dessert wines originated in Europe, where each country developed styles unique to its culture and terroir. Most of the great dessert wines are still produced in Europe, but in the New World winemakers have emulated European dessert wines.

Germany

Dessert wines are made in several German appellations, led by Rheingau, Rheinfaltz, Rheinhessen, Nahe, Mosel, and Franken. Because of significantly different weather patterns every year, the quality of wines varies from region to region in different years. In some years, the weather is foggy enough for the development of *Botrytis cinerea* or cold enough to freeze grapes for Eiswein (ice wine).

Riesling is by far the most prevalent variety used to make dessert wines. German Riesling dessert wines are highly rated because of their balance of fruit, acidity, and sweetness, highlighted by characteristics of soil and place. The most noted dessert wines are classified as Beerenauslese (BA), Trockenbeerenauslese (TBA), and Eiswein. Eiswein is most likely to have the sugar percentage of a BA and less than a TBA. (For more information about German wines see Appendix 2, page 400.)

France

Dessert wines are made in many appellations throughout France, each with its own regional and, in some cases, international popularity.

Sauternes and Barsac

Sauternes is the international star of dessert wines. It competes for best in the world with Germany's TBAs. Wines of Sauternes and Barsac in the southern part of Bordeaux are made of a blend of Sémillon and Sauvignon Blanc. Conditions here are ideal for *Botrytis cinerea*. The cool waters of the Ciron River come in contact with the warmer waters of the Garonne, creating fog that covers the vineyards, but the sun dries out the vines in the afternoon. Because *Botrytis cinerea* develops regularly in these regions, botrytized wines are produced almost every year. Château d'Yquem holds a Premier Cru Supérieur classification followed by several producers in both the Premiers and Deuxièmes Crus classifications. (For information on French classifications, see Appendix 2, page 380.)

The Loire

Because the Loire is much farther north than Sauternes, the weather plays a more dynamic role and causes some poor-quality vintages or no vintage at all for dessert wines. Two principal areas within the Loire produce outstanding dessert wines, both made with Chenin Blanc: Layon and Vouvray.

The sweet wines of the Loire, made from Chenin Blanc, are demi-sec, which is not typically considered a dessert wine, followed by sweet moelleux, and the super-sweet liquoreux. These are highly concentrated wines with high acidity

Château d'Yquem in Sauternes, France, produces esteemed botrytized wine.

that can last for 100 years or more. They should not be consumed until 10 to 15 years after their vintage.

Alsace

Though better known for their dry wines, producers in Alsace make dessert wines from their riper grapes. They are known as Vendange Tardive and Séléction de Grains Nobles. Vendange Tardive, meaning "late picked," are typically made into off-dry wines with high alcohol and richness, the result of the ripeness of the grapes. It is the Séléction de Grains Nobles that make true dessert wines. Within Alsace only three varieties can be made into Séléction de Grains Nobles wines: Riesling, Gewürztraminer, and Pinot Gris. These wines are only made in years when the grapes achieve minimum potential alcohol of 15.1 percent (about 25° Brix) for Riesling and 16.4 percent (about 27° Brix) for Gewürztraminer and Pinot Gris. *Botrytis cinerea* is common among this style of wine.

Italy

Dessert wines are made throughout Italy, and passito is the benchmark style. It is quite common for passito to be made in many regions, in most cases from indigenous grapes.

Northeast

The Veneto is the home to some of the greatest passito wines in Italy. Two well-known wines are Recioto di Soave, which is made from Garganega grapes, and Recioto della Valpolicella, which uses Corvina, Rondinella, and Molinara. Tiny productions of some excellent wines are found and consumed locally, like the Vino Santo of Trentino made of the Nosiola grape.

Friuli produces two dessert wines made from the indigenous grapes, Verduzzo and Picolit. Verduzzo makes a late-harvest wine that varies in sweetness depending on the weather. Picolit is more difficult to manage in the vineyard, but it produces small amounts of excellent botrytized wines.

Tuscany

Tuscany is home to perhaps the most noted dessert wine in Italy, Vin Santo. Vin Santo is a widely produced wine that ranges from dry to sweet and from modest to outstanding quality. It is made in a passito style, from Trebbiano and Malvasia grapes; that is unique to Tuscany. After the dried grapes are pressed, fermentation takes place in old chestnut or oak barrels. New barrels are never used. Fermentation and aging can last for up to six years and the barrels are not topped off, allowing the wine to evaporate and develop an oxidized character.

Other Italian Regions

The passitos from the northwest region are quite good. Caluso Passito from Piedmont is the best known. Sciacchetrà (pronounced *shot-ke-tra*) is a passito-style wine found in the Cinque Terre, a popular tourist destination in Liguria.

In the south, dessert wines are made from Muscat. Sicily is famous for Passito di Pantelleria, a passito made from Muscat of Alexandria (Zibibbo), and Malvasia della Lipari, a dried-grape wine from the same grape variety that is low in alcohol and has aromas and flavors of peaches and apricots.

Austria

Although Austria is better known for its dry wines, varietal dessert wines are some of the country's stars. Many of the best wines are made from Ruländer, Weissburgunder, and Gewürztraminer. The best dessert wines come from the Burgenland. Here Lake Neusiedl forms the same type of fog found in Sauternes and Barsac, which creates conditions well suited for *Botrytis cinerea*. Austrians also produce Eiswein, some dried-grape wines, and late-harvest wines in warmer years. The Austrian wine-classification system closely follows that of Germany, with dessert wines classified as Beerenauslese, Trockenbeerenauslese, and Eiswein.

Hungary

In Hungary the most noted dessert wines, made from Furmint and Hárslevelü grapes, are those of Tokaj, a region in the northeastern part of the country. Tokaji Aszú is the most unusual of the botrytized wines. Sweet must from botrytized grapes is added to finished dry wine, which restarts fermentation. When the wine finishes its second fermentation, alcohol is increased 1 to 2 percent. The wine is then aged.

Sweetness and length of aging are measured by how many puttonyos are added to a 136-liter barrel of wine. A puttony is a tub that holds 25 kilograms of the sweet must. The more puttonyos added, the sweeter the wine. Tokaji Aszú contains from three to six puttonyos, which are indicated on the bottle label. A five- or six-puttonyos Tokaji Aszú is considered one of the best dessert wines in the world. Beyond six puttonyos, wine is Aszú Essencia, which contains even more Aszú paste and is made only in exceptional years.

Canada

Canada produces a range of dessert wines, most notably ice wines. These are made in the region between Lake Ontario and Lake Erie on the northern border of New York State and in the Okanagan Valley in British Columbia. The most desirable variety for ice wine is the French hybrid, Vidal Blanc, followed by Riesling. Late-harvest wines are also produced using both *vinifera* and hybrid varieties.

The United States

Different styles of dessert wines come from different regions of the United States. California and Washington State are known for late-harvest wines with a wide range of sweetness levels. These wines are based on several varieties including Riesling, Sauvignon Blanc, Gewürztraminer, and Zinfandel. Botrytized wines are sometimes found, but ice wines and dried-grape wines are rare. In the colder East and Midwest dessert wines are made from *vinifera* as well as native American and hybrid varieties. New York producers are recognized for their ice wines as well as late-harvest wines. Some of the most noted dessert wines of native American and hybrid varieties come from the grapes Concord, Catawba, Vidal Blanc, Vignoles, and Niagara.

Any fruit can be made into wine, and throughout the United States, well-regarded fruit wines are made from raspberries, apricots, peaches, strawberries, and blackberries. Most are made in a dessert style, and have the distinct aroma and flavor of whatever fruit they are made from.

Other Countries

Late-harvest wines from Sémillon, Sauvignon Blanc, and Riesling were introduced in Australia in the early 1980s by De Bortoli Winery in New South Wales. The botrytized wines of De Bortoli and others echo the botrytized wines of Sauternes. Australia's dessert wines are commonly called stickies.

Every grape-growing country has its own dessert wines, made from the same varieties that are used for the country's dry wines. Dessert wines made in these countries typically follow the established styles of France and Italy with a focus on late-harvest and botrytized wines.

Enjoying Dessert Wines

Buying and Serving Dessert Wines

Dessert-style wines were very popular 40 to 50 years ago, but that is no longer the case. A few examples may be found in your local grocery store, but not the breadth of wines discussed in this chapter. For a wider range of dessert wines, use the Internet to search for a wine shop in your area that focuses on this style. Also, many wineries only sell these small production wines out of their wine shops or tasting rooms. Call the winery directly for availability. If laws permit in your state, you can also order hard-to-find dessert wines on the Internet.

Dessert wines should be stored like any other table or sparkling wine at 55°F to 58°F/13°C to 15°C. Dessert wines, whether white or red, should be served at cool temperatures, usually 50°F to 55°F/10°C to 13°C.

The portion size of a dessert wine is about half of a regular glass of wine, two to three ounces. The wines are so sweet that it would be difficult to drink them like a regular table wine. They are meant to be sipped. The appropriate glassware is typically smaller than a regular table wineglass with a closed rim similar to a Bordeaux glass. A Bordeaux glass works as well, just make sure that you do not pour too much wine in the glass.

Dessert Wine and Food Pairings

The best dessert is a luscious, sensual dessert wine savored by itself. But dessert wines are also a wonderful companion to many foods. The underlying tenet of matching dessert wine with food is: The wine must always be sweeter than the food. The sweeter the dessert, the sweeter the wine must be.

Dessert wines also work with savory foods. Sauternes with sautéed foie gras, for example, is a food and wine match made in heaven. Sweetness is also a good foil for some spicy dishes.

Another rule of thumb is to match wines with the foods of their historic region. Dessert wines with cheeses from the same region, for example, can be a miraculous pairing. Two great combinations are Sauternes with Roquefort and sweet Tokaji with blue cheese. Passito from Northern Italy is commonly served with nut tortes and lightly sweetened cakes. In Tuscany a classic is Vin Santo with biscotti, the dry cookie of the region. Both passito and Vin Santo also work extremely well with regional cheeses. In Germany a BA, TBA, or Eiswein is likely to be served with a fruit strudel or tart, or by itself. Fruit desserts are also served with both the late-harvest wines of the Loire and the Séléction de Grains Nobles wines of Alsace.

In the New World fruit pies, lightly sweetened cookies, cakes, and pound cakes are compatible with a range of dessert wines. Sweet styles of Riesling, Muscat, Gewürztraminer, and Sémillon can all accompany these types of desserts. A red dessert wine, like late-harvest Zinfandel, is best with chocolate.

TOP FIVE FOODS TO PAIR WITH DESSERT WINES

1 Roquefort with Sauternes
2 Crème brûlée with late-harvest Riesling
3 Fall or winter fruit compote and biscotti with a passito
4 Honey-infused cake with a botrytized wine
5 Late-harvest red wine with dark chocolate ice cream

FORTIFIED WINES

Fortified wines require the addition of grape-based spirits to wine, which arrests fermentation and creates an entirely different type of wine. Because they are fortified with spirits, these wines are high in alcohol (15 to 22 percent). Fortified wines can be made in either a dry or a sweeter dessert style. It is the timing of fortification that makes the difference in the level of alcohol and consequently the sweetness of the wine. The earlier in the fermentation stage that fortification occurs, the more sugar will remain in the wine. When a wine is fortified after fermentation is complete, the wine is already dry.

Fortified wines are primarily affected by the blending and aging processes. They are usually made from the local region's indigenous grapes, including a range of red and white varieties.

The British developed fortified wines from Spanish and Portuguese grapes in the 1700s and were the primary consumers. The common fortified wines are sherry, port, and Madeira.

Sherry

Sherry is a fortified wine produced in Andalusia in Southwestern Spain. Factors that characterize sherry include its high alcohol level and oxidation. Dry sherry is made only from the Palomino grape; dessert styles add the variety Pedro Ximenez to the blend. All of the Palomino grapes must be grown in the Jerez Denominación de Origen (DO), which is made up of the area around three towns: Jerez de la Frontera, El Puerto de Santa Maria, and Sanlúcar de Barrameda. Palomino accounts for 95 percent of the grapes used in sherry; Pedro Ximenez makes up the other 5 percent. Palomino grapes thrive in Jerez's albariza soil, which is almost white and consists predominantly of calcium carbonate (chalk) with some clay and sand.

There are five general styles of sherry: fino (including manzanilla), amontillado, palo cortado, oloroso, and cream sherry. Pedro Ximenez (PX) is a fortified wine made from 100 percent Pedro Ximenez grapes that is technically not a sherry but is made in a solera.

Sherry is classically dry with the exception of cream sherry. However, other classically dry styles of sherry (amontillado and oloroso) are commonly sweetened for the international marketplace, including the United States. Variations of the styles identified above, with names like pale dry sherry, medium sherry, and triple cream sherry, are available in the United States. These may be of high quality produced with the same methods as Jerez's sherry, but they may use different grape varieties and can be made in Jerez or outside Spain.

Flor is a naturally occurring yeast that forms on the surface of sherry wine.

Making Sherry

All sherry begins life as a dry white wine. The wine is fermented dry and attains a level of 11.5 to 12 percent alcohol. The producer, usually a large company such as Osborne or González Byass, or a cooperative, prepares the base wine and either keeps it for its own line of sherries or sells it to smaller sherry bodegas (aging facilities). In either case, the most important step follows: converting the wine to sherry.

The base wine is fortified by the addition of grape-based spirits to obtain an alcohol level of 15 percent. This retards spoilage and creates the perfect living environment for **FLOR**, a naturally occurring yeast that, in Jerez's bodegas, typically appears immediately on 15 percent alcohol wine. It covers the surface of the fortified wine, moderating oxidation and adding its own flavor characteristics.

The presence or absence of flor plays a crucial role in determining the style of sherry produced. If flor forms, the fortified wine will become a fino or possibly an amontillado (aged fino). If flor does not develop, the fortified wine will become an oloroso or cream sherry. A palo cortado is created if the flor begins to develop but then stops.

The Solera System

Sherry production employs the solera system of aging. Young wines are blended with older wines; over time these older wines are blended with even older wines, which in turn will be blended with even older wines. A solera is composed of all of these wines from youngest to oldest.

Each style of sherry produced at a bodega has its own solera. Fino, amontillado, palo cortado, or oloroso is added to very old American oak barrels, some more than 100 years old, containing wine of its style that was previously made. The barrels are no more than five-sixths full, leaving plenty of air (known as **ULLAGE**) so the wine oxidizes.

Each aging level within a solera is called a criadera, which means "to raise." There can be as few as one barrel or more than a thousand in a criadera. Each solera contains up to 14 criadera. When there is market demand for sherry, one-third of each barrel in the oldest criadera is drawn off for bottling. One-third of the wine in the next oldest criadera is then transferred into the oldest criadera, and so on through all the levels. When the youngest criadera is drained of one-third of its wine, newly fermented fortified wine is added.

It takes many years to create a solera. Most of the soleras in Jerez were started many decades ago; some are 150 to 200 years old. Since a barrel is never emptied, a sherry contains some wine from every vintage since the solera was created.

The sommelier takes a barrel sample of sherry at Bodegas Osborne in Jerez, Spain.

Sherry Styles

Fino Fermented base wine is fortified with grape-based spirits. In the bodega, naturally formed flor coats the wine that is stored in old oak barrels that comprise the fino solera. Flor moderates the wine's contact with air, controlling oxidation and adding its own flavor characteristics. All finos are dry.

Manzanilla Manzanilla is a type of fino that is only aged in Sanlúcar de Barrameda. The location of the aging facility is the only difference between it and other finos. Manzanilla producers think it is distinct, with a lighter, almost salty character provided by the town's proximity to the sea, and quite different from the finos produced in Jerez de la Frontera or Puerto de Santa Maria. All manzanilla is dry.

Amontillado Amontillado begins as a fino. The wine is moved from a fino solera to an amontillado solera. It is created when the flor dies naturally or alcohol is added to the fino specifically to kill the flor. To kill the flor, the alcohol level must be increased from roughly 15 to 18 percent. Amontillado is considered an aged fino and is oxidized, dry, and turns a darker light brown color. For the American market, a moderately sweetened amontillado is made with the addition of sweet Palomino or Pedro Ximenez wine. It is typically less sweet than a sweetened oloroso and certainly less sweet than cream sherry.

Palo Cortado Palo cortado is made from fortified wine that starts to grow flor and then stops. It is placed in its own solera. The resulting wine falls between an amontillado and an oloroso. A properly made palo cortado has the bouquet of amontillado and the flavor of oloroso. Palo cortado is mostly dry but can be sweetened.

Oloroso Fortified wine that develops no flor is selected for oloroso and placed in its own solera. Oloroso sherries are additionally fortified to a level of 18 to 20 percent. They become very oxidized and turn a fairly dark color. As the wine ages, it begins to evaporate, increasing its alcohol content. Olorosos are dry and have a concentrated nutty character. Most oloroso made for the export market is colored and sweetened by the addition of sweet Palomino or Pedro Ximenez wine. It is less sweet than cream sherry.

Cream Sherry Cream sherry is quite sweet and is made from oloroso with the addition of a percentage of Pedro Ximenez wine. Each bodega has a secret recipe for its own distinctive style of wine. The sweetness of the sun-dried Pedro Ximenez wine combined with the nuttiness of the oloroso makes an exceptional dessert wine.

Pedro Ximenez (PX) Pedro Ximenez begins with 100 percent Pedro Ximenez grapes that are dried on rush mats much as the dried-grape wines of Italy. They are pressed and fermented, then fermentation is stopped with the addition of alcohol. Although not technically a sherry, PX is aged in a solera. The result is a dark, sweet, syrupy fortified wine.

Port

Port is a fortified wine that is always sweet, almost always red, and high in alcohol. Over 300 years ago France and England were at war so the British had no access to French wines. Portugal became the new source of wine for Britain. The English liked the deep, dark Portuguese reds. To ensure that they arrived in England in good condition, brandy was added to the wine, and port was born.

Port is named after Oporto, a city in Northern Portugal at the mouth of the Douro River. The wines are actually aged across the river in the town of Vila Nova de Gaia, where all the major port producers' lodges (aging facilities) are located. Port grapes come from the hillsides up the Douro River. Some of the vineyard sites are as steep as the slopes of the Mosel in Germany and the Côte-Rôtie in the Rhône. Historic vineyards are heavily terraced and are interspersed with newer vineyards on less steep hillsides.

The classic port grape varieties include Touriga Nacional, Tinta Barroca, Touriga Francesa, Tinta Roriz (Tempranillo), and Tinta Cão. The base wine is made in wineries close to the vineyards and then transferred to the lodges in Vila Nova de Gaia for aging and bottling. There are as many as 80 allowed varieties, and many of the base wines come from field blends so the exact mix of varieties may be unidentified. True ports can only come from Portugal. Many producers in other countries make port-style wine from Portuguese varieties or the red varieties that grow well in their area.

Making Port

Grapes are harvested by hand and brought to the winery where they are placed in a tank, or more historically, a lagar. A lagar is a large, flat, open stone or concrete container. It is square or rectangular and only a couple of feet deep. Here vineyard workers traditionally crushed the grapes by stomping on them with their bare feet. Although this technique continues in some places, most lagares or tanks now use mechanical crushing to accomplish the same task as feet—to extract the maximum color and tannins from the skins of grapes.

After crushing, natural yeast causes the must to start fermenting. When the correct levels of alcohol and sweetness are reached, the wine is drawn off from the lagar into a tank that contains grape-based spirits. The spirits kill the yeast, stopping fermentation, and leaving a wine that is sweet and high in alcohol—a port. A port wine has about 10 percent residual sugar and about 20 percent alcohol.

Port Categories

There are two categories of port: wood- or barrel-aged port and vintage port. Wood port is aged in barrels for short or long periods depending on style desired and then fined, filtered, and bottled. Vintage port is only barrel aged for up to two years and then bottled without fining or filtering, leaving a high ratio of solids in the wine.

Vila Nova de Gaia, which sits across the Douro River from Porto, is home to Portugal's port-aging facilities.

Wood Port Styles

Ruby This is a young port with fresh, vibrant color, and red-fruit aromas and flavors. It is typically blended with port from multiple young vintages and is then fined, filtered, and bottled. It is inexpensive.

Tawny Tawny ports are aged in barrel. Over time the wine loses its vibrant red color, turning a tawny brown—hence its name. Tawny wines are fairly young, are made from a blend of multiple vintages, and are inexpensive.

Aged Tawny These wines have been aged a minimum of six years. They are always blended from a combination of many vintages. They are sold as 10-year, 20-year, or occasionally 30- or 40-year tawny. Each of the age designations on the label indicates the average age of the blended components in the wine. As the wines age, they soften, losing their tannins, developing mellower characteristics. The older the designation, the better perceived the quality and correspondingly higher the price. The wine does not continue to age once it is bottled.

Colheita These are wines that have the same character as aged tawny but are made from only one vintage. They are aged for a minimum of seven years and, in many cases, much longer. The vintage is identified on the bottle.

Vintage Port Styles

Vintage This style is considered the premier class of port because it is made from grapes selected in only the best years. A vintage of port can only be declared by the producer after agreement with the Instituto do Vinho do Porto, and this occurs only two to four years out of 10. A young vintage port is big, burly, and robust with heavy tannins. It takes years, even decades, of bottle aging to achieve its best. Because of the high solids, lack of filtering, and extended aging, these wines require decanting before serving to remove heavy sediment.

Late-Bottled Vintage Also known simply as LBV, these wines are made in undeclared years, but are still vintage-dated. They can be aged in wood for up to six years and can be fined, filtered, and bottled, giving the wine a tawny character. Once bottled, the wine will have a much shorter life than true vintage port.

Madeira

Madeira is a Portuguese island in the Atlantic Ocean as well as the name of the fortified wine produced there. Even though the quality of the best of these wines is high, Madeira has limited popularity as a drinking wine and is often used mostly for cooking.

The traditional Madeira grapes are the white varieties Sercial, Verdelho, Bual, and Malvasia (Malmsey). Tinta Negra Mole, a black grape, is the most widely planted variety used to make Madiera, but it produces high-volume, low-quality wines. Madeira comes in a range of styles from dry to sweet, largely determined by the variety that is used. Sercial makes a light, dry style; Verdelho is dry to off-dry; Bual is a sweeter style; and Malvasia makes the sweetest style.

Making Madeira

Like port, Madeira is fortified, but how the wine matures determines its characteristics. The best-quality wines are aged in large casks in attics that absorb the sun's heat. The aging process may last for decades. Younger wines are artificially aged using the estufagem process, in which hot water is circulated around tanks to heat the wine in a saunalike environment. Wine temperatures in the estufa, or heating rooms, can reach 130°F/55°C. This intense aging period of three to six months is equivalent to a period of normal barrel aging for five years.

Other Fortified Wine Styles

Vin Doux Naturel

Vin doux naturel (literally naturally sweet wine) is a sweet fortified wine made in a manner similar to port. A very strong flavor-neutral alcohol is added during fermentation, but it can only make up a total of up to 10 percent of the volume of the wine. By contrast, up to 20 percent of a port's volume can be spirit. The final wine will have an alcohol content of 15 to 18 percent and a range of possible sugar levels. The wines are red or white, but must be made from designated varieties, including Grenache (both red and white subvarieties), Macabeo, and Muscat.

Vermouth

Vermouth is a fortified aromatized wine that can be made of white or red varieties. Botanicals such as herbs, spices, and flowers are added to both dry vermouth (made from white grapes) and sweet vermouth (made from red grapes), giving the wines distinctive aromas and flavors. Each producer uses its own special secret recipe to create these aromatic wines.

Fortified Wine Regions and Appellations

Fortified wines are far less common than dessert wines except in Spain, Portugal, and Southern France. The number of producers and wines outside these three regions is relatively small.

Spain and Portugal

Most of Spain's fortified wines come from the Jerez DO. Málaga is also recognized for its fortified wines. Jerez, Sherry, and Xérès are registered names and may only be used with wines from that DO. Sherries made outside of Jerez can only be called sherry-style. Late-harvest wines are also made in Spain in small quantities, mostly in the country's southern appellations.

Portugal is most noted for its range of fortified wood-aged ports and vintage bottle-aged ports, as well as Madeira. Port is also a registered name, like Jerez, that may only be used with wines from that region. Wines made outside of Oporto are required to use another name entirely or designate the wines as "port-style." Portugal's fortified wines are almost only sweet.

Southern France

Vin doux naturel is produced in several appellations along the Mediterranean coast from the Southern Rhône to the border of Spain. The wines from the Rhône and Languedoc are mostly made of Muscat. Appellations using this grape include Beaumes-de-Venise, Frontignan, Lunel, Mireval, Rasteau, and St. Jean de Minervois. In the Southern Roussillon appellations of Banyuls and Maury, Grenache is used; in Rivesaltes, Grenache, Muscat, or Macabeo.

Other Countries

Fortified wines are produced in Southern Italy, where they are called *vino liquoroso*. Aleatico, a native Apulian grape, is fortified into a portlike wine. Sicily is widely known for Marsala, a fortified wine made from the native white grapes Inzolia and Grillo. It has become mostly used for cooking, but a few producers are attempting to reintroduce high-quality Marsala.

California produces a large volume of fortified wines, mostly in port styles but also some in sherry styles, similar to the classic wines of Spain and Portugal. Most port-style wines in the United States are made from dark red grapes like Zinfandel or Petite Sirah. The Quady Winery in Central California, though, is noted for its port-style wines made from classic Portuguese varieties.

A few port-style fortified wines are made in British Columbia and Ontario, Canada. Fortified wines in both port and sherry styles have long been made in Australia.

Enjoying Fortified Wines

Buying and Serving Fortified Wines

Selecting a fortified wine can be difficult because most stores that sell wine only carry a few brands of fortified wines and usually at the lowest possible price point. Grocery stores and other general retail outlets may sell generic, bulk fortified wines. Use the Internet to find a local wine shop that caters to those interested in fortified wine. The good news is because of their lack of popularity, fortified wines are one of the best wine bargains available. To find classic styles you need to look for sherries from Southern Spain, ports from Portugal, and Madeiras from the island of Madeira. There are also excellent examples of each style made in different locations around the world.

Fortified wines should be stored like any other table or sparkling wine, at 55°F to 58°F/13°C to 15°C. Fortified wines are high in alcohol and will be hot on the palate at high temperatures. Most should be served cool, from 50°F to 55°F/10°C to 13°C. Less expensive wines might be served even cooler.

The portion size for fortified wines is about half that of a regular glass of wine, two to three ounces. The wines are high in alcohol and in some cases so sweet that it is difficult to drink them like a regular table wine. They are meant

TOP FIVE FOODS TO PAIR WITH FORTIFIED WINES

1 Stilton, Roquefort, or Iowa Maytag blue cheese with vintage port

2 Simple tapas of almonds, olives, and sausages with fino or dry amontillado

3 Braised beef, lamb, or pork with dry dark sherry, amontillado, palo cortado, or oloroso

4 Vinegar-based salads or appetizers with fino

5 Dark chocolate with aged tawny

to be swirled and sipped. The appropriate glassware is typically smaller than a regular table wineglass and should have a closed-in rim similar to that of a Bordeaux glass. A Bordeaux glass works as well, just make sure that you do not pour too much wine in the glass.

Fortified Wine and Food Pairings

Fortified wines, both dry and sweet, are excellent accompaniments to food. In Jerez sherries are served through the entire meal. Try a fino or manzanilla sherry with tapas, the little appetizer bites that are found in all of Spain's bars and restaurants, and are now popular throughout the United States. Dry sherries also work exceptionally well with seafood. Probably the easiest combination is to match roasted almonds, salted or not, with any style of sherry. Fino sherries or richer styles also complement mushrooms and onions, particularly in soups. Sweet sherries work with cream, caramel, and nut desserts as long as the wine is sweeter than the dessert.

Traditionally port is served with the famed English blue cheese Stilton. Port is also a perfect combination with dark bittersweet chocolate in any form, from a chocolate-port sorbet to a glass of port alongside a flourless chocolate cake or even a piece of top-quality bittersweet chocolate.

SUMMARY

Dessert and fortified wines are both important wine categories, although they are less familiar to most consumers in the New World than dry table wines or sparkling wines. This is caused in part by small production quantities and lack of availability in New World markets. Dessert and fortified wines are far more popular in Old World countries, where they have a long history and have been an integral part of the dining experience for many centuries.

As consumers become more dining savvy and the popularity of desserts increases, dessert wines will continue to become more highly regarded throughout the world. Fortified wines are likely to continue to struggle for popularity because of their high alcohol content and less-familiar taste profiles.

Beer

Many people know the sound a bottle or can of beer makes when it is opened. Some love the sound because they know what will follow. Many know the refreshment of cold pale lager on a hot day. Others know the comfort of winter ale sipped by a roaring fire on a cold night. Many people know the taste of beer, but they have not all explored the many differences between lagers and ales.

Beer, like wine, is a fermented beverage. Unlike wine, the sugar required for fermentation comes from grain rather than fruit. Whereas wine is produced from many grape varieties, beer uses only a handful of grain types. Like wine, beer is available in a range of colors and flavors, from light and delicate to big and bold, and it pairs well with food.

Beer has been around at least as long as wine. For all of its history, beer has had a connection to hospitality. To refresh the parched traveler, to hoist as part of life's celebrations, to complement great food, or to sip while socializing, beer is the beverage of choice for many.

For most of the 20th century in the United States, beer was viewed as the alcoholic beverage of the blue-collar worker, and the types of beer available were largely limited to golden-colored, light-tasting lagers. However, in the last third of the century, the beer industry focused on increasing the consumer's knowledge of beer and broadened the types and styles available to the consumer. The brewing industry continues to change its perception of the beer drinker from an Everyman to any man or woman who desires an interesting beverage alternative. While increasing awareness, interest, and its many unique offerings, the industry has also focused on keeping beer user-friendly. Beer remains *the* casual adult beverage.

WHAT IS BEER?

Beer is a fermented beverage made from malted grains. The most common grain used for **MALTING** is barley. Malted grains are raw grain kernels that have been steeped in water, allowed to begin germinating, and then kiln-dried. This process converts the grain's starch into maltose and other substances. The production of the sugar maltose—hence the term *malted grain*—is the first step in preparing the starches in the grain for the process of fermentation.

Beer and wine both use fermentation to derive the finished product. However, beer is generally lower in alcohol than wine, typically under 10 percent **ALCOHOL BY VOLUME** (ABV). In the United States, the term *beer* is used generally to define grain-based beverages containing at least 0.5 percent ABV.

A beer brand remains consistent in flavor, year after year, bottle after bottle. The goal for the brewer is to brew the same beer, batch after batch. Wine consumers count on a specific winemaker to make consistently good-tasting wine, with the knowledge that each vintage will be different yet enjoyable. Beer consumers trust that a brewer will brew good-tasting beer, with the added expectation that any bottle of the same product will taste like every other bottle.

Almost all present-day beers contain the same four basic ingredients: malted grain, water, yeast, and hops, a delicate botanical cone resembling a small pinecone that is produced by the flower of a vine. Yet like the winemaker and spirits producer, each brewer can use different combinations of these four components. The resulting beverage reflects differences in the country of origin, the culture, the origin of the ingredients, and the brewer's own preferences. These four basic elements yield dozens of styles and thousands and thousands of brands.

Guinness brewery in Dublin, Ireland

HISTORY OF BEER

Anthropologists and archeologists theorize that grain-based alcohol—the first beer—like other fermented beverages, developed by accident. No one knows for sure, but there is evidence of intentional fermentation as early as the 4th century B.C. History shows that fermented beverages developed independently in cultures throughout the world. For example, North African, Assyrian, Chinese, and Mesoamerican cultures all reference beer in some form. What remain key questions are when did accidental fermentation transition to planned fermentation and, from this, when did actual brewing of beer begin? The development of purposeful brewing seems to have been a gradual process that occurred over centuries rather than at a distinct point in time.

Early pictographs and hieroglyphs suggest that fermented beverages were treated as sacred, both as gifts from above and as religious offerings to gods and goddesses. Sumerian records from the 3rd to 2nd millennia B.C. show that wine and beer were also used for medicinal purposes. The ancient Egyptian medical papers, the Ebers Papyrus, dating from about 1550 B.C., contain hundreds of prescriptions calling, in part, for beer.

For most of recorded history, brewing was considered women's work in the operation of the home or community kitchen. European men finally joined in these brewing responsibilities during the first half of the Middle Ages, as brewing shifted from a home activity to centralized production in monasteries and convents. During this time, beer was not only a hospitable offering for thirsty travelers, but also a sustaining beverage during religious fasts. By 1200 A.D., brewing was a commercial enterprise in Germany, Austria, and England.

The business of brewing became increasingly more important and more lucrative for community governments over the next few hundred years. The Reinheitsgebot, the German purity law of 1516, ordered that the ingredients of beer be restricted to water, barley, and hops. This was an effort to mandate quality. Yeast was added to the law after Louis Pasteur published his famous work with the microorganisms in the 1800s. While the law was repealed in 1987 by the Court of Justice as the European Commission worked to open the market, the law was so important to quality and tradition that many German breweries continue to follow the edict today.

While wine, various distilled spirits, and beer continued to develop in quality and variety throughout the world, beer's strongest foothold remained in Europe. In fact, present-day Belgium, the Czech Republic, Ireland, Germany, and the United Kingdom are the historical homes of most modern beer styles, both ales and lagers. The first record of New World brewing was in the colony of Virginia in 1587. Brewing in North America was well established from the 1600s and its history documented, including the importance of beer on the Mayflower, and throughout the founding and subsequent history of the United States.

Changes to the U.S. Beer Industry

Until the 1850s, most beer in America was locally brewed and served in taverns, pubs, and inns. Frequently, locals could also fill a large vessel at the tavern or pub and take the beer back home.

The advent of railroads and refrigeration made it possible for the more successful local operators, especially those in larger cities, to expand their markets. An increase in mass-marketing and the merging of smaller breweries into giant, corporate enterprises in the late 1800s and into the early 20th century resulted in mass production of a limited number of beer styles in America. World War I, followed by America's experiment with Prohibition, from 1919 until 1933, wiped out most of the remaining smaller breweries and had a devastating impact even on the successful brewing corporations of that time, like Anheuser-Busch, Miller, Coors, and Pabst.

When brewing returned in the mid-1930s, the styles and brands available were more limited in the United States than ever before. Almost every beer produced was a golden lager, with comparatively minor differences in taste. Each brand developed its customer loyalty based more on marketing and advertising than on flavor. The variety of beer styles remained limited until soldiers returning from Europe after World War II brought back new beer-tasting experiences, and the expansion of leisure and business travel in the 1960s and 1970s brought about greater consumer demand for more flavor and styles in American beer. By the late 1970s and into the 1980s, small, community breweries, called microbreweries, and restaurants brewing their own beer, called brewpubs, were opening at a brisk pace to meet this mounting demand for variety.

Because of the escalation of these artisan, or craft, breweries in the last third of the 20th century, the 21st century began with more ale and lager styles available in the United States than at any other time in history. Craft brewing and its segment of the total brewing market continues to grow today, even though the quantity is small when compared with the volume produced by the large, global conglomerate breweries. There is no better time in history to explore—in your glass—the historical underpinnings of the world's great beers and the amazing growth of traditional as well as new, imaginative ale and lager styles.

MAKING BEER

Like winemaking, brewing begins in the field. Barley and other grains, such as wheat, oats, and rye, are sowed and harvested in different countries around the world, and hops are collected from the flowers of tall, climbing vines. Grains have many uses, whether for baking, brewing, feeding livestock, or making fuel. Hops, on the other hand, are grown primarily for use in the brewing of beer. Hops are made available to brewers as fresh, whole hops, refrigerated leaf hops, or as an oil extract.

Most of the grain that the brewer will use needs to be malted to begin to convert the grain's starch to sugars for brewing. The brewing process begins as the brewer accepts the malted grains when they arrive at the brewery.

The Primary Ingredients

Brewing contemporary beer involves a combination of four foundational ingredients: malted grain, water, yeast, and hops. Beer can be brewed using only the first three, though without hops, beer tends to be too sweet on the palate of most consumers. Depending on the desired style, a brewer might also add other ingredients in the brewing process, like spices, herbs, honey, oatmeal, coffee, fruit, or even wood chips, to create a unique offering to consumers.

Grains

Examples of types of barley used in beer production, clockwise from top: two types of crystal malt, two dark roasted malts, unmalted barley, and two light roasted malts
SUSAN STOLTENBERG

Grains serve a number of purposes in brewing, but the primary role is to provide sugar, a nutrient to yeast, which allows fermentation to begin. Barley, with its high starch content, is the preferred grain of brewers. While wheat is the base fermentation grain in some beer styles, even wheat-based beer usually contains barley as well. Rye, corn, rice, or other grains—called adjunct grains—may be used in the brewing process for various contributions to the flavor and texture on the palate, but barley remains the key source of fermentable sugar.

Brewers consider two types of barley for making beer: two-row and six-row. The row count indicates the number of rows of kernels that are on each ear of the plant. Two-row barley costs more per pound than six-row, but most brewers prefer two-row because it yields more fermentable sugar. Two-row barley has a better ratio of starch to husk than six-row barley, and, ultimately, starch is what the brewer wants as the potential sugar source for fermentation.

In addition, barley kernels have a husk that covers and protects the inside of the grain, called the germ, during the malting process. The husks also form a natural filter when liquid is separated from solids later in the brewing process.

Malting and Roasting

Prior to its arrival at the brewery, a maltster—a person who malts grain—steeps the raw grain in warm water, allowing germination to begin. The final step is to kiln-dry the grain, which halts germination. This process begins to convert the naturally occurring insoluble starch into the fermentable sugar, including maltose, and other soluble substances. The finished product is malted barley, or simply malt.

The kilning process involves applying controlled heat. For the bulk of barley malting, the maltster stops the process while the malt is still relatively pale in color. This malt will be the base for fermentation and will yield a golden-colored beer. However, a maltster can also roast unmalted or malted grains to colors ranging from light brown to black.

Brewers use roasted barley and roasted malt in some beer styles to vary the color of the finished product. The addition of roasted grains alters the aroma and flavor of the beer. The darker the roast, the greater its potential for producing dark beers. Likewise, the darker the roast, the more prominent the aroma and flavor of the roasted grain becomes. Some of the darkest of the roasted grains can even contribute bitterness to the taste.

Finally, a maltster can also re-steep malted barley and heat it a second time, at higher temperatures, to crystallize the soluble sugars. This process leads to **CRYSTAL MALT** or caramel (caramelized) malt. These crystal malts are also available in various colors. These crystallized sugars are not fermentable, so when a brewer uses crystal malt, a caramel-like sweetness remains in the beer. The darker crystal malts are used to create a beer of a copper color.

Water

Beer consists primarily of water. The source of the water can contribute to the final flavor of the beer. For example, mountain spring water usually has not had a chance to run downhill through streambeds where it picks up mineral content, and therefore tends to taste purer. Well water and groundwater can pick up minerals, which can affect the taste of the finished beer, for better or worse. Water can also absorb flavors like rust from galvanized pipes or chlorine from water treatment.

Brewers prefer water that is neutral in flavor and unchlorinated. When the local water does not meet the needs of a brewer, he or she can use water from another source or filter the water. Many commercial brewers use a process called reverse osmosis. They run local water through a filtering system that removes undesirable substances and, in fact, all mineral content.

A brewer can then add back desirable elements to filtered water. You may prefer one bottled water over another because of the water's taste. Look at the labels. Is your favorite water pure or does it have some minerals added to affect the taste? A brewer can likewise do the same, based on the beer styles being brewed. Ultimately, the goal for brewers is to use water that does not influence their beer in a negative way. Water should make a positive contribution, or, at a minimum, a neutral contribution to the flavor of beer.

Yeast

Yeast is the third ingredient required to make beer. Yeast feeds on the sugar in grain, generating alcohol and carbon dioxide (CO_2). This is the process of fermentation. The formula for fermentation is:

$$\text{Sugar} + \text{yeast} = \text{alcohol} + CO_2$$

Two basic yeast classes are used to make most beer: top-fermenting yeast and bottom-fermenting yeast. Within these two classes, yeast is further divided into strains. Brewers can choose strains within either class depending on the style of beer being brewed. For example, if a brewer is brewing a beer made with a combination of malted wheat and malted barley, he or she will select a strain of yeast that works best with wheat malts. A third class of wild yeast strains are still used for fermenting some styles of beer.

Top-fermenting yeasts are direct descendants of wild yeast and are the yeast strains that brewers have used for centuries. The classification acquired its descriptive name because the yeast collects at the top of the fermenting liquid. Top-fermenting yeast varieties ferment at warm temperatures, generally between 59°F/15°C and 77°F/25°C. Top-fermenting yeast will not ferment at temperatures below 55°F/13°C.

Bottom-fermenting yeasts, special strains isolated in the 1800s because of their ability to do their work at colder temperatures, begin to sink to the bottom of the liquid as the fermentation process approaches its end. These yeast strains work best between 41°F/5°C and 50°F/10°C, but can ferment at temperatures as low as 34°F/1°C. These yeast strains can also function in the top-fermenting temperature range, but are not as efficient as top-fermenting yeast at the higher temperatures.

Hops

Hops grow on a long, climbing vine. The flower cluster, which looks like small green pinecones, is the part of the plant used in the brewing process. Bitter resins and aromatic oils in the flowers contribute a number of properties to the beer, the most prominent being its aromas and flavors. In addition, hops have a preservative characteristic that helps inhibit spoilage. Finally, a component of hops aids in head retention, meaning that well-crafted beer served in a clean glass will retain its foamy head for some time after pouring.

Hops are available for use in three forms. Although a few brewers use fresh hops for special styles, more frequently brewers use dried, whole leaf hops, which contain intact hop cones. Brewers can also purchase dry, pelletized hops (dried hops powdered and then compressed into small pellets) and liquid hop extract.

Hops are divided into two broad categories: noble hops (or aromatic hops) and high-alpha hops (or bittering hops). Noble hops contribute floral smells and add significant flavor to beer. High-alpha hops contain higher levels of alpha acid and have about twice the bittering potential of noble hops. Whichever type is used, hops contribute bitterness to beer.

Fresh Cascade hops add bitterness to beer.

TOM ZASADZINSKI/CAL POLY POMONA

Beer flavor runs along a continuum from sweet to bitter, with most styles falling somewhere near the center. Without hops, beers would be rather sweet beverages. By using an ingredient that contributes bitterness to the recipe, the brewer makes a beverage that has a more balanced flavor. Some brewers purposely design their beers to be sweeter by using a smaller amount of hops or more bitter by adding extra hops.

The Brewing Process

Brewers can brew at a large commercial operation, like Anheuser-Busch, at smaller operations (microbreweries or brewpubs), or in the home kitchen. While the scale of the equipment is different, and a step or two may be combined in the smaller operations, the process is essentially the same.

Milling

The first step in the brewing process is moving the required amount of malted barley from the grain silo or grain storage area to a milling device, called a roller mill. The roller mill gently breaks the husk of the grain kernels, exposing the meat of the grain, which will allow the soluble sugars inside to come into direct contact with water in the next step. In addition to malted barley, the brewer adds specialty malts, like roasted malts and crystal malts, directly to the mill, depending on the style he or she is brewing. The resultant milled mixture is known as **GRIST**.

Mashing

The brewer transports the grist into a grist hopper, a temporary storage unit that sits above a mash tun, an insulated kettle in which the brewer mixes the grist with hot water, in a process called mashing. The mixture of water and grist in the mash tun is called **MASH** and looks like a giant bowl of hot oatmeal or porridge. The water temperature ranges from a minimum of 113°F/45°C to a maximum of 168°F/76°C, depending on the beer style.

Over the next few hours, the mashing process extracts the malt from the grain and into the water, while finishing the process of converting the grain's starches into fermentable sugars. The brewer controls the temperature to maximize the conversion of starch to sugar, while minimizing the leaching of tan-

Equipment in a new brewhouse: left, brew kettle, and right, mash/lauter tun with grist hopper on top
OWEN WILLIAMS

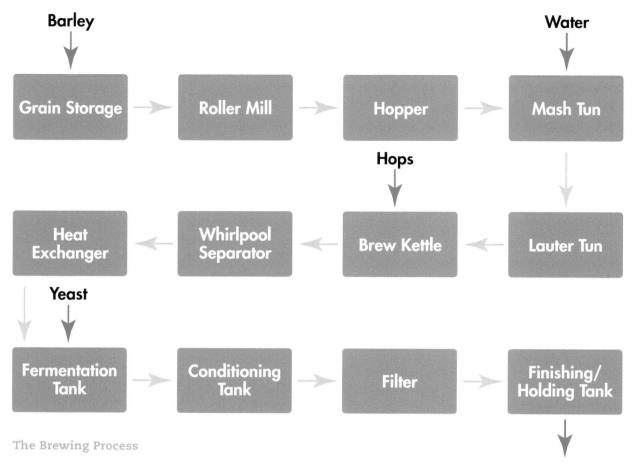

Barley → **Grain Storage** → **Roller Mill** → **Hopper** → **Mash Tun** ← **Water**

Hops ↓

Heat Exchanger ← **Whirlpool Separator** ← **Brew Kettle** ← **Lauter Tun**

Yeast ↓

Fermentation Tank → **Conditioning Tank** → **Filter** → **Finishing/ Holding Tank**

The Brewing Process

nins from the grain husks into the solution, which can lead to off-flavors in the finished product. The finished product is a hot, sweet liquid called **WORT** that is rich in fermentable sugar.

If the brewer is using roasted malts, roasted grains, or crystal malts in the recipe, they are also milled and added to the mash tun. The resulting hot wort contains the aromas, colors, and flavors of the roasted malt and roasted grain. If crystal malt was added, then the hot wort will have taken on the aromas and color of the malt as well as the crystallized sugar that will eventually contribute sweetness to the finished product.

Boiling

Once the mashing process is complete, the brewer must separate the wort from the spent solids, a process called lautering. The brewer draws off the wort from the bottom of a lauter tun, a device similar to a kitchen colander, or from the bottom of a combination mash-lauter tun. The holes allow the liquid to drain off the mash solids.

The hot, sweet wort, a combination of water and grain sugars, is pumped into a brew kettle, where it is boiled. Boiling sterilizes the liquid, helps produce appealing flavors and aromas, and drives off some potentially off-flavors. The boiling time generally lasts approximately one hour.

Adding Hops

The brewer adds the high-alpha, or bittering, hops near the beginning of the boil, and noble, or aromatic, hops closer to the end of the boil. The bittering characteristics of the high-alpha hops affects where the finished product falls on the continuum of sweet to bitter. The boiling of the wort and hops in the brew kettle draws the resin from the hops into the wort, adding bitterness to the beer. However, because hop oil is volatile, much of the potential aroma and flavor from the initial addition of high-alpha hops dissipates into the surrounding environment and does not end up in the beer. This is where the addition of the noble hops comes into play.

Near the end of the boil, the brewer can add noble hops, or even more high-alpha hops, to finish the boiling process. The finishing hops contribute their floral character to the nose and taste of the finished product. With this addition of hops in the last few minutes of the boil, there is enough heat and time to draw the aromatic oil into the wort. However, because the mixture is cooled soon after, the oils remain in solution rather than dissipating into the atmosphere as did the potential aromas and flavors of the hops added at the start of the boil.

Clearing and Cooling

At the end of the boil, the brewer shuts off the heat and a brief settling occurs. The next steps prepare the hot wort for fermentation. The hot wort must be cleared of the spent hops and it must be cooled to an appropriate fermentation temperature.

Depending on the brewery, the hot wort might be centrifuged in the brew kettle or in a whirlpool separator to help clarify the liquid and remove unwanted proteins from the solution. The hot wort may also be passed through a filtering device called a hopback to remove the spent hops.

The hot wort is then pumped through a heat exchanger, which rapidly cools the hot, sugary fluid to a temperature suitable for fermentation. Hot wort that will be top-fermented is cooled to between 59°F/15°C and 77°F/25°C. Wort that will be bottom-fermented is cooled even further, to temperatures between 41°F/5°C and 50°F/10°C.

Fermenting and Conditioning

When the wort is pumped into the fermentation tank, the brewer immediately adds, or pitches, the fourth foundational ingredient, yeast. The recipe and style of beer dictate the type of yeast the brewer will use. For example, if the brewer is making a lager-style beer, he or she will use a bottom-fermenting yeast strain. If he or she is making an ale-style beer, a top-fermenting yeast strain is selected.

This first exposure of the sweet wort to yeast is the beginning of primary fermentation. As the yeast consumes the sugar, it gives off alcohol and CO_2, gradually turning the sweet wort into beer. The alcohol remains in the liquid while the CO_2 generated during primary fermentation is allowed to dissipate into the atmosphere.

Fermentation for top-fermenting yeasts typically lasts three days but may take up to two weeks. Bottom-fermenting yeasts do their work over at least two weeks, with most taking from four to six weeks, and some styles even longer. The longer time for bottom-fermenting beers relates, in part, to the lower fermentation temperatures. Colder temperatures slow the process of fermentation. While the bottom-fermentation method is slower, a benefit is that the extended time also allows a settling process to occur, which produces a clearer and crisper tasting beer than a top-fermenting style.

In addition to top fermentation and bottom fermentation, there is one other category, called spontaneous fermentation. Practiced primarily in a small area near Brussels, Belgium, this method requires the use of open fermentation vessels. This exposes the sweet wort to wild yeast and bacteria in the atmosphere. Once wild yeast and bacteria find their way into the wort, fermentation begins spontaneously. This method is used to produce a class of beer styles referred to collectively as lambic beers.

Secondary Fermentation and Conditioning

Following primary fermentation, the next step can be secondary fermentation or it can be conditioning. If secondary fermentation is used, the beer is pumped out of the primary fermentation tank and off the bed of dead yeast and other insoluble matter that has collected at the bottom of the tank. This prevents unpleasant flavors from developing in the beer. After the live yeast finish converting the sugars, they will settle, leaving a clearer beer.

Brewers may condition the beer in one of three ways. Some small-scale brewers condition the beer in the fermentation tank, while others have separate temperature-controlled conditioning tanks for this process as do many larger operations. For some beer styles, the beer may be cask-conditioned in the serving kegs. The conditioning process varies depending on the style of beer.

In essence, conditioning allows various flavor components to meld. This is similar to what happens after making a sauce in your kitchen. While some sauces are meant to be used immediately, others taste far better the next day. If you make red pasta sauce, right when you finish you can taste the individual bits of onion, tomato, and bell pepper. However, if you refrigerate the sauce overnight, the flavors have blended together to form a wonderfully rich mixture of new flavors. Almost all beers benefit from at least some conditioning time.

For bottom-fermenting beers, brewers use a process known as lagering. Lagering is cold-conditioning at temperatures approaching freezing and the process can last from two to eight weeks, and for particular styles almost a year. This process causes proteins remaining in solution to coagulate and settle to the bottom with the remaining yeast. This settling further clears the beer and also produces a smoother-tasting beer.

For some beer styles, like the more bitter ales, the conditioning stage may include the addition of more hops. The process is called dry-hopping, which involves the addition of uncooked hops directly into the conditioning tank. The process leads to genuine fresh hop aroma and flavor in the finished product.

Filtering and Finishing

While some brewers filter their beer prior to conditioning, most styles are filtered after this time of rest and flavor development. In the past, most filters contained diatomaceous earth as the filtering medium. However, present-day filters use an artificial filtering material. The filter removes yeast and other small particles, leaving a clear finished product. Brewers also produce unfiltered styles, like Hefeweizen, with the yeast still in the bottle or keg, resulting in a cloudy product with flavors and aromas unique to these styles.

Carbonating the Beer

Before the beer is packaged, it must be carbonated. Some bottled and keg beers contain the natural CO_2 produced during a second or third fermentation in the bottle or keg. This is usually achieved by using unfiltered beer and giving the yeast a bit more sugar, called priming sugar, and then sealing the container. This allows the CO2 generated during this final fermentation to remain in solution and, therefore, carbonate the beverage.

In a brewpub, the brewer may use the CO_2 generated during secondary fermentation or during cask-conditioning. This is considered the traditional method of carbonating beer. The brewpub brewer can also inject CO_2 or even nitrogen gas into the serving tank or individual kegs prior to pouring the beer on tap.

However, most commercial brewers force-carbonate their beers for a more consistent finished product. This is done by injecting CO_2 from a tank into a closed system under pressure, like a keg or a holding tank on a bottling or canning line. CO_2 can either be purchased by the brewer from a company selling industrial gases or it can be from a system that captures and purifies the CO_2 from primary fermentation.

Packaging the Beer

The final step in the brewing process is to package the beer for sale and consumption. Commercial brewers package their beers in various sizes of kegs or barrels, in bottles, or in aluminum cans. Smaller commercial breweries might also make cask-conditioned beers available in kegs for bar service. A brewpub can even serve from the keg used for conditioning the beer. In some brewpubs and other restaurants and bars, customers can also purchase beer to go.

BEER STYLES

Beyond the word beer, you may hear a beer drinker comment that he or she prefers ales to lagers, likes hoppy beers more than malty styles, or loves Belgian beers. Much like wine drinkers, beer drinkers use specific terms to describe their favorite beer style.

Beer is divided into two general categories: top-fermenting beer, most commonly known as ale, and bottom-fermenting beer, most commonly known as lager. Numerous styles, with varying colors, aromas, and flavors, can be found within these two broad categories. The yeast type and the related fermentation temperatures create the primary differences between ales and lagers. Ale and lager styles can all be plotted along three continua:

– Light in body ← → Full in body
– Sweet in flavor ← → Bitter in flavor
– Golden in color ← → Black in color

There are more styles of top-fermenting beers than bottom-fermenting because ales, wheat beers, porters, and stouts have a longer history. While brewers lagered beer for hundreds of years to allow some settling to occur, dedicated lager styles did not come about until the 1800s. On production volume alone, the majority of beer brewed worldwide is light lager beers because of their crisp, clean, and refreshing character.

Ales

Ale is a general term describing the world's oldest beers. Historically, the earliest beers were unfiltered and probably dark and cloudy in appearance. Present-day ales are descendants of the ancient beers of the world. The warmer temperature of top fermentation encourages the formation of esters, which are the compounds that give rise to the classic fruity characteristics that contribute to ale's fruity aromas and flavors. While there are dozens of ale styles, most fall into one of three broad categories.

Dark Ales

The various types of porters and stouts are the two largest subcategories of the dark ales, ranging in color from medium-brown to pitch black. These styles have a long history in England, Ireland, and the Americas. The dry bite of malts roasted until they are dark, chocolaty brown, or even black gives these beers a rich, sometimes bitter, character.

While less popular now than they once were, these dark ales still command fierce loyalty among their fans. One of the most recognized brands in the world, Guinness, the popular Irish dry stout, falls into this category. In addition to brown porters, robust porters, and dry stouts, other styles include sweet stout, oatmeal stout, imperial stout, and chocolate stout.

Light Ales

The lighter ales are produced in a wide variety of colors, palates, and strengths, with colors ranging from golden to amber to medium brown. This category contains some styles that are sweet in flavor, or malty, like McEwan's Scotch Ale. Others are more balanced in flavor, like Newcastle Brown Ale, with its additional hint of nuttiness from the roasted grains used to color and flavor this popular brand. At the other extreme are highly hopped and dry-hopped beers, like Stone India Pale Ale from California.

Other styles include cream ales, mild ales, pale ales, bitters, old ales, Irish red ales, and barley wines. This category also includes the golden ales, Abbey ales, and famous Trappist ales from Belgium as well as the Altbiers of German origin.

Wheat Beers

Beers made with a significant portion of wheat, both malted and unmalted, make up the final category of top-fermenting ales. The styles in this category include Belgian witbiers, like Hoegaarden, and the spontaneous-fermentation Belgian lambic beers. The German styles include Southern German Weissbier, or Hefeweizen, with their slightly sour flavor and aromas of cloves or bananas that come from phenols and esters, chemical compounds produced by the way these yeast cultures work with malted wheat. Many of these styles are unfiltered and, therefore, have an interesting cloudy appearance. Other styles include Dunkelweizen (dark wheat), Weizenbock (a stronger Weissbier), and the Northern German wheat-beer style called Berliner Weisse, with its characteristic sourness coming from a unique, bacterial, lactic-acid fermentation.

Lagers

Bottom-fermenting beers are commonly called lagers because of the fermentation and conditioning, or storage method (*lager* in German means "to store") required when brewing at colder temperatures. Because these beers require longer periods of time to finish fermentation, the beers are moved into temperature-controlled conditioning tanks for secondary fermentation and conditioning. The colder bottom-fermentation method, and the longer time for spent yeast and insoluble components to settle, results in products that are dry, clean, and round, but rarely with any of the fruity characteristics of ales. Lager brewing, with a shorter history than ale brewing, has developed fewer styles.

Pale Lagers

Considered the most popular of beer styles, the vast majority of beer brewed worldwide is pale, or light, lager. Pale lager is the classic golden-colored, clear, refreshing beverage topped with a head of white foam that most people think of when they hear the word *beer*. The most common brands in North America, like LaBatt, Budweiser, Miller, Coors, and Corona, are all light lagers. The same can be said for the most prominent brands from Germany, Holland, Brazil, Australia, Japan, South Korea, China, and other countries with significant imports and exports of beer.

However, other styles of pale lagers have similar golden to light amber color. Pilsner (also spelled Pilsener), a more heavily hopped beer than light lager, is a style that originated in the town of Plzeň in the present-day Czech Republic. The brand Pilsner Urquell is the most famous beer of this style and is the original Pilsner-style beer. The category also includes Dortmunder Export, Munich Helles, and Helles Bock, all of German origin.

Dark Lagers

Dark lagers are primarily of German origin, although there are also some classic examples from Austria. These beers range from amber to black in color. Many are double the alcoholic strength of light lagers, and they frequently have flavors that range from sweet or malt-focused, to balanced, but rarely have any hop bitterness. In fact, any perceived bitterness in flavor is most likely from the use of dark-roasted grains in the recipe.

This category includes black lagers, Märzen/Oktoberfest, Munich Dark, Dunkel Bock (dark bock), Doppelbock, Eisbock, and Rauchbier. Märzen-style beers are the most popular in the category and were historically associated with the Oktoberfest in Munich, although paler styles now prevail. However, they continue to be consumed at Oktoberfest celebrations that have sprung up in other parts of the world.

EVALUATING BEER

If you choose to develop your knowledge of beer, you can practice evaluation skills similar to those used by wine drinkers. You can look at the beverage, noting color, clarity, and carbonation. You can smell the beer, evaluating its aromas. You can taste the beer—the start, the finish, and the mouthfeel. Whether tasting the beer alone or with food, you will form an overall impression of every beer and beer style that you try.

Color

Assess a beer's color, clarity, and level of carbonation. Colors range from golden to amber to brown and black. Psychologically, advertising has created in our mind's eye the image of a golden-colored beer in a tall, frosty glass with white, foamy bubbles running down the side as the picture of ultimate refreshment. However, advertising—and color—may not tell the whole story.

While it may be true that a cold, light lager is what you want on a hot day, a similar-looking golden Belgian ale may focus on the malt, have a thicker mouthfeel (more full-bodied), and have double the alcoholic strength of a light lager. Likewise, black lagers look incredibly rich and thick based on their color, but can be as clean and refreshing as a golden-colored lager. For color, simply appreciate it for what it is. As you learn more about beer styles, you will know more about what to expect in terms of color.

Assess clarity, too. Some styles, like an unfiltered German Hefeweizen, should be cloudy, while fully filtered beers should be clear. An unfiltered beer still contains yeast, and yeast in solution will create a cloudy haze in the beer. The lighter the color, the easier it is to evaluate clarity. A Dunkelweizen may be cloudy, but the darker the color, the harder it will be to tell. While most beer styles should be clear, as you learn more about beer styles, you will know more about what to expect in terms of clarity.

Finally, look at the carbonation. Observe the bubbles, their stability, and the thickness and retention of the head, or the foam on top. If you open a bottle or can of beer, and you fail to hear the "hiss" that you expect, you might wonder what went wrong. Likewise, when a beer is poured into a glass for evaluation, you will expect to see bubbles rising to the top, creating a foamy white head.

Some beer styles are livelier than others, but a flaw exists when a beer is absolutely flat, revealing an absence of bubbles. Again, as you experiment with more styles, you may notice subtle changes in liveliness and even in the size of the bubbles.

Aromas, Flavors, and Mouthfeel

Smell your beer before your taste your beer. What can you pick up from the smell alone? Once you taste the beer, can you go back and smell scents you did not smell before? Aroma and flavor—the nose and the tongue—work together.

Familiar terms used to describe the smells and flavors in beer come from the use of roasted malt and include words like *toasted, roasted, nutty, chocolate, coffee, smoky,* and even *burnt.* The use of crystal malt can give beer the aroma and flavor of malty sweetness, toffee, caramel, or rich molasses. The color of a beer might give you hints of what you will smell and taste. For example, the darker a beer, the more likely your nose and tongue will discover aromas and flavors you will describe as chocolate, smoky, or roasted.

Depending on the hops that the brewer uses, the beer may have floral, flowery, spicy, or pungent scents. These aromas and flavors are more likely present when noble hops were added right at the end of the boil. The floral smells and tastes are even easier to identify when the brewer uses noble hops to dry-hop his or her beer in the conditioning process.

Balance

Balanced flavor is created by the contrast of the bitterness in the hops with the residual sweetness of the malted grains. Hops added at the start of the boil are usually high-alpha hops, or bittering hops, and affect the flavor profile of the beer far more than the aroma. This type of hops is used to help many beer styles taste more balanced on the palate. The high-alpha hop resins help bring the sweetness of the malted grain into balance.

Mouthfeel

Beers with a lower hop profile, or those with more malt sugar from the use of crystal malt, may taste sweeter to some. Beers with significant additions of hops can smell wonderfully floral but taste bitter, and even feel oily on the palate to some. The vast majority of styles come across on the palate as relatively balanced, but each of us has differing thresholds for what we call sweet and what we call bitter. Trust your own nose and tongue to tell you what you like.

Beer also has body and how it feels in the mouth is sometimes called mouthfeel. The use of roasted malt, crystal malt, and hop oils as well as alcoholic strength can all affect how a beer feels in your mouth. Some styles are light-bodied, while others are full-bodied. Beers can taste one way when they first hit your palate and another way in the finish as you swallow.

Think about how apple juice feels in your mouth. Now, think of something more full-bodied, like milk. How does the start, finish, and mouthfeel with each beverage differ? While you may have a general preference for mouthfeel, even your own preference can change depending on the time of year. A light-bodied

Coors Light probably seems refreshing after a day of yard work in the summer, whereas a full-bodied Anchor Brewing Christmas Ale might sound great after a day on the slopes of Lake Tahoe in December.

With color, aroma, flavor, and mouthfeel, it helps to pay attention and even take notes if you want to learn more. However, beer is still beer—a beverage for any day and any situation.

ENJOYING BEER

Beer is the easiest and least expensive alcoholic beverage to purchase. You can find it almost anywhere from your local supermarket to sporting venues to amusements parks to casual and fine restaurants. Typically, beer costs less than $10 for a six-pack and is frequently under $2, if your local retailer sells individual bottles.

One of the best places to experiment with different beer styles is in a restaurant with an interesting beer selection. Your beer-appreciating friends all have their favorite locations. Not every beer-friendly place is referred to as a brewpub, bar, tavern, or public house. Frequently, restaurants serving the foods of Belgium, Germany, England, Scotland, and Ireland offer beer styles with historic connections to the country. For example, if you want to try a Belgian witbier, you are more likely to find at least one or two alternatives in a restaurant that offers the cuisine of Belgium.

A brewpub is a combination restaurant and brewery, and it is another great place for a beer lesson or two. These operations sell their own beer on tap. Frequently, they train their staff so that they can help you better understand the styles and what foods they might taste best with. They usually have menus or handouts that explain each of their beers. An added bonus is that many of these operations offer tasting flights consisting of four to six beers in small-portion glasses. Frequently, for less than the cost of a couple of pints, you can experiment with a handful of styles, safely, at one sitting.

Drink beer from a glass for easier evaluation and for an even more enjoyable experience. A glass showcases the appearance of the beer; its potential for a nice, foamy head; and it makes color, clarity, and carbonation easier to assess. Also, pouring beer into a glass allows the beer to present its aromas more easily. Finally, when a beer is poured across the tongue from a glass, the mix of beer with air allows your nose and taste buds to work in concert with one another.

Fresh beer forms a head when poured.
SUSAN STOLTENBERG

Beer and Food Pairings

The discovery of a wonderful beer and food pairing, much like wine and food pairing, is a bit of science, art, psychology, and philosophy, a little trial and error, and a lot of your own opinion. The mouthfeel of ales and lagers spans a continuum from light body to full body, just as the flavor array ranges from sweeter to more bitter and the color spectrum extends from golden to black.

When you are pondering beer and food pairings, consider the foods first. Use the flavors, body, and texture of the foods as the guides for selecting the beers. Think about the dish, and all that goes into making it, and then think about the beer. Start by pairing lighter-bodied ales and lagers with foods that are lighter in flavor and more full-bodied ales and lagers with hearty fare. As you experiment, you will learn more about your likes and dislikes. Get more daring. You may end up with an amazingly tasty contrast, like the classic pairing of a dense, full-bodied porter or stout with the delicate, sweet, mildly briny taste of fresh, raw oysters.

With so many styles of ales and lagers and so many foods, there are obviously endless possibilities. As you begin your beer and food pairing journey, consider the 10 beer styles within five broad classifications and some fine food choices for each that are found in the Beer and Food Pairings list at right.

Use the combinations in the list as a catalyst for your own exploration. Ultimately, follow this basic rule: If you like the beer, and you like the food, and you like the two together, then you have a perfect pairing.

SUMMARY

In general, beer has always been more ordinary and, therefore, more comprehensible than wine. Like wine, it has played a role in religion and worship, medicine, as a social lubricant, as a source of nutrition, and as a way to enhance the enjoyment of a quality of life. Other than water, beer has been portrayed as *the* quencher of thirst. It remains the adult beverage with the widest appeal, and today a broad variety of ales and lagers are available around the globe.

Evaluating wine, especially while consuming it in public, is a time-honored tradition and rarely draws a stare from a casual observer. Unfortunately, evaluating beer in public can still cause your neighbor to gawk, unless you happen to be consuming in an establishment known for its beer selection. Like the growing number of beer styles, with time, this will change.

While it may not seem as natural as doing so with wine, engage all of your senses while enjoying your beer. Appreciate your ale or lager, with or without food and with recognition of your location and the season—are you at a baseball game eating a foot-long hot dog on a warm July evening or surrounded by friends in December in the lodge after a day on the ski slopes? Whether in a plastic cup or the appropriate glass, swirl, look, sniff, taste, swallow, and ponder. Make beer exploration another path on your journey of lifelong learning.

BEER AND FOOD PAIRINGS

Dark Ales

PORTER Roasted lamb or a peanut butter- or espresso-based dessert

IMPERIAL STOUT Fresh-sliced Parmesan or well-aged cheddar cheese

Lighter Ales

AMBER ALE Burgers or barbecues that use sweeter sauces

INDIA PALE ALE Piquant Tex-Mex, Thai food, or a spicy curry dish

Wheat Beers

BELGIAN WITBIER Steamed, buttery mussels or a citrus-based dessert

AMERICAN OR GERMAN HEFE-WEIZEN Sushi, herbed cheese, or Key lime pie

Pale Lagers

PILSNER Spicy buffalo wings, grilled chicken, or grilled bratwurst

ASIAN LAGER Smoked fish, sushi, or fried calamari with garlic aioli

Darker Lagers

VIENNA-STYLE LAGER Spicy Mexican food or tomato-based pizza

DOPPELBOCK A rich, roasted game dish or German chocolate cake

20

Distilled Spirits

The word *spirit* carries with it the allure of its origins in alchemy, an ancient practice that combines science, art, spiritualism, and mysticism. Alchemists, building on the philosophy of the ancient Greeks, recognized four elements that constituted matter: earth, air, water, and fire. A fifth element, ether, was thought to be the essence of life. Because alcohol is obtained from plant materials, they believed that by extracting the vapors from plants they were capturing the essence, or spirit, of life. So the term *spirit* became associated with distilled alcohol.

DISTILLED SPIRITS have long been hugely popular, not only for the wonderful drinking experience but also because of the relative ease of producing and handling them. Spirits are fairly simple to make, last indefinitely even after the bottle has been opened, tolerate variable temperatures, and are easy to transport.

WHAT IS A SPIRIT?

A spirit, or liquor, is an alcoholic beverage made by distilling fermented grains, fruits, vegetables, or other plant materials that concentrates and purifies the alcohol and captures the aromatic essence of those substances.

Although all spirits are created by distillation, not all spirits are the same. As with wine and beer, each country or region creates a type of distilled beverage that reflects its land and its people, based on whichever ingredients are most readily available. In Scotland whisky is made using peat and barley. In France, Spain, and Italy, local wine is distilled into brandy. In Holland barley, corn, and rye provide the base for gin, while in America corn is used to make bourbon. Scandinavians use potatoes; Mexicans, agave; and the people of various Caribbean islands, sugarcane. In addition, native yeasts, local water, and flavorings change the character of each beverage, resulting in numerous variations, while master distillers and blenders manipulate the ingredients to create distinctive spirits.

HISTORY OF SPIRITS

The earliest references to alcoholic beverages appear as long ago as the 4th century B.C. in China, Egypt, Mesopotamia, and the Near East. Many fermented beverages developed accidentally as fruits that were overly ripe, fermented naturally. However, these early drinks were not distilled beverages.

Distillation was first described by the Greeks, who used large covered pots to distill perfume, flavorings, and medicines. In the 8th and 9th centuries A.D., Muslim alchemists became the first to distill alcohol in covered pots while conducting experiments to extract an "elixir of life" from a variety of substances. They were responsible for developing the pot still, a type of still that is sometimes used today.

Muslims introduced distillation to Spain. From there it spread throughout Europe where it gained a foothold during the Middle Ages. Early distillers thought that alcohol contained the essence of life, and expressed it by naming spirits "the water of life" in their many languages:

- Latin: Aqua vitae
- Scandinavian: Aquavit
- French: Eau-de-vie
- Scottish Gaelic: Uisge beatha
- Irish Gaelic: Usquebaugh

Spirits for Health and Pleasure

By the 12th century, spirits were being made in significant quantities. At this point, they were valued for their medicinal qualities and were believed to cure a variety of ills. They were often prescribed to stimulate circulation or to help with digestive problems, a remedy that is still used today. Imbibing spirits in social situations was limited to the rich.

By the13th century, alcohol distillation was commercialized and spirits were more accessible to greater numbers of people. In the 15th century, spirits became a drink of pleasure as well as a tonic. Each country developed its special form of spirit based on its local agricultural crops. In addition, monks were producing all kinds of liqueurs, such as Chartreuse and Bénédictine. Spirits were immensely popular, and because they did not spoil readily and could be stored easily in kegs, they were shipped around Europe and to and from the New World.

By the 18th and 19th centuries, spirits were cheap and readily available thanks to the invention of the column still, which made mass production possible. With greater accessibility to spirits, addiction became a serious problem in urban environments of Europe, and movements formed to control the social problems related to excessive drinking. In the United States during this time, spirits were plentiful, as farmers could make more money from corn by producing whiskey than by selling it as a food crop. By the 19th century, mixing spirits with other beverages to make cocktails gained a following. The sophisticated enjoyed serving cocktails before dinner, and the Martini, Manhattan, and Gin and Tonic first appeared on the scene.

Spirits in the 20th Century

During the 20th century, the fortunes of spirits rose and fell with various trends. In 1919 Congress bowed to pressure from Temperance Movement forces and enacted the Volstead Act, which outlawed the manufacture, transportation, and sale of alcoholic beverages (see Chapter 2, Prohibition, page 12). Prohibition was a dismal failure and the law was repealed in 1933. The 1930s heralded an era of glamorous nightclubs and sophisticated beverages as epitomized by Hollywood, but World War II put an end to this lifestyle.

At the end of the war, people renewed their interest in distilled spirits. In the United States during the 1950s, vodka began to take center stage away from whiskeys because it was easy to mix with fruit juices, carbonated beverages, and flavorings to create sweet drinks. Liqueurs were also used to make mixed drinks, such as the Brandy Alexander and the Grasshopper.

In the 1960s and 1970s, hard liquor fell out of favor while beer and wine consumption increased, but the 1980s saw a renewed interest in dining and drinking along with a resurgence in vodka sales. By the 1990s, young people were rediscovering the pleasures of classic cocktails. There was renewed interest in premium-grade single-batch spirits, led by Scotch and then bourbon, and gin took the place of vodka as the preferred drink of the sophisticated.

Perhaps the biggest change, however, was the consolidation of the distilled-spirits industry. Large distillers bought up small producers by the hundreds. For example, in Ireland only three large companies remain from the thousands of small businesses that once produced whiskey. Although there are now only a few distilleries, each produces many brands. Each company jealously guards its labels to maintain the distinctive characteristics of the spirits that were made by the original small distilleries.

Moonshine

Moonshine, an illegal homemade alcoholic beverage, has been with us since people first discovered that they could produce distilled spirits at home inexpensively and avoid paying taxes on their liquor. They often distilled alcohol outside by the light of the moon, hence the name moonshine.

Nowhere did making moonshine take hold as it did in the early American frontier, where corn was cheap and easy to grow and early settlers balked at government restraints. During the 1700s in the backwoods, homemade whiskey was used to trade for food, clothing, and other goods.

Illegal production of moonshine mushroomed during Prohibition when all legal commercial alcohol production ceased in the United States. Gin was especially easy to make because it does not require aging. People used whatever alcohol was available, including deadly wood alcohol (methanol). They added juniper berries and spices, and soaked the brew for a few days, often in their bathtubs, and the term *bathtub gin* was born.

While it may be easy to produce a spirit, moonshine comes at a price. Without using a properly designed still, it is difficult to create an alcoholic beverage that does not have impurities or an alcohol content that is too high. Any form of alcohol is toxic at high levels or if it is improperly distilled. Because moonshine is often poorly made, it may be crude and harsh, and drinking it can cause sickness, blindness, or even death.

MAKING DISTILLED SPIRITS

Distillation is a process that uses evaporation and condensation to extract and concentrate the alcohol in a fermented liquid. Although it seems simple, it is a complex procedure because the distiller must understand the nuances of the base ingredients, yeast, flavorings, and other substances in the fermented liquid, both those wanted and unwanted in the finished product, in order to create a beverage worth drinking.

The following factors contribute to the unique character of each recipe:

- the base—grain, fruit, or vegetable
- yeast
- water source
- distillation techniques
- flavoring and coloring
- blending
- aging

The raw materials that form the base of the fermented liquid add the primary flavor components to a spirit. Distillers usually use grains or fruits, but sugarcane, agave, potatoes, sugar beets, other roots, and other plant materials are also made into distilled beverages. The yeast strain that the distiller selects for fermenting the base affects the flavors and helps each distiller create a uniquely flavored beverage, and additional herbs, spices, and other botanicals may be used to further enhance the aromas and tastes of the spirit.

Agave is only one of the many types of plant materials used to make distilled spirits.

Water can also give a spirit much of its individuality because it imparts the flavor of its environment to the beverage. Does it come from a spring located in a granite outcropping or does it flow over peat? Is it natural water or distilled to remove minerals and other substances? All of these factors lead to very different aromas and flavors in the finished liquor.

Making distilled spirits requires three basic steps:

1 Fermentation
2 Distillation
3 Finishing

Fermentation

All distilled spirits originate from fermented liquid, which is obtained by adding yeast and water to grains or fruits. The yeast converts the sugar in the grains or fruits to alcohol, yielding a fermented mixture with an alcohol content of 5 to 12 percent. This serves as the base for the distilled spirit.

Before fermentation can begin, fruits are crushed to release their sugar, making it available to the yeast. Grains must first be milled to expose the starch

Pot stills, based on ancient design, are used to produce some kinds of distilled spirits.

inside their kernels. They are then mashed, which involves mixing the grains with water and cooking them, which enables enzymes to convert the starch to sugar. This mash is now ready for fermentation. Some spirits are made from malted grains. Malting involves soaking grains in water until they germinate, which produces some of the enzymes needed to break down starch into sugar. The grain is kiln-dried to stop plant growth. At this point, it can be fermented.

Distillation

Distillation concentrates the alcohol, flavors, and aromas in a spirit, and may occur one or more times. Distillation is based on the fact that water and alcohol evaporate at different temperatures, and can be separated by heating. The boiling point of ethyl alcohol (ethanol), the type we can drink, is 173°F/78°C; the boiling point of water is 212°F/100°C. When fermented liquid is heated to 173°F/78°C in a still, the alcohol, along with other compounds, vaporizes and separates from the liquid water. The alcoholic vapors and aromatics are collected, and when cooled they condense to a liquid state, called **DISTILLATE**. Because some of the water has been removed, the distillate has a higher concentration of alcohol and fewer flavor components than the original liquid. A fermented liquid with 5 to 10 percent alcohol content can be concentrated into a 50 to 70 percent alcohol distillate. Each time the alcohol passes through the still it becomes more concentrated and has fewer impurities, a process known as rectification.

Heads, Hearts, and Tails

Fermentation produces substances that are unpleasant to taste and smell and, in some cases, may even be toxic. In addition, components of yeast, botanical substances, and even distillation create unwanted by-products. In the fermented liquid, these by-products form three levels of concentrates that evaporate at different temperatures during distillation: heads, hearts (main run), and tails.

Heads contain toxic forms of alcohol, such as methanol (wood alcohol) and acetone, that must be extracted from the distillate. They evaporate first because they have a lower boiling point than ethanol and water, and are captured as liquid and removed.

Hearts are the portion of the distillate composed of ethanol, flavor components, and water. When the temperature reaches 173°F/78°C, the desirable alcohol condenses and is reserved for the alcoholic beverage.

Tails contain both desirable and undesirable elements, including **FUSEL OILS** and **CONGENERS**, which condense at higher temperatures than the heads and hearts. Fusel oils are heavy alcohols that lend oiliness, spiciness, and heat to the spirit. However, they can be harsh and unpleasant tasting can cause stomach upset, hence the term *rotgut;* and may be deadly in sufficient quantities. Fusel oils add character to some spirits, such as whiskey, but are undesirable in clear spirits, such as vodka. Congeners are chemical compounds that give a beverage its characteristic taste, aroma, and body. Some distilled spirits, gin for example, contain many different added congeners, but they are removed from other spirits, such as vodka.

Because heads, hearts, and tails evaporate at different temperatures, skilled distillers control these elements to create unique flavor profiles in their beverages. During distillation, they remove impure and unwanted elements, while capturing desirable fusel oils and congeners and adding them to the main run.

The Still

All spirits are processed in a still to concentrate the alcohol, to obtain flavors and aromas, and to remove unwanted elements. Two types of stills are used: the **POT STILL (BATCH STILL)**, and the **COLUMN STILL (CONTINUOUS STILL)**.

Pot Still

The pot still, also called a batch still, has not changed significantly since its development in the 8th century. It consists of a large, enclosed pot, usually copper, with a broad rounded bottom. Fermented liquid is poured directly into the pot and is heated to vaporize the alcohol. The vapors collect in a narrow, tapered neck. The neck connects the pot to the worm, a coiled metal tube, where the vapors are cooled and condense back into a liquid. Pot distillation leaves some water and flavor compounds in the vapor along with the alcohol, so it is customary when using a pot still to double-or triple-distill.

Pot Distillation

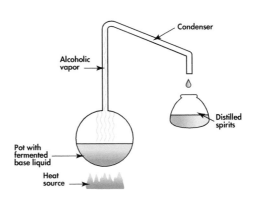

Pot distillation is a slow process because it takes time for the liquid to reach the temperatures needed to extract the heads, hearts, and tails. However, it produces the finest liquors because the distiller has a great deal of control over its flavor components. Pot-distilled spirits are produced in small batches, so it is possible to achieve complex mixtures of flavors and aromas. Although pot distillation produces excellent liquors, it is an expensive process because it is time consuming and liquors cannot be mass-produced in these stills.

Column Still

The column still, also known as a continuous still, consists of at least two tall, narrow metal columns. Steam enters the first column from the bottom and fermented liquid enters slowly from the top. The rising steam removes the desirable alcohol from the liquid and carries it to the second column. The heads, which have a low boiling point, collect at the top where it is cooler and are siphoned off. The tails, which have a high boiling point, collect at the bottom where it is hotter. The main run alcohol condenses in the middle, where it is extracted and then condensed. This is a faster process than pot distillation because heads, hearts, and tails can be separated at the same time.

Distillers like column distillation because it proceeds continuously and does not require much monitoring. Therefore, it reduces time, labor, and costs. They have less control over the composition of the beverage than in pot

Continuous Distillation

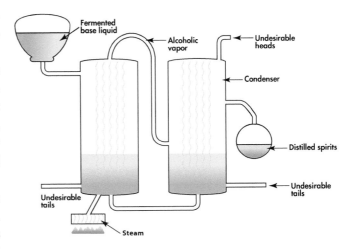

stills but obtain greater consistency from batch to batch. Because the alcohol is distilled continuously, the distiller can extract higher concentrations of alcohol to obtain a neutral flavored distillate with 90 to 95 percent alcohol content. However, the process can remove many of the congeners that give spirits their character.

Alcohol Content and Purity

Alcohol content is usually identified as alcohol by volume (ABV) and is designated as a percentage of alcohol in relation to water. The higher the ABV of the distilled liquid, the purer and more neutral it is. Chemically pure alcohol is 100 percent ABV or 200 proof (see Proof, at left). Distilled at 95 percent ABV (190 proof), **NEUTRAL SPIRITS** are odorless, colorless, and flavorless. Although they are typically made from grains, any plant material that can be fermented can be used to create them. These form the base of other types of spirits, including blended whiskey, vodka, gin, liqueurs, and bitters. Bottled spirits usually contain between 35 and 50 percent ABV, although the ABV can be lower or higher.

Finishing

When the distillate leaves the still, it is a colorless, rough, and harsh liquid. Distillers employ numerous techniques to enhance the finished product. They can flavor the alcohol with herbs, spices, fruits, and other essences, such as citrus peels, and sometimes even include a bit of wine or sherry. Depending on the type of spirit and the regulations governing that spirit, color may be added. Because it is made from natural sugar, caramel is frequently used to produce a consistent golden or amber color or to adjust the flavor. In addition to flavoring and coloring the beverage, aging and blending are also used to alter the distillate and create the final spirit.

When the spirit is ready for bottling, final adjustments are made to the color and flavor. Then it is cut with water to obtain the desired proof, usually to 40 percent ABV (80 proof). The difference between 80 proof (40 percent ABV) and 60 proof (30 percent ABV) is simply the amount of water that was used to cut the alcohol content.

Aging

Spirits are separated into two categories: white goods and brown goods. In order to retain their clarity and purity, white goods are not aged. These clear spirits include gin, vodka, light tequila, light rum, and eaux-de-vie. Brown goods are aged in oak, which imparts tawny color, mellows and smooths the alcohol, and adds complex flavors, such as vanilla, spice, floral, and wood. Dark rum; some tequila; all types of whiskey, including Scotch and bourbon; and grape-based brandy are brown goods. During aging, water and alcohol evaporate from the barrel, further concentrating the spirit.

PROOF

Proof, a term that is used only in the United States, indicates the alcohol content of a bottled spirit. Proof number is double the actual percent ABV.

100 Proof = 50% ABV

80 Proof = 40% ABV

60 Proof = 30% ABV

The term derives from the old method of testing the quality of a given spirit. A drink was poured onto gunpowder and then lit. If the flame sputtered out, the alcohol content was too low. If the gunpowder burst into flame, the alcohol content was too high. But if the gunpowder burned slowly, it was proof that the beverage contained the right amount of alcohol.

Blending

Blending is an art that requires the blender to achieve the right flavor balance in the beverage and maintain its consistency from one year to the next. Blending enables each distillery to develop singular recipes for each brand it produces. When blending, the distiller combines different types of spirits or the same spirit taken from different lots, and can add ingredients, such as sugar, spices, herbs, wine, sherry, and other flavorings and essences.

GRAIN-BASED SPIRITS

Grain-based spirits can be produced from any grain, but barley, corn, wheat, and rye are used most often in Western countries. Although these spirits come from a common source, grain, each grain selected by the distiller imparts a noticeably different character to the spirit. For example, wheat is smooth and rye is strong. In addition, the combination and percentage of different grains used, distillation techniques, blending recipes, aging, and flavor additives differentiate each spirit.

Whiskey

Whiskey is made primarily from barley, which may or may not be malted, as well as from corn, rye, or wheat. Whiskey is almost always aged in barrels, but the type of barrel and the length of time it matures help create distinctive characteristics. Whiskey ranges from light and lean to bold and deep, depending on the way it is distilled, blended, and aged. Each whiskey region also has its own unique geography, which impacts the final style of whiskey much more than some other types of spirits. See Table 20.1 for a summary of whiskey styles.

Scotch Whisky

Scotch whisky acquires its unique flavor from malted barley. To be called Scotch whisky, it must be distilled and aged in Scotland. It is matured a minimum of three years and is cut to a strength of 40 to 43 percent ABV (80 to 86 proof) before bottling. There are two types of Scotch whisky: malt whisky and grain whisky.

MALT WHISKY is made from malted barley that is double- or triple-distilled in a pot still to 70 percent ABV. Some distillers dry the malt over peat fires (see Scotch Whisky and Peat, page 336), which give the whisky a strong, smoky character. As mandated by law, malt whisky must be barrel aged for three years, but it is more common to age it from 10 to 25 years. It matures in previously used bourbon or sherry barrels or in new oak barrels, each of which imparts different characteristics to the finished whisky.

GRAIN WHISKY is made from a combination of raw grains, usually wheat or corn, with 10 to 15 percent malted barley, distilled in a column still to 95 percent ABV. Grain whisky is generally less flavorful than malt whisky because it is composed of a blend of grains.

WHISKEY OR WHISKY?

How do you know which is the correct way to spell this beverage? In the United States and Ireland, whiskey is spelled with an "e". In Scotland, Canada, Japan, and New Zealand, whisky is spelled without an "e". In this text, we use whiskey (whiskeys) except when we refer specifically to Scotch or Canadian whisky (whiskies).

Single malt Scotch is highly prized by connoisseurs.

SCOTCH WHISKY AND PEAT

Peat is a layer of partially decayed plant matter that covers the ground in wet areas of Scotland, as well as other boggy areas throughout the world. Peat is an inexpensive and easily collected fuel source. In the Highlands, the Scots use peat fires to dry malted barley. These highly regarded whiskies obtain distinctive, smoky flavors from the peat fires.

Blended whisky can combine dozens of different batches of both malt and grain whiskies. The distiller selects each whisky based on the flavor components that its unique qualities will contribute to the blend. The ratio of malt whisky to grain whisky in a blend affects the flavor and the cost, with the price increasing as the percentage of malt whisky rises.

Scotland has six distinct whisky-producing regions. Five, the Highlands, Speyside, Islay Island, Cambeltown, and the Islands whisky region, are noted for malt whiskies. The sixth, the Lowlands, produces primarily grain whisky. Between and within each region, there is a great deal of individuality in whisky styles because of the broad range of environments.

Premium Whisky

Some of the best malt whisky is reserved for single malt Scotch. These highly flavored and distinctive whiskies are produced using the batch process in a pot still by a single distillery. Single malt Scotches are produced mostly by small distilleries, primarily in the Highlands, Speyside, Islay Island, and Orkney in the Island whisky region. They are produced in a broad range of styles, and each distillery is noted for its specific brand. Single malt Scotches have a small market share but are highly prized by connoisseurs. Expensive whiskies are very complex with older and rarer malts used in the blend. They have darker and heavier flavors than regular Scotch. The age of the youngest whisky included in the blend is indicated on the label.

Irish Whiskey

Malted and unmalted barley forms the base of Irish whiskey, which is pot distilled twice and then distilled a third time in a column still. However, unlike Scotch, peat is not involved in the drying process. Irish whiskey is aged for a minimum of three years in used casks, often former bourbon barrels, and it is almost always blended. Irish whiskey is lighter than Scotch. Famous Irish distilleries, such as Jamesons, have been purchased by large corporations that produce dozens of brands, all with special recipes and styles.

American Whiskeys

U.S. law strictly regulates whiskey production. It defines the grains that can be used in each type of whiskey and in what percentages, establishes the alcohol level in the distillate and the final proof of the bottled spirit, and regulates aging and additives. By law, all American whiskeys are made from grain mash, distilled to 90 percent ABV or less, reduced to 62.5 percent ABV, aged in wood, and bottled at a minimum of 40 percent ABV (80 proof).

Most American whiskeys are made from **SOUR MASH**. To make sour mash, some of the residue, including the yeast, from one batch of mash is reserved and added to a new batch of mash, just like the process of making sourdough bread. Many distillers prefer sour mash because it gives the whiskey consistent flavors.

Bourbon

The quintessential American whiskey, bourbon received its name from Bourbon County, Kentucky, where it was originally produced from corn, an abundant crop in the American colonies. By law, it must include at least 51 percent corn, but it is often made from a closely guarded recipe using 70 to 80 percent corn and 10 to 15 percent malted barley, with the remainder rye or wheat. Bourbon is distilled in a column still, followed by a second distillation in a pot still to less than 80 percent ABV (160 proof). It must mature at no more than 62.5 percent ABV (125 proof) for a minimum of two years in new, toasted American oak barrels, although four years is more common, and sometimes it is even aged as long as 10 years. By law, distillers may not color, flavor, or sweeten bourbon. Bourbon is usually bottled at 80 proof. It has a deeper color and stronger vanilla notes than Scotch.

Barrel-proof bourbon has become quite popular. As bourbon ages, water evaporates from the barrel, concentrating the alcohol. This highly alcoholic bourbon is not cut with water, but is bottled directly from the barrel at levels that are well over 100 proof.

As with Scotch, there is a trend toward premium bourbons. Rather than the process of blending selections from several barrels, single barrel bourbon comes from an individual barrel that is specifically selected for its outstanding quality. These are bottled one barrel at a time. Small-batch bourbon is composed of the whiskey from several outstanding barrels that are blended to create an even finer bourbon. Premium bourbons are often whiskeys that have been aged longer and generally bottled at a higher proof than other bourbons.

American distillers are proud of their small-batch and single-barrel whiskies.

Tennessee Whiskey

Tennessee whiskey is a smooth whiskey made from at least 51 percent corn that is filtered through sugar maple charcoal to remove some of its congeners, a process that can take as long as 10 days. It is distilled at less than 80 percent ABV and aged for at least two years. The maple adds a hint of sweetness to the whiskey.

Rye Whiskey

Rye whiskey is made with at least 51 percent rye, is diluted to less than 80 percent ABV, and is aged for at least two years in new toasted barrels. These aggressive, strong whiskeys are generally used for blending to give "backbone" to other whiskeys in the blend. They are usually dry, grainy, and spicy.

American Blended Whiskey

Blended whiskeys in the United States are made of a combination of straight grain whiskey, bourbon or rye for example, and neutral spirits or light whiskey (high-alcohol whiskey). The neutral spirits reduce the aggressiveness of the grain whiskey. The distiller combines dozens of different whiskeys to achieve the desired blend, but the law requires that at least 20 percent comes from straight grain whiskey. The taste of the final blended whiskey varies greatly depending on the proportion of grain whiskey to neutral spirits included in the mix. These are generally bottled at 80 proof. Blended whiskeys may have dozens of styles, including unaged, if dictated by the house recipe.

Table 20.1 Whiskey Styles

WHISKEY	GRAINS	DISTILLATION	ABV	AGING	BLENDING
Scotch: malted whisky	Malted barley Usually dried over peat fire	Pot still Double- or triple-distilled	70%*	3–25 years Old sherry or bourbon barrels or new oak	Single malts Blended whiskies
Scotch: grain whisky	Wheat or corn 10% -15% malted barley	Column still	95%*	3 years minimum	Blended whiskies
Irish whiskey	Malted and unmalted barley	Double distilled, pot still Third distillation, column still	80% maximum*	3 years minimum	Blended whiskeys
Bourbon	51% corn, minimum Malted barley, rye, wheat	First distillation, column still Second distillation, pot still	80% maximum*	2–10 years New, toasted American oak	Single barrel Small batch Blended whiskeys
Tennessee whiskey	51% corn, minimum	Distilled then filtered through sugar maple charcoal	80% maximum*	2 years, minimum New toasted oak	
Rye whiskey	51% rye		80% maximum*	2 years, minimum New toasted oak	Blended whiskeys
Canadian whisky	Corn, malted barley, wheat, rye	Column still	40% minimum **	3 years, minimum Old toasted oak	Blended whiskies

* Distilled ABV

** Bottled ABV

Canadian Whisky

Canadian whisky is made from a combination of corn, malted barley, wheat, and rye. Unlike other whiskeys, no one grain dominates. It is column distilled and aged for at least three years, usually longer, in old toasted oak barrels. Canadian whiskies are blended with batches of different characters and ages, and are typically smoother and milder than other grain-based whiskies. Rye contributes spicy flavor to Canadian whisky, but its aggressiveness is tempered by using small amounts in proportion to other grains and by extended aging in old barrels.

Gin

Gin's defining characteristics come from juniper berries and, to a lesser extent, other botanicals. Distilled-gin production begins with a neutral grain spirit, rather than a mash, that the distiller usually purchases from another producer. A column still is used to concentrate the alcohol. It is then redistilled with juniper berries, herbs, and spices. Compound gins, which are considered lower quality than distilled gins, are not redistilled with botanicals. Instead, the juniper berries and other flavorings are added to the distillate.

Because water from natural sources can impart unwanted flavors to the gin, demineralized water is used. To retain its clarity, gin is usually not aged. If the distiller decides to age gin, it is matured for a much shorter time than whiskey, anywhere from a few months to at most three years.

In addition to juniper berries, the best gins contain as many as 15 botanicals, such as lemon and orange peels, herbs, dried fruits, and spices, most commonly coriander, cardamom, and anise. The distiller has an important role in selecting the ingredients that are used in the distillery's recipes. Which ingredients are selected and how they are combined determine the complexity and uniqueness of each gin.

Gin is described as lean, racy, bright, clean—a spirit with character. It has complex undertones because it incorporates botanicals. Of course, juniper is the primary flavor and aroma in gin, but it also can have licorice, peppers, lemons, cucumbers, roses, oranges, and a myriad other flavors, all of which should be subtle so they do not overwhelm the juniper. Newer styles of gin tend to be lighter with more botanicals and spices, whereas traditional styles are stronger with heavier juniper aromas.

Juniper berries are the dominant flavor in gin.

Infusing Flavors

Distillers use one of three methods to infuse flavors into gin:

Cold Compounding Botanicals are crushed and steeped in alcohol for a few days. The infused liquid is added to the distillate before the gin is bottled. This method is used for the lowest-quality compound gins.

Essential Oil Method Botanicals are crushed and cooked to remove the essential oils. The oils are combined with alcohol and concentrated. The oil mixture is added to a neutral grain spirit, which absorbs the flavor and aroma components. This liquid is usually redistilled with the gin.

Gin-head Distillation Neutral spirits are vaporized and passed through a basket containing juniper berries and botanicals. As a result, they absorb the flavors from the ingredients. The flavored vapors are condensed and become gin. This method is used for the best gins.

Types of Gin

Several styles of gin are produced, but the most common are London Dry and Genever. London Dry gin, or English Dry gin, uses a 96 percent neutral grain spirit that is made primarily from corn, with smaller amounts of malted barley and other grains. It is distilled in a column still, so it has few congeners, and then redistilled to 95 percent ABV (190 proof) in a pot still with juniper berries and other botanicals. American dry gin, a version of London dry, is made from 100 percent neutral spirits, which are distilled to a lower alcohol level, 40 percent ABV (80 proof), than other dry gins, resulting in a less flavorful product.

Genever, also known as Dutch or Holland gin, uses a base of malted barley, corn, wheat, and rye. Although column stills are used for most gins, Genever is distilled two or three times in a pot still to a low proof (37.5 percent ABV), which allows it to retain many of its congeners. During the final distillation, juniper berries, spices, and herbs are added. Genever is sometimes aged and is fuller-bodied, more malted, and heavier than dry gin.

Vodka

Vodka has always been a favorite liquor because it has a smooth, neutral flavor. Its origins are highly contested, with Russians and Poles both laying claim to its discovery. The majority of vodka is made from grains, with a few exceptions that have a base of potatoes, beets, molasses, and more recently grapes.

Most vodka has no distinct character, aroma, or taste, and is colorless. This high level of purity comes from distilling the mash in a column still, which extracts most of the congeners, and then charcoal-filtering the distillate to remove any remaining aromas and flavors. Vodka is rectified to a very high alcohol level, about 96 percent, more so than other types of grain spirits. Because vodka is flavor-neutral, its most important element is water, which imparts its qualities to the final product. Vodka is not aged and is bottled at 80 to 110 proof.

Depending on the style of vodka, it may be creamy, sweet, or grainy, but all vodka is characterized by a crisp, clear, smooth taste. Because it is basically flavorless after distillation, vodka lends itself to the addition of almost any spice, herb, or fruit. Blood oranges, caraway, cherries, chocolate, peppers, cinnamon, limes, lemons, nutmeg, and raspberries are just a few of the ingredients that can be used to flavor vodka. The producer must identify flavor additives on the label. Flavored vodka has become very popular in the United States in the past few years.

RUM AND TEQUILA

Most spirits are made from grains or fruits, but rum and tequila are different. These spirits originated in tropical climates where native plants, sugarcane and agave, were readily available and fermented naturally. European conquerors realized that the fermented product from these plants could be distilled, and rum and tequila were created. The Europeans introduced these new spirits to other countries as they traveled the world, and they quickly gained popularity.

Rum

Rum originated on Caribbean sugar plantations with the discovery that the byproducts of sugar production could be fermented and distilled. Rum is made from sugarcane juice, cane syrup, or molasses, the residue that remains after sugarcane is boiled to crystallize the sugar. Molasses is the preferred base because it provides an ideal home for yeasts; ferments easily; has a strong, distinctive flavor; and is inexpensive.

Rum is produced by allowing a mash of molasses, yeast, and distilled water to ferment from 12 to 48 hours or more. The type of yeast used with the molasses contributes significantly to the final taste. Rum can be single-, double-, or triple-distilled.

Light-bodied rums are made from molasses that ferments from 12 to 20 hours. They are distilled to high alcohol levels in column stills and usually are charcoal filtered. These two processes result in highly purified rum with minimal flavors, aromas, and color.

Dark, full-bodied rums ferment a minimum of one to two days, often longer, and are pot distilled. Because of this, they retain congeners, which give them more flavor and aromatics than light-bodied rums. Dark rums obtain color and flavor as they mature in used American whiskey casks, and caramel or molasses may be added to provide even more color and flavor.

Blending is a critical step in rum production because it establishes the characteristics desired by each producer. A master blender combines rums from different stills, different purities, and different vintages.

Rum production is not usually as highly regulated as the production of other spirits, and the laws vary greatly from one country to the next. Maturation can last from eight months to five years and ABV at bottling ranges from 38 to 50 percent (76 to 100 proof).

Rum Classifications

Rum is produced in four classifications, which are based on the length of maturation:

1 Clear, white, or silver rum is aged in stainless steel tanks or oak barrels for one year and has clear color and subtle flavor.
2 Amber or golden rum is matured for at least three years and has medium body and smooth, mellow flavors.
3 Dark rum is aged for five years or longer and develops rich, caramel flavors and is smoother and mellower than amber rum.
4 Añejo, or aged, rum is superior-quality rum blended from several outstanding vintages or batches, with the date of the youngest vintage printed on the label. Liked aged Scotch, aged rums have become very popular.

Rum is distilled worldwide in places as diverse as South Africa, Russia, and Central and South America. However, most rum is distilled in Caribbean countries, such as the Virgin Islands, Martinique, Barbados, Jamaica, Trinidad, and especially Puerto Rico.

When you buy rum, it is very important to read the label because production and styles change from country to country. The label indicates the country of origin, alcohol content, name of the distiller, where it was blended, and whether it was made from molasses or cane juice. Knowing the country where the rum is made gives you an indication of its style. For example, Puerto Rican rum is usually light, Jamaican and Bermudan are usually dark, and Haitian is golden. Common flavor and aroma descriptors for rum include honey, spice, vanilla, banana, caramel, butterscotch, tobacco, and leather.

Tequila and Mezcal

Tequila's origins date to the Aztecs, who drank pulque, naturally fermented agave juice, during religious ceremonies. Spanish conquistadors applied grain-distilling processes to the pulque, and mezcal was born.

Tequila and its cousin, mezcal, are made from the hearts of agave plants that grow in limited regions of Mexico. Agave hearts are cooked, which give tequila and mezcal their intense browned flavor, and then they are mashed, mixed with water, and fermented.

Tequila is primarily produced in the Jalisco region, but four other states are allowed to make tequila. Of the 400 species of blue agave, only the particular variety *Agave tequilana Weber* var. *azul* can be used in tequila. The agave mash is double-distilled in either a pot or column still. Column-distilled tequilas are cleaner and blander than those that are pot distilled. Some tequilas are not aged, but those that are matured do not remain in barrel for as long as other spirits because the wood flavors would overwhelm the agave. Tequila is intense with a complex aftertaste. It has notes of sweet agave and can be floral, earthy, fruity, and smoky.

Mezcal is a rustic beverage made by small producers in Central Oaxaca. Any species of agave, except the blue agave used in tequila, is allowed. The hearts are roasted in ovens heated by charcoal fires, which gives it a smoky aroma. It is double-distilled: first in a small clay pot, and then in a large metal pot. Most mezcal is not aged.

Tequila Classifications

The Norma Oficial Mexicana Tequila (NOM), under the jurisdiction of the Mexican government, designates the regions and zones where blue agave can be grown and regulates how tequila is produced, bottled, and labeled. Each authorized distillery has its own NOM number, which tells the buyer that the tequila is authentic. An unnumbered bottle indicates that the product is not genuine tequila. The regulations define two qualities of tequila:

1 One hundred percent agave tequila uses only blue agave juice and natural yeast. These are the only tequilas that can state "100 percent Agave" on the label. They must be distilled, aged, and bottled in Mexico.

2 Mixto combines 51 percent blue agave juice with 49 percent diluted molasses or cane sugar and commercial yeast. Mixto must be distilled and aged in Mexico but can be bottled elsewhere. Distillers ship a large percentage of mixto to the United States for bottling, mainly in California.

Whether it is 100 percent agave or mixto, all tequila is classified into five categories. From lowest to highest, they are:

Blanco (Silver) The original style, silver tequila is clear and unaged.

Joven Abocado (Gold) Gold tequila is unaged, but it may be colored or flavored with some aged tequila or caramel.

Reposado Reposado is aged from two to 12 months in oak barrels, which results in a mellow tequila.

THE WORM

The tequila worm has been immortalized in many a Hollywood Western when the cowboy downs his shot of tequila, worm and all. However, tequila never has a worm in it. The worm, in fact, is a moth larva that is found growing on some of the agave species used to produce mezcal. Mezcal containing worms may be labeled *"con gusano."* Some say that drinking good tequila is like swallowing liquid velvet, whereas drinking mezcal is like downing barbed wire.

Añejo To earn this designation, tequilas are aged for a minimum of 12 months, and usually not more than four years, in old bourbon or cognac barrels. Añejo tequila has richer, deeper coloring and more complex flavors than the other classifications.

Extra Añejo Tequilas are aged for a minimum of three years, usually longer, in oak barrels. They are dark, complex, and smooth.

FRUIT-BASED
SPIRITS

In wine-producing countries, it was a small leap from making wine to distilling wine to create a stronger beverage—brandy. It did not take long for distillers to realize that if brandy could be made from grapes, distilled spirits could also be made from other fruits, such as apples, pears, and cherries.

Fruits and other plant materials, such as seeds and stems, make excellent beverages because congeners in the fruit contain flavor components that are easily absorbed by alcohol. Grains can be stored, so it is possible to make grain-based spirits year round. However, the production of fruit-based spirits depends on the seasonal availability of fruit and must be initiated shortly after the fruits ripen. As with grains, fruits must be fermented to obtain the alcohol needed for distillation.

Artistic bottles are part of the pleasure of buying tequila.

Brandy

Brandy is produced from wine made from neutral-flavored white grapes. Although French brandies are noted for superior quality, the United States, Spain, Italy, Germany, Mexico, Greece, Israel, South Africa, and South America are also brandy-producing countries. Brandy is aged to obtain color, mellowness, aroma, and flavors. When matured in new oak, brandy acquires more color but can be bitter; whereas in old oak, it picks up less color but is softer. Caramel can be added to brandy to boost its color, and it is generally bottled at 80 proof.

Cognac

Cognac, the most famous brandy, is produced in Cognac, north of Bordeaux on the Atlantic coast of France. The Cognac appellation is divided into six growing regions or crus, which indicate the quality and character of the grapes. Grande Champagne and Petite Champagne are considered the best crus and are often listed on Cognac labels. Rigorous laws govern production in Cognac. First, brandy labeled Cognac must be produced in one of the Cognac crus, and it must be distilled by March 31 in the year following harvest. Second, 90 percent of the grapes must consist of Ugni Blanc, a neutral grape, Folle Blanche, and Colombard, and 10 percent can include any other white variety. Third, it must be double-distilled in an alembic Charentais still, a small, long-necked copper pot indigenous to the region, to a maximum of 70 percent ABV.

There are also stringent requirements that govern the minimum aging time for each designation or category of Cognac. However, most Cognacs are aged longer than required. The designation, which identifies the age of the youngest brandy used in the blend, must appear on the label. The designations are:

Very Special (VS) This is the lowest designation and the least expensive Cognac. It must age for at least 2½ years but typically ages for five years.

Very Superior Old Pale (VSOP) Required to age for 4½ years but is usually aged for seven to 10 years.

Extra Old (XO) Usually matures from 15 to 25 years, although the law requires only 6½ years of aging.

Grande Reserve Usually aged for about 50 years, although the law does not mandate an aging requirement for this designation.

All Cognac is first aged in new French oak to pick up color and flavor from the wood. Cognac selected for long aging programs may be transferred to old oak barrels to control the color and to keep it from picking up bitter flavors from the wood.

Blending is a crucial component of Cognac production. A master blender combines different batches, ages, and crus to create complex and unique Cognacs, each with its own personality. During blending, caramel, sugar, and oak extract can be added to enhance the taste. Cognac is usually bottled at 40 percent ABV (80 proof). It ranges in color from golden to deep reddish-brown and can taste fruity and flowery, or toasty and rich.

Armagnac

The world's oldest brandy, Armagnac is produced in Gascony in Southwestern France. Like Cognac, its production is highly regulated and distillation must be finished by March 31 in the year after the grapes are harvested. Ten varieties of grapes are permitted, but Ugni Blanc, Baco 22A, Folle Blanche, and Colombard compose the bulk of Armagnac. The grapes must be grown in one of three designated regions: Armagnac-Ténaréze, Haut-Armagnac, and Bas-Armagnac, which is recognized for producing the highest-quality grapes.

Armagnac is single-distilled in an alembic Armagnacais, a specialized type of column still indigenous to the region. In this still, some of the heads and tails are retained, which gives the distillate floral and fruity aromas. It is aged in new French oak for a short period and then transferred to old oak barrels. However, Armagnac is often aged for 10 years, much longer than most Cognac. Longer aging gives it a rich mahogany color, develops intricate flavors and aromas, and results in Armagnac having stronger flavors than most Cognac.

Most Armagnacs are blended, but exceptional batches are reserved and sold as single-vintage and single-vineyard Armagnacs. The law permits the addition of caramel. Armagnac is reduced to 40 percent (80 proof) before it is bottled, and the label identifies the age of the youngest brandy in the blend.

American Brandy

Brandy has been made in California since the 1800s, but until recently its quality could not compare with that of French brandy. California wine producers realized that they could expand their winemaking skills to brandy production, and they are now creating brandies that can compete with the finest brandies from France. The laws governing brandy production in California are not as strict as European regulations, so producers have much greater leeway. The law states that grapes included in California brandy must be grown in California, but it does not specify varieties. Thompson Seedless, Colombard, Ugni Blanc, and Folle Blanche are the primary grapes chosen for brandy production.

California brandy must be distilled in the state. It is often distilled in column stills, but small boutique distilleries are leading a movement toward alembic pot stills. By law, California brandy must be aged for a minimum of two years; however, distillers may age premium brandy for up to 10 years. It is first aged in new oak to add color and flavor, then it is matured in used brandy or bourbon casks, which temper the wood's influence. American brandy is lighter, smoother, and more fragrant than Cognac or Armagnac.

Eau-de-Vie (Fruit Brandy)

Eaux-de-vie, or fruit brandies, are produced from fruits other than grapes. Drinking a glass of eau-de-vie is like eating a piece of fruit. Distillation captures the essence of the fermented fruit and imparts it to the alcohol. Fruit brandies are typically double-distilled in pot stills to about 70 percent ABV. Unlike grape-based brandy, which obtains color from barrel aging, eaux-de-vie are not barrel aged, so they retain their clear color. They are usually bottled at 80 to 90 proof. Eaux-de-vie are dry and unsweetened and are distilled at a higher alcohol content than liqueurs.

The best-known eau-de-vie, Calvados or apple brandy, is an exception because it is aged. It is made from a combination of sweet, tart, and bitter apple varieties that are fermented to produce cider. The cider is double-distilled in either a pot still, producing a complex brandy that ages well, or a column still, resulting in a fresh brandy with less complex flavors. Unlike the typical unaged fruit brandy, Calvados is matured for a minimum of two years in French oak barrels that have previously been used for sherry or port. However, most Calvados is aged for much longer, and those that are over 20 years old are highly prized.

Familiar fruit eaux-de-vie include Poire Williams, made from pears; Kirsch, made from cherries; Framboise, made from raspberries; and Cassis, made from black currants.

Pomace Brandy

Grape pomace—the stems, seeds, pulp, and skins left over after pressing grapes for wine—has been used for centuries to make beverages. These solids contain small amounts of sugar, juice, and flavor components that can be fermented and distilled to make a high-alcohol drink. Traditionally, these beverages made use of the by-products of winemaking and were rarely aged. They produced pomace brandy, a strong, fiery, raw wine substitute that was given to peasants in place of more expensive wine.

Commonly known by its Italian name, *grappa* (*marc* in France), pomace brandy was made from any combination of red and white grapes, and little thought was given to the varieties added to the mix. Today, producers are creating fine grappa and marc, which are mellower and have a range of flavor profiles. Producers use carefully selected varieties, vintages, or vineyards to create artisanal spirits that reflect the grapes' personalities. Pomace brandy is often aged in oak to add color and mellow harshness. Flavoring pomace brandy with a variety of ingredients to give it individuality and character is becoming popular.

FLAVORED ALCOHOLS

Alcohol is good at absorbing the characteristics of solid ingredients, like herbs, spices, flowers, nuts, seeds, and fruits, so spirits lend themselves well to flavoring. These ingredients can be either soaked in alcohol or distilled with alcohol to produce liqueurs or bitters. Gin can technically be considered a flavored alcohol because it is distilled with herbs, spices, and other botanicals, and flavored vodka is very popular. However, these spirits are not categorized as liqueurs because they are not sweetened. It is sugar that is the defining ingredient of a liqueur.

Liqueurs

Liqueurs, or cordials, were probably the first widely consumed distilled spirits. They were originally used as medicines and tonics by monks, who developed special recipes for their liqueurs by adding spices and herbs to cover up the taste of the raw alcohol and make them more palatable.

To be classified as a liqueur, a beverage must contain three ingredients: alcohol, sweetener, and flavoring. The base is usually a pure 190 proof alcohol, but sometimes brandy, rum, or whiskey is used. The sugar moderates the acidity of some kinds of fruits, softens the bitterness of herbs, and heightens the aromas. Sugar content for liqueurs generally ranges from at least 20 percent up to 35 percent. It is easy to confuse crème liqueurs and cream liqueurs. Crème liqueurs contain 40 percent sugar and are sweet, heavy, and dense. Cream liqueurs are made with real cream or a cream substitute and are thick and mild. Flavoring for both kinds of liqueurs can be obtained from fruits, nuts, seeds, beans, or herbs.

Distillers employ three methods to extract flavors from botanicals:

Maceration Botanicals are soaked in alcohol until the alcohol absorbs their flavors. The simplest method, it is best used with delicate fruits, such as raspberries.

Percolation In a process similar to percolating coffee, alcohol is pumped over botanicals, and the flavors are infused into the alcohol as it slowly drips over them. The process repeats until the flavors are strong enough. This method works well with beans and pods, such as vanilla and cocoa.

Table 20.2 Common Liqueurs

LIQUEUR	ORIGIN	BASE	PRIMARY FLAVORINGS
Amaretto	Italy	Neutral spirit	Apricot pits
Baileys Original Irish Cream	Ireland	Irish whiskey	Chocolate, vanilla, cream
Bénédictine	France	Cognac	Herbs
Chambord	France	Neutral spirit	Black raspberries
Chartreuse	France	Neutral spirit or brandy	About 130 plants and herbs
Cherry Heering	Denmark	Brandy	Cherries
Cointreau	France	Neutral spirit	Orange peels from many varieties
Crème de Cacao	Origin unknown	Neutral spirit	Cacao and vanilla beans
Crème de Cassis	France	Neutral spirit	Black currants
Crème de Menthe	Origin unknown	Neutral spirit	Mint
Curaçao	Holland	Neutral spirit, sometimes brandy	Laraha (bitter oranges) orange peels
Drambuie	Scotland	Scotch	Heather honey, herbs
Frangelico	Italy	Neutral spirit	Hazelnuts
Galliano	Italy	Neutral spirit	Herbs
Grand Marnier	France	Cognac	Orange peels
Jägermeister	Germany	Neutral spirit	Herbs
Kahlúa	Mexico	Neutral spirit	Coffee beans
Limoncello	Italy	Neutral spirit	Lemons
Midori	Japan	Neutral spirit	Yubari melons
Nocino	Italy	Neutral spirit	Green walnuts
Pernod	France	Neutral spirit	Anise
Sambuca	Italy	Neutral spirit	Elderberry flowers, anise
Sortilège	Canada	Canadian whisky	Maple syrup

Distillation Aromatic components, such as flowers, berries, and seeds, are extracted and concentrated in a pot still. This is the preferred method for dried botanicals because it draws out their flavors.

Liqueurs are often clear when they are bottled, but some are produced with vibrant colors, such as greens, reds, blues, and oranges. Liqueurs generally obtain these colors, from the natural ingredients that make up the primary flavors. If artificial colors or flavors are used, which is very unusual, the label must say "artificial" or "imitation." If an aged spirit, such as whiskey, forms the base, the liqueur will pick up its amber color. Before they are bottled, liqueurs are filtered to remove any bits of the flavoring ingredients. Liqueurs are generally bottled at a lower alcohol content than spirits, from 35 to 40 percent ABV (70 to 80 proof). However, a few are bottled at higher proofs.

Distillers can be more creative with liqueurs than with most other types of spirits. Any imaginable combination of botanicals can be used to develop signature liqueurs. Table 20.2 lists a representative sample of the hundreds of liqueurs on the market. This offers a glimpse at the great diversity of these beverages.

Bitters

Bitters are rooted in early medicinal beverages that were distilled with bitter botanicals, such as wormwood, chamomile, orange peel, and almond. Often bitters are made from the bark, roots, stems, and seeds of plants, which are more astringent than the flowers or fruits. They typically contain considerably less sugar than liqueurs. Bitters today are usually complex blends of herbs and spices that are used as aperitifs to prepare the digestive system for eating, or as digestifs to stimulate digestion after a meal. They are considered to be stomach settlers and hangover cures. Campari is the most highly recognized and popular bitter. Table 20.3 contains examples of some better-known bitters.

EVALUATING SPIRITS

When evaluating spirits, we use the same structure that we use for wine or beer: look, smell, taste. Although the procedures are similar, there are general guidelines to follow when tasting spirits. First, they must be evaluated by category: gin with gin, Scotch with Scotch, apple brandy with apple brandy. You cannot assess two different types of spirits at the same time, gin with apple brandy for example, because their characteristics are so different. Second, because of the amount of alcohol in each beverage, palate fatigue sets in easily. No more than six beverages should be evaluated at one sitting. Third, whereas wine should be nosed deeply, spirits should be inhaled delicately. Finally, you should only sample spirits straight, not in cocktails, because the flavored ingredients in mixed drinks alter the aromas and flavors of the distilled beverages.

Table 20.3 Common Bitters

BITTER	ORIGIN	BASE	PRIMARY FLAVORINGS
Amaro	Italy	Neutral spirit, sometimes wine	Herbs, roots
Angostura Bitters	Trinidad	Neutral spirit	Gentian, other herbs
Campari	Italy	Neutral spirit	Herbs, chinotto (bitter Italian citrus fruit)
Cynar	Italy	Neutral spirit	Artichokes
Fernet-Branca	Italy	Neutral spirit	40 herbs, spices, other plants
Gammel Dansk	Denmark	Neutral spirit	Herbs
Peychaud's Bitters	U.S. (Creole origins)	Neutral spirit	Gentian, other herbs

Color

All spirits are colorless following distillation. Clear spirits, such as vodka, remain colorless because they are not barrel aged or are only aged for a short time in old barrels that do not leach much color. While it may seem as if color evaluation is not important in clear spirits, their clarity should be assessed. They should be bright, fresh, and translucent, not hazy.

Colored spirits are evaluated on the depth and clarity of their color. In barrel-aged beverages, like whiskeys and brandies, color is an important evaluation component. Aged spirits can be tawny, amber, or copper, even deep mahogany. The depth of the color indicates the length of barrel aging and the age of the barrel used. The longer the spirit rests in the barrel, the deeper its color becomes. The age of the barrel also has an impact, with younger barrels imparting more color to the spirit than older barrels.

Liqueurs come in every color imaginable, depending on the flavoring ingredients, base spirit, and production methods. Most often colored liqueurs reflect the ingredients used. For example, crème de menthe is often green. Sometimes, however, the color has no relationship to the ingredients. Curaçao, which is made from bitter-orange peels, can be orange, blue, green, or red. Additionally, some liqueurs are clear, like most grappas for example, while others, like cream liqueurs, are opaque. Colored or clear liqueurs should appear translucent with no sign of haziness or particles from the flavoring ingredients. Because they contain a high percentage of sugar, many will appear viscous in the glass. Cream liqueurs should be opaque, smooth, and creamy.

Aromas, Flavors, and Mouthfeel

Some spirits are distilled to such a high level of purity that they are odorless, but most spirits have distinct and powerful aromas and flavors. Some are the by-product of distillation. Aromas and flavors of the base ingredient, whether barley, agave, or grapes, remain after distillation is finished. Some spirits also pick up aromas and flavors from botanicals that may be added during or after distillation. Finally, barrel aging adds aromas and flavors to spirits. They are more intense if the beverage is aged in a new barrel, and less intense if an old barrel is used. The unique tastes of each type of spirit should be immediately recognizable. For example, gin should always have the scent and taste of juniper berries, one of its major components.

The taste of most spirits is subtle, though complex, and depending on the type of spirit, may contain notes of coffee, toffee, juniper berries, smoke, agave, honey, herbs and spices, fruits, or flowers to name a few. Liqueurs also have delicate aromas and sweet undertones, and can include additional aromas and flavors of citrus, nuts, distilled alcohol, or chocolate.

The final part of evaluating a spirit is assessing its mouthfeel. Spirits should be smooth, creamy, and warm in the mouth. Although hot on the palate because of the high alcohol content, spirits should not be bitter. They should have a long finish on the palate. Liqueurs should present a sweet sensation in the mouth without being cloying. Aged spirits in all categories are mellower and have more complex, nuanced aromas than unaged spirits.

ENJOYING SPIRITS

Spirits are versatile. They can be served straight out of the bottle or mixed with other beverages; on their own or with food; over ice, at room temperature, or diluted with water. Cocktails or mixed drinks combine distilled beverages with any number of other ingredients, including liqueurs or bitters, sweeteners, juices, soft drinks, milk or cream, and other flavorings. While many cocktails, such as the Martini, have become standards, there are always new combinations being developed that come and go with changing tastes.

Unlike wine, which can age well in the bottle, once a spirit is bottled it does not improve. Therefore, storing spirits until they are older is of no value. After they are opened, spirits oxidize and may turn color or become cloudy. However, this does not impact quality, so spirits can be kept indefinitely. The exception is tequila because the agave flavor in tequila begins to lose its potency when it is exposed to air. Store spirits, tightly closed, in an upright position in a place that is cool and dark.

There is a shape and size of glass for every beverage. If money is no object, you can purchase a different glass for every type of spirit that you plan to serve. However, the glasses in Table 20.4 will meet your basic needs.

If you frequently serve more specialized spirits, such as brandies or liqueurs, you can add glasses designed for these beverages. Brandy snifters have fallen out of favor for serving brandies and whiskeys because the large, open bowl disperses aromas. Cognac glasses, which have a smaller opening, are preferred because they concentrate aromas, and cordial or sherry glasses can also be used. Table 20.5 lists some specialized spirits glasses.

Table 20.4 Basic Spirits Glasses

GLASS TYPE	GLASS SHAPE	SERVING SIZE	BEVERAGE TYPE
Cocktail	V-shaped bowl with long stem	4–6 oz.	Martinis, Manhattans, Cosmopolitans, and other mixed drinks that are served neat
Collins	Narrow, straight-sided, flat-bottomed	10–14 oz.	On the rocks and mixed drinks, especially tropical drinks
Highball	Straight-sided, flat-bottomed	10–12 oz.	On the rocks and mixed drinks
Old-fashioned	Straight-sided, flat-bottomed	8–10 oz.	On the rocks and mixed drinks
Shot	Small, flat-bottomed	1.5–2 oz.	Neat

Serving Spirits

Whiskey

Blended whiskeys are good starting points for beginning whiskey drinkers because they are softer and smoother than straight whiskeys. Whiskey is a fine base for simple or complex mixed drinks, or it can be served on the rocks, with a splash, or neat. In order to savor the individual characteristics of single malt Scotch and small batch bourbons, they should be served neat at room temperature.

Examples of Whiskey Cocktails
BLENDED WHISKEY Manhattan, Seven and Seven, Whiskey Sour
BOURBON Eggnog, Mint Julep, Old-Fashioned
SCOTCH Rob Roy, Rusty Nail, Scotch and Soda

Gin

Because they are heavy, Genever-style gins do not mix well and are best served neat or on the rocks. Lighter English dry gins blend well in cocktails, including the famous Martini, but they are also good over ice.

Examples of Gin Cocktails
Gimlet, Gin and Tonic, Martini, Sloe Gin Fizz, Tom Collins

Vodka

Because it is a neutral alcohol, vodka is the most versatile spirit. It can be flavored, used in cocktails, served neat, or poured over ice. Vodka is best if it is served icy cold, and it can even be frozen overnight for a particularly crisp drink. Drink premium vodka neat or on the rocks.

Examples of Vodka Cocktails
Bloody Mary, Martini, Salty Dog, Screwdriver, White Russian

Table 20.5 Specialized Spirits Glasses

GLASS TYPE	GLASS SHAPE	SERVING SIZE	BEVERAGE TYPE
Brandy snifter	Short-stemmed with large, open bowl	6–8 oz.	Some liqueurs served over ice
Cognac	Medium-stemmed with tulip-shaped bowl	6 oz.	Cognac and Armagnac, brandy, single malt Scotch, bourbon, and aged tequila
Cordial	Small, short-stemmed	2 oz.	Bitters and some liqueurs, brandy, whiskey
Irish coffee mugs	Short-stemmed with handle	8–10 oz.	Hot beverages
Margarita/coupette	Stemmed with large, open bowl and broad rim	12 oz.	Margaritas, daiquiris, fruity cocktails

Spirit Categories

WELL Low-priced house-brand
    spirits

CALL High-quality brand-name
    spirits

PREMIUM, TOP-SHELF
    Highest-quality and -priced
    brand-name spirits

Serving Terms

HIGHBALL Served with ice and soda

JIGGER Two-ounce measure

NEAT, STRAIGHT, OR
    STRAIGHT UP Chilled and
    served without ice

ON THE ROCKS Poured over ice
    cubes

SHOT 1.5 ounces served neat in a
    shot glass

UP Chilled over ice, then strained
    and poured into a glass

WITH A SPLASH Served with a
    small amount of water

WITH A TWIST Served with a small
    slice of lemon rind, twisted to
    release lemon oil

Rum

Light rum has little flavor, so it is perfect for mixed drinks, especially those with fruit or cola. Because it has strong, sometimes spicy flavors, dark rum is not as well suited for making cocktails. To enjoy it at its best, drink it neat or in rum punches.

Examples of Rum Cocktails

Daiquiri, Mai Tai, Mojito, Piña Colada, Rum and Coke

Tequila

Aged 100 percent agave tequila is best when it is slowly sipped, while young tequila is best for mixed drinks. Tequila is usually served in a shot glass, but a cognac glass enhances the aromas of aged tequila. Serve it at room temperature because chilling masks its aromas. In Mexico sips of tequila are often alternated with sips of sangrita, a mixture of tomato and orange juice with salt and chiles. Of course, there is also the classic tequila shot with salt and lime.

Examples of Tequila Cocktails

Margarita, Tequila Fizz, Tequila Slammer, Tequila Sour, Tequila Sunrise

Brandy

Fine brandy—whether Cognac, Armagnac, or American—is designed to be savored straight up in a tulip-shaped cognac glass, which focuses the aromas. Although best when it is served neat, brandy can also be used in mixed drinks. Eau-de-vie and grappa are excellent aperitifs. They should be slightly chilled but not iced, and are good either neat or mixed with sparkling water or tonic.

Examples of Brandy Cocktails

B&B, Brandy Alexander, Hot Apple Toddy, Singapore Sling, Tom and Jerry

Liqueurs and Bitters

Liqueurs are perfect after-dinner drinks when chilled and served alone, poured over crushed ice, or flamed. Liqueurs are excellent substitutes for dessert. They are meant to be sipped slowly in small amounts, so are usually served in two-ounce cordial glasses. Typically, bitters are served in small glasses before or after a meal. It is unusual to find cocktails made from bitters.

Examples of Liqueur-based Cocktails

B-52, Black Russian, Grasshopper, Kir Royale, Midori Sour

Spirits are the most versatile category of alcoholic beverages. They can be made from almost any plant material, including grains, fruits, seeds, and roots. They absorb flavors, colors, and sweeteners and can be mixed with untold ingredients to create distinctive cocktails.

Whether the drink is a classic cocktail or an innovative flavored vodka, consumers have renewed their interest in spirits in the past couple of decades. Ultrapremium crafted beverages, including small-batch bourbon and single malt Scotch, have gained a following, and rum, vodka, and other spirits flavored with fresh, natural ingredients are popular. Bartenders are experimenting with ever more unusual mixed drink combinations, but classic cocktails continue to appeal to many consumers. A relatively new trend is low-alcohol, sweet, fruity distilled spirits. Whatever the changes in taste, there is little doubt that people will continue to enjoy the pleasures of drinking distilled spirits.

Spirits and Food Pairing

In general, spirits are meant to be savored by themselves before or after a meal because their alcoholic strength can overpower food. However, mixed drinks are often served with appetizers before a meal, particularly salty snacks. Liqueurs sometimes accompany dessert because their sweetness complements sweet food, but more often they take the place of dessert or are one of its ingredients. Although wine and beer are most often the preferred beverages with a meal, some people enjoy pairing spirits with food. The food must be flavorful enough to stand up to the alcohol: Rich or spicy dishes are often the best choices. Suggested foods to pair with specific types of spirits follow.

Aquavit Anchovies, blinis, caviar, creamy cheeses, gravlax, hearty food, herring, Indian food, Mexican food, spicy foods, Szechwan food, Thai food

Armagnac Apples, duck, figs, foie gras, prunes

Bloody Mary Corned beef, eggs, Mexican food, oysters, shellfish, smoked fish

Brandy Chocolate, custard, fish, flamed dishes, fruit, gravy, ice cream, marinades, rabbit, sauces, soufleés

Calvados Apples and apple desserts, bread pudding, cheeses, cream sauce

Cognac Asian food, foie gras, roast duck, roast squab

Eau-de-Vie/Pomace Brandy Coffee, fruit sorbets, ice cream

Gin Clams, crab, English cheeses, oysters, pork, salmon, seafood, shellfish, spicy or salty foods, spring rolls, Szechwan dumplings

Liqueurs Coffee, ice cream

Margarita Mexican food, spicy food

Mojito Caribbean food, spicy fish, spicy food

Rum Desserts, rum cake, sweet fruits, tarts

Scotch Braised, smoked, or stewed meat, dark chocolate, game meats, oysters, smoked fish

Tequila Chiles, chips and salsa, guacamole, salty food

Vodka Anchovies, blinis, caviar, creamy cheeses, gravlax, herring, Indian food, Mexican food, oysters, pickled vegetables, spicy and salty foods, Szechwan food, Thai food

Whiskey Beef, cheeses, duck, foie gras, Irish food, Japanese food, mushrooms, roast pork, salmon, smoked fish, smoky food, spicy food, stews

21 Purchasing Alcoholic Beverages

Purchasing alcoholic beverages can be daunting. Not only do you have to deal with three main beverage categories, wine, beer, and spirits, but today you have a global economy with more producers, more brands, more style adaptations, and more products confronting you. Producers are addressing consumers' desires and entertaining new ideas that would not have been considered only a few years ago.

If you want to be a sophisticated buyer, minimum skills are required in order to successfully navigate the wealth of choices in any retail store or restaurant. These include the ability to read and interpret beverage labels, shop for alcoholic beverages in different kinds of retail stores, and knowledgeably order alcoholic beverages in restaurants.

For the food-service industry professional, a basic understanding of beverage-label terms, price categories, and appropriate service techniques is also essential to provide an enjoyable experience to every customer.

UNDERSTANDING
BEVERAGE LABELS

The look of a label, its design, drives the buying decision for some consumers. But to become a knowledgeable alcoholic beverage buyer, you must be able to read and understand the information on the label. The label tells you who produced the beverage; its variety, class, or type; and its alcohol content, along with other information. This information provides clues about the style of the beverage contained in the bottle.

In the United States, the Alcohol and Tobacco Tax and Trade Bureau (TTB) regulates and authorizes all domestic and international alcoholic beverage labels. (The specific TTB guidelines for alcoholic beverage labels can be found in Appendix 1, page 376).

Reading Wine Labels

Wine labels are more difficult to understand than beer and spirits labels for several reasons. First, there are dozens of varietal wines and hundreds of wine blends available in the marketplace. The grape varieties in the wine are not always identified consistently on labels, which can be very confusing. Second, the TTB requires more detailed information on a wine label than a beer or spirits label. Finally, Old World wine labels are even more complicated than New World labels because of language differences and because the grape variety, in many cases, is not printed on the label.

Although all of the information on a wine label is helpful, mastering key pieces of information enables you to understand the kind of wine you are buying. The following sections outline general guidelines for interpreting labels. Appendices 2 and 3 (pages 380–447) include a series of Old World and New World labels that will help you to thoroughly understand wine-label information.

Reading Old World Labels

French, Italian, and Spanish wines are commonly identified by appellation, but grape variety or both appellation and variety can also be used. French Alsatian and Vin de Pays (VDP) wine labels identify the wine by variety, and German wines are almost always identified by grape variety.

Most European appellations are noted for particular grape varieties or specific regulated blends. Knowing which appellations produce which varietals or blends is the most difficult part of understanding European wine labels. As an example, in France Chardonnay is most prominently cultivated in Burgundy. The label only indicates the specific appellation in Burgundy where the grapes were grown and does not mention that the wine is made from Chardonnay grapes. However, if the wine is from Burgundy and it is white, it is almost guaranteed to be Chardonnay—virtually the only white grape grown in the appellation.

As sales of European wines increase in the United States and other New World markets, Old World producers or their American importers are beginning to print the grape variety on appellation-only labels destined for the United States, making it easier for the English-speaking consumer to know what is in the bottle.

This is the information you want to look for on an Old World label (see Reading a French Wine Label and Reading an Italian Wine Label, page 356):

NEW EUROPEAN UNION WINE REGULATIONS

New European Union (EU) wine laws already established will be fully implemented in 2012. The new regulations are intended to make European wines more competitive in the international marketplace.

Wines of Origin are labeled Protected Designation of Origin PDO or Protected Geographical Indication PGI. PDO will be the equivalent of AOC in France and DOC/DOCG in Italy. PGT will be the equivalent of VDP in France and IGT in Italy.

The system works in parallel with established national designations (ex. AOC and DOC) and the national designation may be used in place of the EU designation. Base requirements have been established. As an example, PDO requires 100 percent of the grapes to come from a specific region.

Wines without Origin formerly classified VDT will be labeled as "vin," "vini," or "wine." Wines can be identified by varietal and may also identify the vintage. Wines can still just be labeled "red," "white," or "rosé."

Reading a French Wine Label

1   *Old vines*
2   *Selected vineyard of Burgundy*
3   *Estate bottled*
4   *Vintage*
5   *Estate name*
6   *Appellation*
7   *Quality classification with appellation*
8   *Name and address of producer*
9   *Alcohol content*
10 *Net contents*
11 *Country of origin*

Reading an Italian Wine Label

1   *Proprietary name/castle name*
2   *Varietal Brunello (Sangiovese), appellation Montalcino*
3   *Quality classification*
4   *Vintage*
5   *Estate name*

Name of the Wine (Brand) The name can be the estate where the wine was made, a local village, or a proprietary name. The brand name has a prominent position on the label. The brand name is usually not the name of the producer. The producer is printed on the bottom of the label.

Appellation All European wines have an appellation, which is listed on the center of the front label. Learn the major appellations of France, Italy, Spain, and Germany. Other European countries follow the same model. For classified wines (see Appendices 2 and 3, pages 380–447), only specific grape varieties are permitted in each appellation. Once you know an appellation, you can determine which grape varieties are likely to be included in the wine. European wine quality is largely based on appellation, which has been the focal point for hundreds of years.

Vintage This indicates the year the grapes were grown. In Europe the weather can be variable from year to year, changing the quality and taste of a wine in a particular appellation. The vintage is one indication of the characteristics and quality of a wine. A knowledgeable salesperson and wine publications are good sources of vintage information.

Alcohol Level Alcohol level can be a good indicator of climate and grape-growing and winemaking philosophy and style. Alcohol levels below 14 percent can indicate the use of traditional techniques demonstrated by lower sugar levels at harvest and crisp, acidic wines (see Chapter 4, Viticulture, page 40, and Chapter 5, Viniculture, page 70). Alcohol levels 14 percent or higher can indicate the use of modern techniques with riper grapes producing soft, fruity wines. In the European Union 0.5 percent is the allowable tolerance from the stated alcohol level. When 13.5 percent is printed on the label, the actual alcohol level in the wine will be between 13 and 14 percent. This is a more stringent requirement than the U.S. allowances.

Reading New World Labels

New World labels are easier to understand than Old World labels because they identify the variety and are printed in English. New World labels are created in either a traditional style or a designer style. It is easy to locate important information on a traditional label. Designer labels have the same key pieces of information, but they are often presented in a format that makes them less apparent to the reader (see sample labels, page 358–359). Important information on a New World label includes:

Name of the Wine (Brand) The name can be the brand name or the producer. An established brand name offers the assurance of quality, but a new brand that was created to sell the wine may generate interest and excitement. It can be of equal quality but has not yet developed a firm reputation.

Grape Variety The name of the grape, like Chardonnay or Pinot Noir, is printed on the front of the label if the wine is named for a grape. In the United States, a varietal wine, like Chardonnay, only requires 75 percent Chardonnay in the bottle. The back label may list the grapes that compose the other

Reading a Traditional California Wine Label

1 *Brand name*
2 *Varietal designation*
3 *Appellation*
4 *Vintage*
5 *Method of production*
6 *Name and address of winery*
7 *Web site (not required)*
8 *Alcohol content*
9 *Net contents*
10 *Sulfites declaration*
11 *Marketing information*
12 *Government warning statement*

25 percent. If the wine is a blend, the label may have a proprietary name or may simply state: White Wine or Red Wine. The back label may identify the grapes included in the blend. Australian labels typically add this information to the label.

Alcohol Content As with Old World wines, alcohol content can indicate ripeness, the techniques used, and the style. Alcohol levels 14 percent or higher can indicate that riper grapes were used and that the wine is likely to be fruitier. Wines below 14 percent alcohol are likely to be crisp and more acidic. The stated alcohol level does not necessarily indicate the accurate alcohol content of the wine in the bottle. The TTB permits alcohol levels to vary by 1.5 percent on wines with a printed alcohol content of 14 percent and below, and 1 percent in wines with stated alcohol over 14 percent.

Geographic Area The location or appellation, such as the American Viticultural Area (see Chapter 4, Appellations in the United States, page 51), is where the grapes were grown, not where the wine was made. A more specific location generally indicates (although not always) that the wine was made with specific grapes.

CORRA

(1)— CORRA

(2)— CABERNET SAUVIGNON

(3)— NAPA VALLEY

(4)— 2006

2006 • NAPA VALLEY • CABERNET SAUVIGNON

CORRA, A CELTIC DEITY REPRESENTING PROPHECY, APPEARS IN THE FORM OF A CRANE. AS THE LIGHT OF HER SHADOW CASTS A RAY OF PROMISE, THE EXPERIENCED HANDS CRAFTING CORRA DEVELOP THE POTENTIAL AND PROMISE OF YOUNG WINEMAKERS. —(11)

Celia Welch Masyczek

CELIA WELCH MASYCZEK, WINEMAKER

(5)— PRODUCED AND BOTTLED BY CORRA
(6)— NAPA, CALIFORNIA
(7)— WWW.CORRAWINES.COM
(8)— ALC. 14.5% BY VOL.
(9)— 1.5L
(10)— CONTAINS SULFITES

GOVERNMENT WARNING: (1) ACCORDING TO —(12) THE SURGEON GENERAL, WOMEN SHOULD NOT DRINK ALCOHOLIC BEVERAGES DURING PREGNANCY BECAUSE OF THE RISK OF BIRTH DEFECTS. (2) CONSUMPTION OF ALCOHOLIC BEVERAGES IMPAIRS YOUR ABILITY TO DRIVE A CAR OR OPERATE MACHINERY, AND MAY CAUSE HEALTH PROBLEMS.

Reading a Designer California Wine Label

1 *Proprietary name*

2 *Appellation with table wine designation*

3 *Alcohol content*

4 *Identification of grapes used to make wine (not required)*

5 *Marketing information*

6 *Method of production*

7 *Name and address of winery*

8 *Government warning statement*

9 *Sulfites declaration*

10 *Net contents*

11 *UPC code (not required)*

12 *Web site (not required)*

CALIFORNIA RED TABLE WINE

④— Middle Sister's blend of zinfandel, merlot and cabernet sauvignon.

⑤ Did you know that birth order is commonly believed to have a profound and lasting effect on psychological development? And that the middle sister has a greater chance of having a special wine named just for her? A sassy blend of our three favorite red varieties perfect for sipping before, during and after our favorite family meals. Some people are just born lucky. Now give me back my blouse.

www.middlesisterwines.com —⑫
We donate a dime a bottle to causes women care about.

⑥— VINTED AND BOTTLED BY MIDDLE SISTER —⑦
SANTA ROSA, CALIFORNIA

⑧— **GOVERNMENT WARNING:** (1) ACCORDING TO THE SURGEON GENERAL, WOMEN SHOULD NOT DRINK ALCOHOLIC BEVERAGES DURING PREGNANCY BECAUSE OF THE RISK OF BIRTH DEFECTS. (2) CONSUMPTION OF ALCOHOLIC BEVERAGES IMPAIRS YOUR ABILITY TO DRIVE A CAR OR OPERATE MACHINERY, AND MAY CAUSE HEALTH PROBLEMS.

⑨— **CONTAINS SULFITES**

⑩— 750 ML

⑪

Reading Beer Labels

The TTB requires beer labels to provide the same information as wine and spirits labels, except that the alcohol-content information is not mandatory but rather is allowed. Big brewers tend to have straightforward, traditional labels, whereas smaller microbreweries commonly develop creative labels for personalized brands (see Reading a Beer Label, at left). The most important information on a beer label is:

Brand Name Quality and consistency are closely tied to the brand.

Class Designation This tells you whether the beer is a lager, ale, wheat beer, stout, and so on. Each class of beer has different and distinguishing characteristics.

Alcohol Content Although alcohol content is not required by law, most beers indicate the level of alcohol.

Marketing Information The back label is often used effectively to identify ingredients and brewing methods, and sometimes to tell a story about the brand.

Reading a Beer Label

1 *Neck band with brand name and class designation*
2 *List of ingredients*
3 *Alcohol content (ABV)*
4 *Net contents*
5 *Name and address of bottler*
6 *Class designation*
7 *Brand name*
8 *Government warning statement*
9 *UPC code (not required)*
10 *Marketing information*

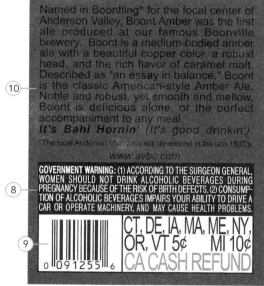

Named in Boontling* for the focal center of Anderson Valley, Boont Amber was the first ale produced at our famous Boonville brewery. Boont is a medium-bodied amber ale with a beautiful copper color, a robust head, and the rich flavor of caramel malt. Described as "an essay in balance," Boont is the classic American-style Amber Ale. Noble and robust, yet smooth and mellow, Boont is delicious alone, or the perfect accompaniment to any meal.

It's Bahl Hornin' *(It's good drinkin').*

The local Anderson Valley dialect, developed in the late 1800's.

www.avbc.com

GOVERNMENT WARNING: (1) ACCORDING TO THE SURGEON GENERAL, WOMEN SHOULD NOT DRINK ALCOHOLIC BEVERAGES DURING PREGNANCY BECAUSE OF THE RISK OF BIRTH DEFECTS. (2) CONSUMPTION OF ALCOHOLIC BEVERAGES IMPAIRS YOUR ABILITY TO DRIVE A CAR OR OPERATE MACHINERY, AND MAY CAUSE HEALTH PROBLEMS.

CT, DE, IA, MA, ME, NY, OR, VT 5¢   MI 10¢
CA CASH REFUND

0 09125 5 6

Reading Spirits Labels

Many brands of spirits have been made for decades and their labels are traditional and easy to read (see Reading a Spirits Label, at right). The important information on a spirits label is:

Brand Name The brand name is tremendously important because it represents a specific quality to the buyer.

Class Designation The class of spirit printed on the label—vodka, Scotch, brandy—identifies the type of spirit being purchased or ordered. The brand also conveys that information to the buyer.

Country of Origin Origin is an additional indication of quality. Many spirits were developed in specific countries or regions, and quality is tied to that location, such as Scotch from Scotland, gin from Holland or England, and bourbon from Kentucky.

Alcohol by Volume ABV indicates the potency of the spirit. Alcohol levels vary tremendously from one type of spirit to another.

Reading a Spirits Label

1   *Brand name/producer*
2   *Class designation*
3   *Name and address of producer*
4   *Net contents*
5   *Alcohol content (ABV)*
6   *Marketing information*
7   *Government warning statement*
8   *UPC code (not required)*

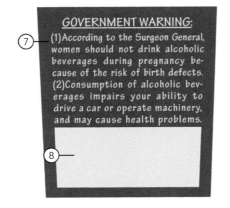

BUYING ALCOHOLIC BEVERAGES IN RETAIL STORES

Many types of outlets sell wine, beer, and spirits, each with a different focus. Availability in each type of store varies by state, but in general alcoholic beverages are sold in liquor stores, supermarkets, department stores, drug stores, specialty stores, big-box stores, convenience stores, state stores, and at wineries, breweries, and distilleries. In addition, online businesses are becoming major sources for wine, beer, and spirits purchases. Each of these outlets has its own strengths and weaknesses in terms of selection, price, and customer service.

Selection

In the United States, each state controls how and where alcohol is sold (see Chapter 2, State Laws, page 16). You will need to determine the exact requirements and restrictions in your state. State alcoholic-beverage control boards maintain Web sites with this information.

With this in mind, some generalizations can be made about outlets that sell alcoholic beverages. In the past retail outlets tended to focus on only one type of alcoholic beverage. For example, liquor stores sold spirits and wine shops sold wine. While some outlets, most notably specialty shops, maintain a narrow focus, many retail stores are now selling all types of alcoholic beverages. In general, the larger the store, the wider the selection of popular brands that is stocked. Some premium brands may only be available in specialty shops.

Regional differences in availability also exist. European wines are commonly sold on the East Coast, but a broad selection is harder to find on the West Coast, where Californian or Australian wines are more prevalent. Stores in the Midwest and the South may carry more brands of bourbon, while in California you are more likely to find a greater selection of tequila.

Whether they sell wine, beer, spirits, or some combination of the three, specialty stores offer the widest selection of beverages from major brands to handcrafted premium brands. Each shop has a different focus based on the owner's preferences and interests. Some emphasize wine, others focus on high-end spirits, and some larger stores carry every type of alcoholic beverage. Some focus on wine, beer, or spirits from a particular country or region, such as France or Britain, often selling the foods from those countries as well.

Supermarket offerings often depend on the size of the store. They generally sell the greatest variety of beer and focus on brand-name spirits at moderate prices. Wine departments range from those that carry a relatively limited selection of inexpensive wines to those that feature large, well-stocked departments with excellent wines.

In many areas liquor stores, which traditionally sold primarily spirits and beer, now often include some wine in their inventory. Drug stores may carry a limited selection of major-brand alcoholic beverages at competitive prices, and convenience stores might sell only a few items for the grab-and-go customer. Although they typically offer a limited range of brands, big-box stores have become some of the largest retailers of alcoholic beverages.

Wineries, breweries, and distilleries sometimes sell their products directly in their tasting rooms, which gives the customer the opportunity to sample the beverage before making a purchase. Frequently, specialty handcrafted beverages are made in limited quantities and are only available for purchase on-site.

Internet retailers offer the greatest selection of wine, beer, and spirits, but they can only be purchased in states into which alcoholic beverages can legally be shipped. Internet retailers sell widely available brands as well as beverages from regional and international producers that may not be available in the local market. Again, each state has laws governing shipping alcoholic beverages.

Shopping at a specialty store, such as Wally's, the highest-rated shop in Los Angeles, provides wide choices and excellent customer service.
STEVE WALLACE

Price

Each state's alcoholic beverage control board establishes its own pricing guidelines. Most states allow for some form of competitive pricing, but in others the licensing agency controls the price no matter what type of retail outlet sells the beverage (see Chapter 2, Control and Competitive States, page 18). Alcoholic-beverage control states like Pennsylvania have only one wholesaler and retailer, the state itself.

Generally the larger the store or chain, the lower the price charged for wine, beer, or spirits. High sales volume allows chain retailers to negotiate lower wholesale prices from producers and distributors, and therefore, they can set lower profit margins to offer lower retail prices. Big-box stores offer some of the lowest prices because sales volumes of the products they carry are extremely high. Chain drug stores and supermarkets always price competitively and use extensive discounting and sales to increase customer traffic. Independent specialty shops generally have higher prices because of increased selection and low average sales volume. In addition, they tend to carry a greater number of high-end beverages.

Pricing is generally competitive for wine, beer, and spirits purchased online, but shipping adds an additional cost, although some Internet retailers include free shipping with a minimum purchase. In general the most expensive place to purchase alcoholic beverages is at a winery, brewery, or distillery. The producers do not want to undercut the retailers who are selling their products. Also, the beverages are often premium brands, made with high-quality products that can command higher prices.

Finally, beverage pricing is being impacted by online sites, like wine-searcher. com, which can inform the consumer about who is selling a particular wine at what price. Prices can be compared locally, by state, or even nationally. Retail pricing is affected because consumers can look for the lowest price for the identical product.

Customer Service

A skilled salesperson can be a great help in choosing the right beverage. A salesperson should demonstrate general knowledge of the product, whether wine, beer, or spirits, and particular knowledge of the products available in his or her store. He or she should be able to answer questions about style, characteristics, value, food considerations, and if appropriate, mixed drinks and cocktails. Be wary of a salesperson who recommends only those beverages at the top of, or beyond, your price range, or who only quotes beverage magazine ratings. A salesperson should be more interested in winning your long-term business than making an immediate sale.

Knowledgeable sales staff are usually found at high-end stores and independently owned shops where the focus is on customer service. Although you will probably pay more at a specialty shop, you generally receive better customer service and find a better selection than at many chain retailers. In many cases, the best service is at a winery, brewery, or distillery, where you can speak to someone with firsthand information about the product, and sometimes you may even talk to the person who made the beverage.

Finding Additional Information

Besides the direct assistance you can receive from a salesperson, there are a number of places to go for information that will help you make beverage selections. Most stores that sell alcoholic beverages have shelf talkers. This is a small card placed in front of a beverage that includes a description of its characteristics, awards won, or ratings from beverage industry professionals and writers. Specialty shops often publish newsletters that are available in-store and as e-newsletters that are full of beverage reviews.

Dozens of magazines, newspapers, and newsletters geared to specific alcoholic beverages offer helpful information about new products, tasting and evaluating beverages, and purchasing beverages. The Internet also contains a wealth of information. There are hundreds of sites that offer educational materials, commentaries from experts, and blogs from interested consumers who offer third-party endorsements of specific beverages. The Internet can also steer you to books, magazines, and other print publications.

Purchasing Wine

Before you venture into a store to buy wine, you need to know your budget, the occasion, the style, and whether you plan to drink the wine soon or age it.

It may also be helpful to shop in a store that has a wide selection of wines and a helpful sales staff.

Budget

Before shopping for a bottle of wine, know how much you want to spend. Break it down to a per-bottle amount whether you are buying a couple of bottles or a case or two. If you are purchasing multiple bottles, you may want to consider cost averaging: splurging on one bottle while being budget conscious on another.

Your focus will be quite different when choosing between a bottle under $10 and one at $50. At the lower price point, your options might be more limited, but it may also force you to try a new varietal, brand, or region. As your purchasing price point increases, your options increase as well. Keep in mind that price is not always an indication of quality. Some of the best wines we have tasted have been bargains. There is nothing more enjoyable than finding a low-priced wine and after tasting it discovering that it is worth a bit more than you paid.

Occasion

What is the occasion or purpose for the wine you are buying? Is it for a simple dinner with burgers or pizza, or is it for an extraordinary meal with special guests? Are you looking for a wine to go with a particular dish? What you drink with poached fish may be different from what you serve with grilled steak. Are you going to drink it yourself, or is it a special gift? Are you trying to expand your knowledge with something unfamiliar, or are you looking for your favorite style of wine? The answer to each question will help you make your selection.

Style

Your first buying decision may be as simple as deciding between a white wine and a red wine. As your knowledge and sophistication develop, you should start paying attention to other style factors. Do you prefer big, powerful, fruity wines, or do you like wines with restrained fruit or nonfruit characteristics? Do you like wines with high acid levels like Riesling, Sauvignon Blanc, or Sangiovese? Do you prefer wines made from cool climates or warm climates? Do you like oak flavors and aromas from new oak barrels and heavy barrel toasting, or do you prefer wines with less oak character? Each factor will help you select a particular style of wine. It can also guide you to trying something new and different.

Aging Potential

Are you buying a wine to drink within the next few weeks or months, or do you plan to cellar it for future consumption? If you are serious about becoming a wine collector, you must familiarize yourself with important factors before you begin making purchases of wines to age. Variety, total acidity, pH, tannins, and alcohol level all affect a wine's aging potential. Considering such factors will help you decide if a wine can be aged. Some of this information is unlikely to be printed on the label, so you will have to research the wine before buying it.

Total acidity and pH are frequently available on the producer's Web site under technical information or with a call to the winery.

The producer and appellation are also important elements in determining how well a wine ages. Most New World wines are made for immediate consumption. Only those that are made in a traditional style are likely to have a long life. Old World wines from great producers are still the best bet for cellaring.

A knowledgeable salesperson can be very helpful when you are making a decision about aging wines because he or she may understand the finer points about the wine's quality. If you plan to age wines, keep in mind that you must have a proper storage facility with temperatures of no more than 62°F/14°C and at least 70 percent humidity.

Purchasing Beer

The decision on which beer to buy is generally based on brand, class, and price. Major domestic brands, like Budweiser and Miller, and international brands, like Heineken, dominate beer sales, but a range of smaller craft brewers are in regional, and sometimes national, distribution. Local brewers may also make and sell beer of exceptional quality in their local markets. Many of the major domestic brewers have purchased smaller craft producers, so you may not always be able to tell who actually owns a brand without some investigation. Producers' Web sites, local or trade press, or beer tastings are the best ways to learn more about available brands.

Knowledge of the different classes of beer is also extremely helpful. Within the major classes of ales and lagers, there are many subcategories of beer, each with its own unique characteristics (see Chapter 19, Beer Styles, page 321).

The common brands of beer are available in almost every kind of retail outlet. Craft brews must be purchased in specialty shops, brewpubs or breweries, and online.

Beer is generally the most affordable alcoholic beverage. The large national brands typically have the lowest cost, while craft beers are priced the highest. Beer is usually sold by six-pack, 12-pack, or 24-bottle cases. The larger the pack, the lower the price per bottle is likely to be. A six-pack of a major brand sells for under $8, a 12-pack averages $15 or less, and a 24-bottle case goes for $25 or less. A 22-ounce bottle of craft beer costs from $4 and up. You can also purchase beer in quarter kegs, half kegs, and full kegs, which are good for large groups.

Beer quality begins to deteriorate immediately after it is brewed and bottled. Beer stored under constant refrigeration is best. In a retail store, beer that is not refrigerated should be turned over quickly. Ideally beer should be consumed within six months of bottling, even sooner if possible. Some brewers date-stamp their beer so you know how fresh it is, but most do not. To ensure freshness you should buy beer from a reputable retailer who uses proper storage and handling with continuous turnover of its beer stock.

Purchasing Spirits

When you purchase spirits, important considerations include quality level and brand name. The brands for each spirit type are generally categorized into quality levels by price with the higher-priced selections generally of higher quality. As with wine and beer, you can find small distillers that specialize in one type of high-end beverage, grappa for example.

Which brand and quality level to purchase depends largely on how the spirit will be served. The highest-quality spirits should be served on their own, on ice, or with water, so their taste can be savored. Some cocktails, like classic Martinis, also require high-quality spirits because they depend on the distinct flavor characteristics of the primary spirits—in the case of Martinis, gin and vermouth. Most cocktails, however, contain highly flavored ingredients that mask the taste of the spirit. It is almost impossible to differentiate between brands, so generic, low-priced spirits may be adequate.

Like beer, moderately priced brand-name spirits are found in almost any kind of retail store. However, for premium beverages you have to shop at specialty stores, the distilleries, or online. At the low end, a 750-milliliter bottle may cost under $20. The average price range is $20 to $50. Some premium-brand spirits, such as small-batch bourbons, single malt Scotch, and Cognac, can sell from $50 to several hundred dollars. Spirits are often sold in bottles that are works of art, which adds to the cost. Spirits last indefinitely, even after they are opened.

ORDERING ALCOHOLIC BEVERAGES IN RESTAURANTS

In restaurants wine, beer, and spirits add to the total dining experience. Spirits create a wonderful start and end to a meal and, occasionally, can be a complement to food. Beer is generally a more casual beverage that goes particularly well with a broad range of everyday foods, while wine is served with every kind of meal. One of the advantages of ordering beer or spirits is the simplicity of their service. Wine service, on the other hand, requires a long-established formal process that requires both the server and the guest to follow specific steps.

Ordering Wine

If you are dining in an upscale restaurant in a major city or resort, like San Francisco, New York, or Las Vegas, you are likely to receive very sophisticated wine service from professional sommeliers. However, most restaurants do not offer that level of service, so it will increase your enjoyment if you know how to review a wine list, select and order wine, and evaluate proper service.

In most restaurants wine is sold by the bottle or the glass. Flights of wine are also becoming increasingly popular in wine bars and upscale restaurants. Service is different for each of these options.

Wine Lists

Restaurant wine offerings vary widely from those that feature only one house white and one house red to those that have two or three dozen wines or more. Fine-dining establishments often have very extensive wine lists with hundreds of choices listed in a book.

To make all but the most basic decision, you must be able to understand a wine list. There are two styles of wine lists: A standard wine list is structured either by varietal or by country and appellation; a progressive wine list is organized by wine style from light to medium to heavy body. In Europe standard lists are usually organized by appellation rather than varietal, while lists that segregate wines by varietal are more common in the United States and other New World countries. Progressive wine lists are becoming more popular in the United States but still constitute a small percentage of wine lists.

Whether standard or progressive, the list should provide the varietal or blend, vintage, producer, appellation, and the price for each wine offered. When you receive the wine list, the first considerations are generally price, wine style, and what food you are going to eat. If price is important, start reading the wine

Frequently Asked Questions About Wine Service

How many bottles should I plan to order for dinner? The average diner drinks two 6-ounce glasses of wine. You should plan on a half bottle per person.

I am with a large party. How far can I stretch a bottle of wine? A bottle normally serves four 6-ounce glasses, but it can serve up to eight guests, each with a three-ounce pour. This allows enough for a toast and a sip or two, but it is not enough to accompany a meal.

My table has just finished a bottle of wine and has ordered a different wine. Should the server bring new glassware? Absolutely. Every guest should receive a new glass for every new wine. The server should pour a small amount of the new wine for the host's approval.

Is it acceptable to bring your own bottle of wine to a restaurant? Yes, in many cases. In some states restaurants that do not have a wine license can allow you to bring your own wine. These restaurants are called BYOB (Bring Your Own Bottle). If the restaurant provides appropriate glassware and directs the wine service, the host should add an appropriate gratuity to the bill. Many restaurants that have wine lists also allow you to bring your own bottle of wine. Most charge a corkage fee ranging from $5 to $50. Generally the more sophisticated the restaurant the higher the corkage fee.

Here are some rules to follow:

- Do not bring a commonly available inexpensive wine.
- Do not bring a wine that is likely to be on the restaurant's wine list.
- Bring a wine that is special, unique, or interesting.
- Call in advance for approval or inform the manager or maître d' upon arrival.
- If multiple bottles will be served during the meal, purchase at least one bottle from the restaurant's wine list.
- Pour a glass for the server, manager, or chef if you have brought a special wine. They are likely to appreciate the opportunity to try it, and you might find the corkage fee waived.

STANDARD WINE LIST

Chardonnay

Alamos Mendoza, 2008	$20
Robert Young, Alexander Valley, 2006	$40
Patz & Hall, Dutton Ranch, Russian River, 2005	$42

Sauvignon Blanc

Brander Sauvignon Blanc, Santa Ynez V., 2008	$24
Pascal Jolivet, Sancerre 2008	$28

Riesling and Gewürztraminer

Chateau Ste. Michelle "Eroica" Riesling, 2008	$33
Fess Parker White Riesling, Santa Barbara, 2008	$21
Navarro Gewürztraminer, Anderson V., 2009	$24

Other White Wines

Champalou, Vouvray, 2009	$25
Pazo Senorans Albarino, Rias Baixas, 2009	$31
Bridlewood Viognier, Reserve, Santa Barbara C., 2007	$25

Pinot Noir

Hitching Post, Santa Rita Hills, 2005	$40
Dog Point Vineyard, Marlborough, 2005	$45
Vincent Girardin Burgogne Cuvée, 2004	$34

Cabernet Sauvignon

Frog's Leap, Napa V., 2006	$52
Louis Martini Monte Rosso Vineyard, Sonoma C., 2004	$55

Syrah

Eberle Steinbeck Vinyard, Paso Robles, 2005	$31
Bernard Burgaud Cote-Rotie, 2004	$60
Groom Shiraz, Barossa V., 2002	$50

Other Red Wines

Savignola Paolina Chianti Classico, 2006	$28
Viña Tondonia, Rioja Crianza, 2002	$52
Howell Mountain Zinfandel, Old Vines, 2006	$30

PROGRESSIVE WINE LIST

Fruitier Whites - *easy sipping with light foods*

Chateau Ste. Michelle "Eroica" Riesling, 2008	$33
Champalou, Vouvray, 2009	$25
Pazo Senorans Alarino, Rias Baixas, 2008	$31

Medium, Drier Wines - *for foods with a kick!*

Brander Sauvignon Blanc, Santa Ynez V., 2008	$24
Planeta La Segreta Bianco, Sicilia, 2009	$20

Full, Rich, Dry Wines - *great with chicken and fish*

Patz & Hall, Dutton Ranch, Chardonney, 2005	$42
Bridlewood Viognier, Reserve, Santa Barbara C., 2007	$25
Navarro Gewürztraminer Anderson V., 2009	$24

Sparkling Wine - *for celebrating with great friends!*

Roederer Estate Brut NV, Anderson V.	$30
Piper-Heidsieck Brut Champagne, NV	$42
Borgo Magredo Prosecco, NV	$24

Medium Body Reds - *more delicate, yet full flavor*

Savignola Paolina Chianti Classico, 2006	$28
Dog Point Vineyard Pinot Noir, Marlborough, 2005	$45
Vina Tondonia, Rioja, Crianza, 2002	$52

Full, Spicy Reds - *great with our burgers!*

Banfi Rosso di Montalcino, 2006	$29
St. Francis Merlot, Sonoma C., 2007	$27

Big Reds - *full flavor wines for all flavor dishes*

Howell Mountain Zinfandel, Old Vines, 2006	$30
Louis Martini Cabernet Sauvignon Monte Rosso, 2004	$55
Groom Shiraz, Barossa V., 2002	$50

Dessert Wines

Banfi Rosa Regale, Acqui, 2008	$7.50
Quady Essensia Orange Muscat, California	$6.00
Galleano 3 Friends Port, Cucamonga	$6.00

list from right to left, looking at the price first to determine which wines fall within your price range.

Wine Markup

Wine markups vary in different restaurants, but the standard markup ranges from two to four times the wholesale cost of the wine. A restaurant that pays $10 for a bottle of wine sells it for $20 to $40. The restaurant must mark up the wine to cover the costs of service, glassware, and overhead, and also allow for some profit. Higher markups should come with better service, better glassware, and a better wine selection.

Wines by the glass usually have a higher markup than wine by the bottle to cover the risk of not selling all the wine in the bottle. The markup is likely to be three to five times the wholesale price.

If you are by yourself, it may be prudent to order wine by the glass. If two or more people are ordering, a bottle will provide the better value, but all guests will have to agree on the wine.

Selecting Wine

There are two approaches to selecting wine to go with a meal: Choose a wine first and then order food courses to complement the wine, or decide which food courses to order and then select a wine to pair with the food. Most restaurant owners develop the food menu first and then create a wine list to complement the food. That is also the typical way diners order food and wine.

If your server asks for your drink order as soon as you arrive at the table, order an aperitif, a cocktail, a glass of wine, or water first, and review both the food

Most frequently used wine glasses, from left:
All-purpose, Sparkling, Burgundy, Bordeaux
TOM ZASADZINSKI/CAL POLY POMONA

menu and the wine list before ordering a bottle of wine. If the restaurant offers half bottles, which contain two 6-ounce pours of wine, think about ordering two different types of wine in half bottles. Doing this enables you to make a better match between the wine and food courses. You also have the option of ordering a different glass of wine with each course.

A well-informed staff member can suggest interesting wines based on your budget and food choices and may be able to recommend wines that are not on the list that will meet your needs. If the staff's immediate response is to suggest only the more expensive wines on the list, you may prefer to depend on your own skills to choose your wine.

Ordering a Bottle of Wine

When ordering wine by the bottle at a table-service restaurant, whether it is a three-star restaurant with a sommelier or a more casual establishment, the procedures for opening a bottle of wine adhere to a standard sequence. Less formal restaurants may not include all of the steps, but more formal restaurants should provide more exacting procedures. The important steps are listed in Table 21.1.

Here are key points to follow when ordering and being served a bottle of wine:

1 Ask questions about wines that are unfamiliar but look interesting.
2 Remember the important information from the list—producer, variety, vintage, and appellation—and make sure the wine that is brought to the table is exactly the same as the one you ordered.
3 After the wine has been opened and a taste has been poured for you, swirl the glass to aerate the wine, which intensifies its aromas. Smell the wine, take a sip, and approve the wine unless you think there is something wrong with it.

Table 21.1 Sequence of Wine Service When Ordering a Bottle of Wine

SERVER RESPONSIBILITY	HOST RESPONSIBILITY
Present wine list to host	Accept list
Ask if there are questions, make recommendations	Ask questions, get clarification on wines
Accept, restate order for clarification	Place order
Place correct glassware on table	
Present bottle (ensure producer, brand, vintage, appellation/vineyard correct to wine list)	Approve accuracy of selection
Open bottle (cut foil, remove cork or screw cap, clean top of bottle)	Observe process
Pour 1 oz. in host's glass	Swirl, smell, taste, approve, or reject
Serve guests	
Serve host last	Propose toast (if appropriate)
Place bottle of red wine on table, place bottle of white wine in ice bucket or on table according to host's preference	

Table 21.2 Proper Serving Temperatures for Wine

WINE CATEGORY	WINE STYLE	FAHRENHEIT TEMPERATURE	CELSIUS TEMPERATURE
Red wine	Light, fruity	50°– 55°	10°–13°
	Medium-bodied	56°–60°	13°–16°
	Full-bodied	60°–64°	16°–18°
White wine	Crisp, light, fruity	48°– 54°	9°–12°
	Full-bodied	54°– 59°	12°–15°
Rosé		50°– 54°	10°–12°
Sparkling wine		43°– 50°	6°–10°
Dessert wine		43°– 47°	6°–8°
Dry sherry		43°– 47°	6°–8°
Sweet sherry and port		54°– 60°	12°–16°

If you reject it, you should be able to say specifically what is wrong, for example, "It is oxidized" or "It is corked." You can ask your server to open another bottle. You should not reject a wine because you do not like its taste. However, if you are honest with your server, he or she is likely to allow you to make another choice.

4 If a white wine is too cold, let the bottle sit on the table to allow it to come up to the proper temperature before it is served. Table 21.2 shows proper serving temperatures for different types of wine.

5 If a red wine is too warm, request an ice bucket with a slurry of ice and water. Place the bottle up to its shoulder in the ice bucket for five minutes to bring the wine down to proper serving temperature. Then set the bottle on the table.

6 Make sure the server does not fill the glass to the rim. Overpouring is the most common problem with poorly trained servers. Ask the server to fill the glasses about one-third full, allowing room to swirl the wine.

7 A wine ordered with a particular course should be poured before the food arrives.

Ordering Sparkling Wine

Sparkling-wine service is similar to table-wine service with the following variations:

1 Sparkling wine must remain ice cold, so it should be brought to the table with an ice bucket containing enough ice and water to immerse the bottle up to its shoulder.

2 The server removes the foil, releases the cage from the bottle, and gently removes the cork. The sound of the cork release should resemble a sigh, and no wine should foam out of the bottle.

3   Sparkling wine should be served in a flute, a tall cylindrical glass designed to help maintain carbon dioxide and showcase the bubbles or (bead) in the wine.

4   After the wine is poured, the remainder should be kept in the ice bucket.

Ordering Wine by the Glass

Ordering wine by the glass is extremely common and popular with many restaurant guests. At casual restaurants, wine is sometimes poured from a jug or box into a carafe, which is placed on your table, or you might be given two or three selections and the wine will be poured into a glass before it is brought to the table.

At more sophisticated restaurants, a good wines-by-the-glass program offers several choices from a list and the wine is poured from the bottle at the table. Just as with wine by the bottle, service for wine by the glass follows a sequence of steps, which is outlined in Table 21.3.

Here are some key points to follow when ordering and being served a glass of wine:

1   A standard wine-by-the-glass pour is five to six ounces.

2   Request to see the bottle to ensure that you are receiving the wine you ordered. The most frequent discrepancies between what you ordered and what is being poured are with vintage, appellation, or vineyard designations. There can be significant differences in quality between vintages of the same wine or between different appellations or vineyard designations from the same producer.

3   The wine should be fresh. A very busy restaurant typically serves multiple bottles of wine by the glass each day, preventing its wines from becoming tired and stale. If a bottle is more than one day old, ask the server to open a new bottle. There are systems that control freshness by injecting nitrogen gas into the bottle to displace oxygen. Some restaurants use this as a marketing tool, and it does ensure freshness, but most restaurants simply cork the bottles and put them in the refrigerator overnight to help retard oxidation.

4   If the glass is too small to swirl the wine, ask for a larger glass.

Table 21.3 Sequence of Service When Ordering Wine by the Glass

SERVER/BARTENDER RESPONSIBILITY	GUEST RESPONSIBILITY
Present wine by glass list to guest	Review list
Ask if any questions, make recommendations	Ask questions about wine offerings
	Place order
Place correct glassware in front of guest	
Present bottle (ensure producer, brand, vintage, appellation/vineyard correct to wine list)	Approve accuracy of selection
Pour 1 oz. for guest	Swirl, smell, taste, approve, or reject
Complete wine pour	

Wine Flights

Wine flights are a great way to sample several different kinds of wines at one time. You can try something new without spending a lot of money on a bottle that you might not like. Flights are normally composed of small pours of three or four wines that have a particular theme. For example, a flight might consist of Pinot Noirs from different countries or appellations, or it might pair a different style of wine with each course of food that you ordered.

The process is the same as ordering and being served wine by the glass. However, the wines are preset, so you do not have to decide which wines you want to order.

Ordering Beer

Because beer is the least expensive choice of alcoholic beverages in restaurants, it can be an affordable alternative if wine prices are excessive. It is a great thirst quencher by itself or is a good accompaniment to food. As with wine, you want to match the style of beer to the type of food that is ordered. (See Chapter 19, Beer and Food Pairings, page 327, for recommendations.)

In a restaurant your beer choices can be limited to a few national brands, or you may be presented with a list of several dozen beers that includes international brands and craft beer. Beer is offered on tap (draft) or by the bottle. Draft beer is sold by the 12-ounce glass, the 16-ounce pint, the 24- or 32-ounce schooner, and the pitcher. Bottles usually contain 12 ounces, although craft and imported beer is often sold in 22-ounce or 750-milliliter bottles.

If you order a bottle, the server often brings a frosted glass along with it. If it is a light lager, that may be fine. However as with wine, the aromas and flavors of many beers are masked if they are served too cold or poured into a cold glass. Ask for an unchilled glass. When you order beer on tap, the server brings the glass from the bar to the table. Whether it is poured into the glass from the tap or a bottle, it should have a foamy head. If it is flat, request a new bottle or glass.

Ordering Spirits

Spirits are generally ordered before a meal in the form of a mixed drink or cocktail or, perhaps, an aperitif to stimulate the appetite. At the end of a meal brandy or liqueur often replaces dessert, or a digestif might be served to aid digestion. Spirits are rarely served with a meal because their high alcohol content is often in conflict with food. Spirits are usually poured or mixed into a cocktail at the bar and brought to your table. (For recommendations on serving spirits with food, see Chapter 20, Spirits and Food Pairing, page 353.)

Bartenders are familiar with the many types of spirits and know how to mix hundreds of cocktails. Unless you have a great deal of experience ordering drinks, you want to rely on their expertise. Your decision should take into account the price you want to pay and the quality of the spirit.

Spirits fall into three basic quality and price categories: well, call, and premium. A fourth, super premium, may be found in some restaurants. Well, the least expensive option, is what the restaurant uses to make a drink if you do not ask for a particular brand. Call is a quality brand and premium is a more exclusive brand. If you order a vodka Martini, the bartender will use whichever brand of vodka the restaurant has purchased as its well vodka. Generally, this a generic brand, although some establishments select a high-quality call or premium spirit as the house well brand. If you ask for a Smirnoff Martini, the bartender will make it using Smirnoff vodka and charge a higher call price. If a Grey Goose Martini is ordered, the bartender will make it using Grey Goose vodka, charging the even higher premium price.

Some restaurants highlight specialty cocktails and have trained mixologists who prepare them. Most restaurants offer a wide selection of many kinds of spirits, but some specialize in particular spirits like single malt Scotch or tequila. They also may offer dozens of choices of their specialty in a wide price range and carry some of the best and most expensive brands in the marketplace.

Serving wine by the glass has become an important component of restaurant wine service.
WILFRED WONG

SUMMARY

Purchasing and ordering wine, beer, and spirits in a retail store, over the Internet, or in a restaurant is easy, but as with any consumer product, you must take the time to familiarize yourself with the range of choices available and the appropriate handling of each beverage in order to successfully navigate the retail shelf or menu.

The place to start is learning how to read and interpret beverage labels, so you know that the bottle contains what you want to buy. Each time you taste and explore different beverages, you add to your knowledge of wine, beer, and spirits. As you gain more confidence, do not hesitate to take charge to get the beverage and service you want, particularly in a restaurant.

TTB Requirements and Regulations for Alcoholic Beverage Labels

In the United States all wine, beer, and spirits labels are controlled and processed by the Alcohol and Tobacco Tax and Trade Bureau (TTB). The TTB authorizes domestic and international labels before they are placed on bottles for sale. Individual states also have jurisdiction over alcohol labeling. They can add, but not take away, any requirements on a label approved by the TTB. What follows are the specific TTB requirements on alcoholic beverage labels in the categories of wine, malt beverage (beer), and spirits.

Requirements for Domestic and Foreign Wine

Numbers in the following text refer to numbers on labels.

Brand (1) The name used to identify the product. Commonly it is the producer's name but it can also be a proprietary name. The brand may not mislead the consumer as to appellation, origin, or any other characteristic.

Vintage Date (2) The year the grapes were grown, not when the wine was bottled.

- If a vintage date is shown, an appellation of origin, smaller than a country, must also be shown.
- 85 percent of the wine must be from the vintage year if the wine has a state or county appellation or foreign equivalent.
- 95 percent of the wine must be from the vintage year if it came from a specific American Viticultural Area (AVA) or foreign equivalent.

Varietal or Other Designation (3) Wines are not required to have a grape-variety designation. Wines may simply be identified as white, red, rosé, or dessert wine.

- If identified by varietal, such as Chardonnay or Cabernet Sauvignon, 75 percent of the named grape variety must be used in the wine, and 75 percent of the named grape variety must come from the appellation stated on the label.
- *Labrusca* grape varieties, like Concord, require only 51 percent of the named variety to identify that wine by varietal name.
- Wines from other countries, particularly in Europe, can be identified by appellation rather than by grape variety. They can also be identified by combining variety and place, proprietary name, and legendary name.

Appellation of Origin or AVA (4)

- The name of the place where the grapes were grown. It can be a country, state, county, or a specific area defined by the TTB based on its characteristics, called an AVA in the United States.
- The foreign equivalent, which also identifies the wine by appellation rather than by variety. Each wine-producing country has its own system for defining and naming specific appellations.
- 75 percent of the appellation-named wine must come from that appellation.

Estate Bottled Appears on the label only if 100 percent of the wine came from grapes grown or controlled by the winery, and the wine was made and bottled at the winery in a continuous process. The winery must be within a specific appellation or AVA.

Producer or Importer Name and Address (10) The name of the producer or the company importing the wine from a foreign country. It must include the address (city and state) of the winery or principal place of business.

– The name of the winery that is responsible for the contents in the bottle must appear on the label. In most cases, the listed winery produces the wine. However, a winery may use its name on a label for a wine that was produced by someone else.

– Along with the producer's name, there must also be a statement clarifying the winery's role in making the wine **(5)**:

– Grown, produced, and bottled by – Equivalent to "Estate Bottled" (see above). The wine came from grapes grown or controlled by the winery, and the wine was made and bottled at the winery in a continuous process. The winery must be within a specified AVA, or

– Produced and bottled by – A minimum of 75 percent of the wine in the bottle was produced and bottled by the named winery, or

– Vinted and bottled by – Does not indicate the origin of the grapes or who made the wine. The wine was cellared or stored at the named winery, or

– Cellared and bottled by – Means exactly the same as vinted and bottled by.

Net Contents The total volume of wine stated in metric units, ranging in size from 50 milliliters to three liters (can be blown into glass). Oversized bottles must be made to hold an even number of liters. The standard bottle size is 750 milliliters (25.4 ounces).

Alcohol Content (9) Alcoholic strength is indicated by percentage of alcohol by volume (ABV). In the United States, regulations require that there be no more than 1.5 percent difference from the stated amount when alcohol content is 14 percent or below. There can be no more than 1 percent difference from the stated amount when it is above 14 percent.

Country of Origin For imported wines, a statement indicating country of origin is required.

Government Health Warning Statement (8) The standard consumption-risk warning on the back label of every bottle of alcohol.

Net contents blown into glass not shown on label.

Numbers above refer to text on label requirements.

Declaration of Sulfites (6) A statement "Contains Sulfites" for any wine with 10 or more parts per million (ppm) of sulfur dioxide is required if the wine is destined for interstate commerce.

Back Label Additional marketing information **(11)**, which has few restrictions, may be printed on the back label. It may include information on vineyards, varieties, processing, and flavor characteristics, as well as website **(7)**, telephone number, and promotional information.

Although the TTB has established these criteria for international labels, many European requirements are more stringent.

Requirements for Beer (Malt Beverages)

Numbers in the following text refer to numbers on label.

Malt beverage is the term used by the TTB to include all classes of beer as well as flavored malt beverages. The following information is required on an approved TTB malt beverage label:

Brand (1) The name used to identify the malt beverage. Commonly it is the brewery's name but it can also be a proprietary name. The brand may not mislead the consumer as to type of malt beverage or its origin.

Class (7) The largest class of malt beverage is beer, followed by the two subclasses of beer: ale and lager. Types of ale or lager, like wheat beer, porter, and stout, are allowed as the identifying class on a label.

Bottler or Importer Name and Address (2) The name of the brewery (producer) or the company importing the product from a foreign country. It must include the address of the brewery or principal place of business.

Alcohol Content (3) The TTB "allows" (but does not require) breweries to state the alcohol content of beer. Increasingly more producers identify ABV. Some states require that alcohol content is printed on the label.

Net Contents (5) The total volume in the container must be stated in English units (ounces, pints, etc.).

Country of Origin For imported malt beverages, a statement indicating country of origin is required.

Draft/Draught Indicates the beverage has not been pasteurized. Terms such as *draft brewed* or *draft beer flavor* can only be used if the label includes a statement that the beverage has been pasteurized.

Light/Lite/Low-Carb Indicates a lower calorie content than a "standard" product. The calories, carbohydrates, protein, and fat content must be printed on the label. Low-carbohydrate can only be used if the beer contains no more than seven grams of carbohydrates per 12-ounce serving.

Government Health Warning Statement (4) The standard consumption-risk warning on the back label of every bottle of alcohol.

Back Label (6) Additional marketing information, which has few restrictions, may be printed on the back label. It may include ingredients, processing, and flavor characteristics, as well as promotional information.

Numbers above refer to text on label requirements.

Requirements for Spirits

Numbers in the following text refer to numbers on labels.

The following information is required on an approved TTB spirits label:

Brand (1) The name used to identify the product. Commonly it is the producer's name but it can also be a proprietary name. The brand may not mislead the consumer as to type of spirit or its origin.

Bottler or Importer Name and Address (2) The name of the producer in the United States or the company importing the product from a foreign country. It must also include the address of the production facility or principal place of business.

Class or Type of Spirit (3) Class of spirit (gin, vodka, rum, etc.). Each class must meet specific ingredient and processing requirements for that class.

Alcohol Content (4) Alcoholic strength is indicated as ABV. In addition, proof, which is double the ABV, may be included.

Net Contents (5) The total volume of product stated in metric units, ranging in size from 50 milliliters to 1.75 liters.

Country of Origin (6) For imported spirits, a statement indicating country of origin is required.

Government Health Warning Statement (7) The standard consumption-risk warning on the back label of every bottle of alcohol.

Back Label (8) Additional marketing information, which has few restrictions, may also be printed on the back label. It may include information on ingredients, processing, and flavor characteristics, as well as promotional information.

Numbers above refer to text on label requirements.

Old World Wine Regions

Almost all Old World countries produce wines, but the four countries identified in this appendix produce the largest amount of wine that is available in the international market. They are commonly referred to as the most prominent countries in the Old World because they have instituted standards for grape growing and winemaking and have long standing reputations.

FRANCE

French Wine Classifications

French wine law defines four quality levels, listed from lowest to highest: VDT, VDP, VDQS, and AOC.

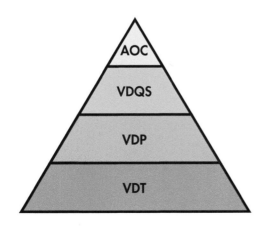

French Wine Classifications

Vin de Table (VDT)

VDT is the lowest quality level. These are everyday drinking wines that are identified simply by the words *rouge*, *blanc*, or *rosé* on the bottle. They are only sold in France.

Vin de Pays (VDP)

VDP is the next highest level. The requirements are less restrictive than those used for VDQS and AOC wines, but they still include yield, minimum alcohol, and production techniques, among other criteria. These wines must be made from approved grapes and are associated with specific geographic areas (appellations). Growers are permitted to use nontraditional grapes that are not approved for AOC wines of that region.

Vin Délimité de Qualité Supérieure (VDQS)

VDQS is considered an intermediate step between VDP and AOC wines. They are more tightly controlled than VDP wines. Regulations identify the designated production area, varietals, alcohol content, yields, and vineyard and winery practices. The wine is subject to analysis and tasting by a panel of experts. VDQS wines receive a seal of approval from the region's wine growers' association. Only a limited number of VDQS wines are made.

Appellation d'Origine Contrôlée (AOC)

AOC wines, or controlled appellation wines, are the highest quality of French wines and must adhere to strict regulations. The grapes must come from a specific parcel of land that is recognized for its outstanding terroir as well as its history and tradition. The wine must reflect the characteristics of its terroir, and only certain traditional varieties are allowed in each appellation. In some appellations all the wines are 100 percent variety-specific. In others the wines must be blended, and some AOCs permit both 100 percent varietal and blended wines. In addition to allowed varieties, regulations establish maximum yield, alcohol level, vineyard and winemaking practices, and sometimes aging criteria. The wines are subject to analysis and tasting by a panel of experts.

Appellation Levels

Appellations are divided into three levels from the general to the specific (see French Appellation, Subappellation, and Village Areas, below). The first level is a regional AOC, Bordeaux for example. These encompass a wide area and generally produce generic red and/or white wine. Grapes can come from any vineyard in the regional Bordeaux AOC. These are the lowest-quality level of controlled-appellation wines. They are named for the region, in this example Bordeaux.

The second level is a specific subappellation that covers a large defined area within the regional AOC. The grapes must come from the subappellation. Médoc is an AOC within Bordeaux. Only grapes that are grown in the Médoc can be used in the wine. The wines are named for the subappellation, in this example Médoc.

Finally, the best villages in an area may receive their own AOC. These produce the finest appellation wines made from grapes grown within a specific village. The Margaux AOC, which is located in the Médoc, is an example of a village AOC. The grapes must be grown in designated Margaux vineyards. Not all villages are classified appellations. The vineyards within the village must have unique characteristics to be recognized as an appellation. The wines are given the village name, in this example Margaux.

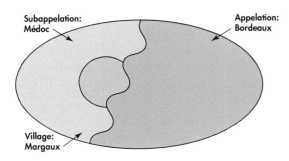

French Appellation, Subappellation, and Village Areas

Additional Classifications

Several of the regional appellations have added their own quality classifications to the AOC requirements. Also, the best vineyards in an appellation are frequently identified as Grand Cru or Premier Cru. Additional classifications include:

CLASSIFICATIONS IN ALSACE

1 Alsace or Vin d'Alsace AOC
 – This includes the entire region.
2 Alsace Grand Cru
 – The designation identifies 51 specific vineyards.
 – The specific vineyard is named on the label.

CLASSIFICATIONS IN BORDEAUX

These apply to a property within an appellation. There is no single classification that covers all the appellations. Most classifications are revised periodically.

MÉDOC CLASSIFICATIONS

1 1855 Classification
 – The top châteaux (estates) of the Médoc are organized into five crus or growths. The first growth is called Premier Cru Classé (First Growth), the last Cinquième Cru Classé (Fifth Growth). The wines are ranked by price, price equating to quality.
2 Cru Bourgeois (established 1920)
 – This identifies a quality category just below the five Crus Classés.
 – It contains three levels: Cru Bourgeois, Cru Supérieur, and Cru Exceptionnel.
3 Cru Artisans Classification (revised 2002)
 – This is reserved for small family wineries. The family must grow, make, and sell the wines themselves.

SAUTERNES-BARSAC CLASSIFICATION (established with 1855 Classification)
 – The 27 sweet white wines of Sauternes and Barsac are classified into two crus: Premier Cru and Deuxième Cru.
 – Only one wine, Château d'Yquem, has been designated a Premier Cru Supérieur.

GRAVES CLASSIFICATION (established 1953)
 – Wines are classified by district and color, red, white, or both. There is only one level.

ST. ÉMILION CLASSIFICATION (established 1955)
 – This classification has very stringent requirements, including the requirement that half the vines must be

over 12 years old.

- There are two categories: Premier Grand Cru Classé and Grand Cru Classé. Château Cheval Blanc and Château Ausone have been given special status.

CLASSIFICATIONS IN BURGUNDY
CÔTE D'OR
1 Regional AOC
- The name includes the region and the wine type, red or white.
- *Example: Bourgogne Rouge, Bourgogne Blanc*

2 Specific Subappellation AOC
- The name includes the region and the area within Burgundy.
- *Example: Bourgogne Hautes Côtes de Nuits*

3 Village AOC
- The name includes the area and the village.
- *Example: Nuits-St-Georges*

4 Premier Cru (First Growth)
- The grapes come from specific vineyards within a designated village that have special characteristics. These are the second best sites.
- If all the grapes come from one vineyard, the name includes the village followed by Premier Cru and the vineyard.
- *Example: Aloxe-Corton Premier Cru Les Fournières*
- If the grapes come from several cru vineyards, the name includes the village and Premier Cru.
- *Example: Aloxe-Corton Premier Cru*

5 Grand Cru (Great Growth)
- The grapes come from the very best vineyard sites within a designated village.
- The name includes the vineyard plus Grand Cru.
- *Example: Le Montrachet Grand Cru*

CHABLIS
1 Petit Chablis AOC
- The grapes can come from any vineyard in Chablis.
- A level below Chablis, the grapes come from the least desirable vineyards, usually located at the tops of slopes.
- The label states only Petit Chablis.

2 Chablis AOC
- The grapes come from any vineyard in Chablis, usually those located on the slopes.
- The label states only Chablis.

3 Premier Cru
- The grapes meet the same criteria as Côte d'Or.
- The name includes Chablis, Premier Cru, and the vineyard.
- *Example: Chablis Premier Cru Montmains*
- If the grapes come from more than one designated vineyard, a vineyard name is not used.
- *Example: Chablis Premier Cru*

4 Grand Cru
- The grapes meet the same criteria as Côte d'Or. Only seven vineyards qualify.
- The name includes Chablis, followed by Grand Cru and the vineyard name.
- *Example: Chablis Grand Cru Blanchot*

CÔTE CHALONNAISE
1 Regional AOC
- The grapes come from any vineyard within Côte Chalonnaise.
- The name states only Bourgogne Côte Chalonnaise.

2 Village AOC
- The grapes come from vineyards in the village area.
- Only the village name is used.
- *Example: Mercurey*

3 Premier Cru
- The grapes come from a defined village and estate.
- The name includes the village and the estate.
- *Example: Mercurey Clos Voyen*

4 Bouzeron AOC
- Aligoté is the only authorized grape and wine.
- Bouzeron is the only name used.

MÂCONNAIS
1 Regional AOC
- The grapes can come from any area within Mâcon.
- Usually only the name Mâcon is used.
- If Mâcon Supérieur is used, the wine requires 0.5 percent more alcohol.

2 Mâcon-Villages AOC
- The grapes come from any village within Mâcon.
- Wines have the generic name, Mâcon-Villages.

3 Mâcon-Villages plus Village Name AOC
- The grapes come from a specific village.
- The name includes the region and the village name.
- *Example: Mâcon-Lugny*

4 Individual AOCs

- The grapes come from a named appellation consisting of designated vineyards within specified villages.
- The AOC name is used.
  Example: Pouilly-Fuissé
- The AOC name can also include the name of the vineyard.
- *Example: Pouilly-Fuissé Les Cray*

BEAUJOLAIS

1 Beaujolais AOC

- The general appellation includes three levels: Beaujolais, Beaujolais Supérieur, and Nouveau Beaujolais.
- These names are used to identify the wines.

2 Beaujolais-Villages AOC

- These wines are blends of grapes from several villages.
- They are identified by the generic village designation.
- *Example: Beaujolais-Villages*

3 Cru Beaujolais

- This is the highest quality level in Beaujolais.
- A cru refers to a village rather than a vineyard. There are 10 crus.
- The village name is used; the regional designation, Beaujolais, is not.
- *Example: Moulin-à-Vent*

CLASSIFICATIONS IN THE RHÔNE

1 Côtes du Rhône AOC

- The grapes can come from the entire region but are usually from the south.
- The name states only Côtes du Rhône.

2 Côtes du Rhône-Villages AOC

- The grapes come from any village within the region.
- The minimum grape maturity requirement is higher than those of regional AOC wines.
- The name states only Côtes du Rhône-Villages. The name of the village may not be used.

3 Côtes du Rhône-Villages with Village Name

- 19 villages may use the village name.
- *Example: Côtes du Rhône-Villages Rasteau*

4 Cru Wines

- These are the premier wines.
- There are 15 cru appellations.
- Only the estate name is used.
- *Example: Hermitage*

French Wine Labels

Two classes of French wine are sold in the international marketplace: AOC and VDP.

AOC

Here are examples of French AOC wine labels from the major regions that demonstrate regional differences and highlight difficult-to-understand pieces of label information.

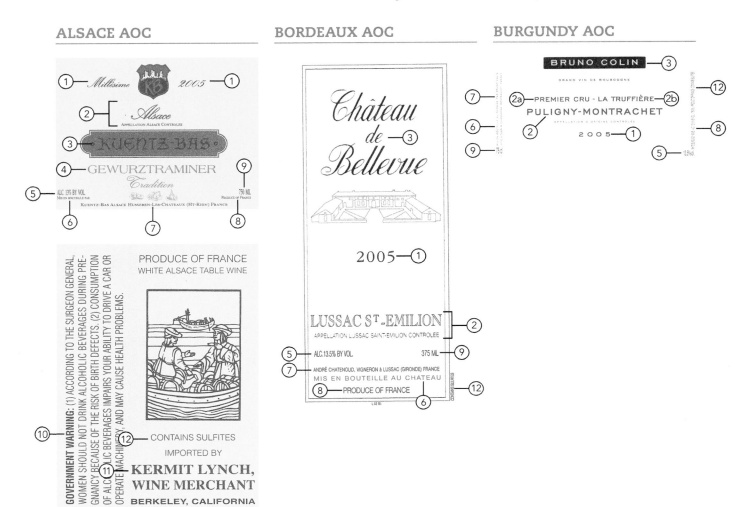

ALSACE AOC

BORDEAUX AOC

BURGUNDY AOC

CHAMPAGNE AOC

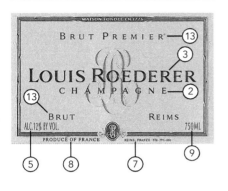

THE LOIRE AOC

THE RHÔNE AOC

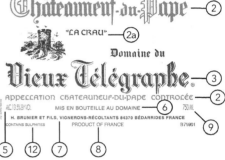

1 Vintage
2 Appellation
  a Vineyard designation
  b Vineyard name
3 Name of producer or estate
4 Varietal (only Alsace AOC)
5 Alcohol content
6 Estate bottled by producer
7 Name and address of producer
8 Country of origin
9 Net contents
10 Health warning statement
11 Name and address of importer
12 Sulfites declaration
13 Style
14 Proprietary blend of producer

Other back labels for European wines generally add information required or desired specifically for the American/international market as shown on the first label.

VDP

VDP wines are sold as varietal wines. In the American marketplace, both French-style labels and, more frequently, American-style labels are found on VDP wines.

VDP LABEL

③— MAS GABINÈLE

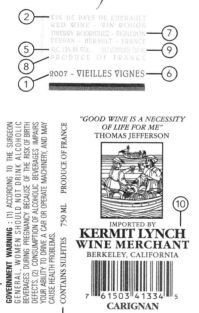

Carignan —④

② — VIN DE PAYS DE L'HERAULT
RED WINE · VIN ROUGE
THIERRY RODRIGUEZ · VIGNERON — ⑦
VEYRAN · HERAULT · FRANCE
⑤— ALC. 13% BY VOL.   NET CONTENTS 750 ML — ⑨
⑧— PRODUCE OF FRANCE
2007 - VIEILLES VIGNES — ⑥
①

1 Vintage
2 Classification of wine (VDP) from village within Languedoc
3 Estate name
4 Varietal
5 Alcohol content
6 Old vines
7 Name and address of producer
8 Country of origin
9 Net contents
10 Name and address of importer
11 Health warning statement
12 Sulfites declaration

VDP LABEL FOR AMERICAN MARKET BY AMERICAN PRODUCER

③—
THE EARTH FRIENDLY WINERY™

④— PINOT NOIR
①— 2009

17% LESS GLASS RESULTING IN 14% LESS CARBON EMISSIONS

THE EARTH FRIENDLY WINERY

The Earth Friendly Winery·   DENNIS MARTIN

PINOT NOIR · VIN DE PAYS D'OC — ②
⑤— PRODUCT OF FRANCE   ALC 13.5% BY VOL
IMPORTED & BOTTLED
⑥— BY FETZER VINEYARDS
HOPLAND, MENDOCINO
COUNTY, CALIFORNIA
750 ml CONTAINS SULFITES

1 Vintage
2 Classification and French appellation
3 American winery brand
4 Varietal
5 Country of origin
6 Importer of wine

France

UNITED KINGDOM

Corsica

BELGIUM

Reims

Paris

CHAMPAGNE

Strasbourg

ALSACE

Seine

ATLANTIC
OCEAN

Tours

THE LOIRE

Loire

BURGUNDY

JURA

SWITZERLAND

SAVOY

Lyon

ITALY

Bordeaux

BORDEAUX

Dordogne

Garonne

SOUTH-WEST

THE RHÔNE

Rhône

Alps

Nice

PROVENCE

P y r e n e e s

LANGUEDOC-
ROUSSILLON

*Mediterranean
Sea*

SPAIN

N

Legend:
- Bordeaux
- The Loire
- Champagne
- Alsace
- Burgundy
- The Rhône
- Provence
- Corsica
- Languedoc-
  Roussillon
- South-West
- Jura
- Savoy

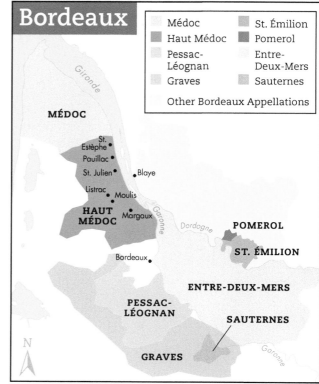

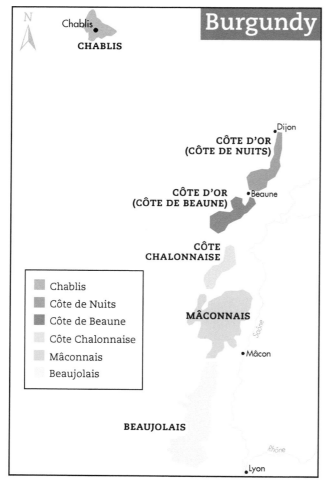

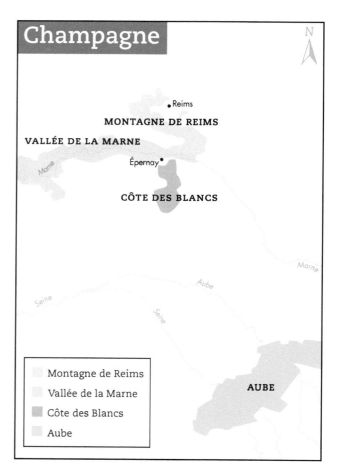

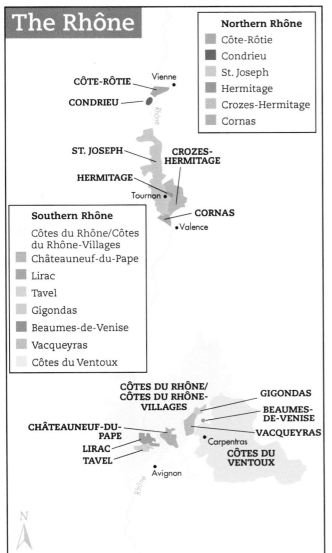

The Rhône

Northern Rhône
- Côte-Rôtie
- Condrieu
- St. Joseph
- Hermitage
- Crozes-Hermitage
- Cornas

CÔTE-RÔTIE — Vienne
CONDRIEU

ST. JOSEPH — CROZES-HERMITAGE
HERMITAGE
Tournon — CORNAS
Valence

Southern Rhône
- Côtes du Rhône/Côtes du Rhône-Villages
- Châteauneuf-du-Pape
- Lirac
- Tavel
- Gigondas
- Beaumes-de-Venise
- Vacqueyras
- Côtes du Ventoux

CÔTES DU RHÔNE/CÔTES DU RHÔNE-VILLAGES — GIGONDAS
BEAUMES-DE-VENISE
CHÂTEAUNEUF-DU-PAPE
VACQUEYRAS
Carpentras
LIRAC
TAVEL
CÔTES DU VENTOUX
Avignon

Languedoc-Roussillon

N

BLANQUETTE DE LIMOUX
MINERVOIS
CÔTES DU LANGUEDOC
MUSCAT DE FRONTIGNAN
Carcassonne
Mediterranean Sea
CORBIÈRES
CÔTES DU ROUSSILLON

Pyrenees
CÔTES DU ROUSSILLON-VILLAGES
BANYULS/COULLIOURE
SPAIN

- Banyuls/Coullioure
- Côtes du Roussillon
- Côtes du Roussillon-Villages
- Corbières
- Blanquette de Limoux
- Minervois
- Côtes du Languedoc
- Muscat de Frontignan

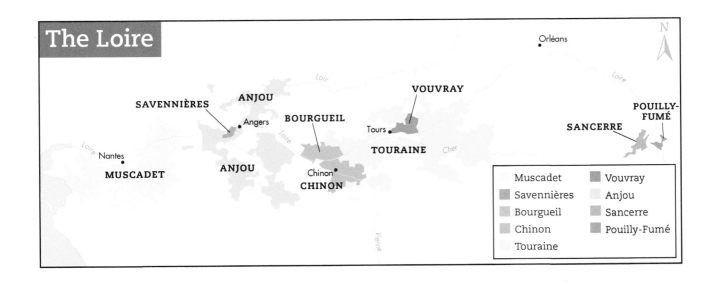

The Loire

N
Orléans
Loir
Loire

VOUVRAY
SAVENNIÈRES — ANJOU
BOURGUEIL
Angers
Tours
POUILLY-FUMÉ
SANCERRE
TOURAINE
Cher
Loire
Nantes
MUSCADET
ANJOU
Chinon
CHINON

Vienne

- Muscadet
- Savennières
- Bourgueil
- Chinon
- Touraine
- Vouvray
- Anjou
- Sancerre
- Pouilly-Fumé

ITALY

Italian Wine Classifications

Italian wine law defines four classifications, listed from lowest to highest: VdT, IGT, DOC, and DOCG.

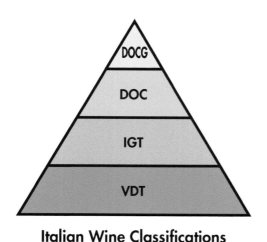

Italian Wine Classifications

Vino da Tavola (VdT)

This includes the lowest quality of table wines produced and sold. These wines are often sold as bulk wines or are used for blending.

Indicazione Geografica Tipica (IGT)

This classification is similar to Vin du Pays in France. Indigenous and international varieties are permitted and controls on production methods within this group are very flexible.

Denominazione di Origine Controllata (DOC)

Regulations establish varieties, zones, growing guidelines, including yields, and production techniques, especially aging. Wines are identified by the region, specific zones within the region, or the variety. A DOC designation does not guarantee quality. It only provides assurance of the grapes' origins and the producers' compliance with the regulations. As of early 2011, there are well over 300 DOCs.

Denominazione di Origine Controllata e Garantita (DOCG)

This is the most rigorous designation. The law spells out which varieties can be used in the wine and where the vineyards are located as well as growing conditions and production techniques. The wines must be made from grapes harvested from vineyards with lower yields than those permitted in the other classifications. The quality of DOCG wines is guaranteed based on a series of chemical tests and sensory evaluation by a tasting panel. When these wines are bottled, a numbered seal is applied over the corks. As of early 2011, there were 49 DOCGs.

Italian Wine Labels

Three classes of Italian wine are sold in the international marketplace: IGT, DOC, and DOCG.

IGT

IGT wine labels are very easy to read and understand because most IGT wines sold in the international market come from the larger, well-known regions and are made from familiar Italian varieties, or even international varieties.

DOC and DOCG

DOC and DOCG wines follow the same labeling guidelines. Each wine can be labeled by grape variety, appellation, grape variety with appellation, proprietary name, or legendary name. Most wines in this category are identified by the variety, appellation, or a combination of the two. Wines with proprietary or legendary names frequently list the grape variety on the front or back label, but this is not required. Each DOC and DOCG zone sets its own requirements. Wines in these categories display either DOC or DOCG on the front label.

IGT WINE

(4) CUM LAUDE

(1) 2006

(3) CASTELLO BANFI
MONTALCINO

da vigneti

(6) *Cabernet Sauvignon Merlot Sangiovese Syrah*

(4) CUM LAUDE

Toscana (2)

(2d)

2006 - VINTAGE RED WINE

Cum Laude is a cuvée that is produced "with honors". A harmonious blend of Tuscan and International red grape varieties. Aged 14 months in oak barriques. Unfiltered. Succulent nuances of blackberry fruit and soft tannins that will mature with Ph.D. dignity.

CASTELLO BANFI
MONTALCINO

1ST WINERY in the WORLD
Recognized for Exceptional
Environmental, Social & Ethical Responsibility,
& Leadership in Customer Satisfaction
(18)

(13) ESTATE BOTTLED
(10) BANFI
MONTALCINO, ITALY
(14) IMPORTED BY
BANFI VINTNERS
OLD BROOKVILLE, NY
(11) 750 ML · ALC. 13% BY VOL.
(15) PRODUCT OF ITALY
(17) CONTAINS SULFITES

GOVERNMENT WARNING: (1) ACCORDING TO THE SURGEON GENERAL, WOMEN SHOULD NOT DRINK ALCOHOLIC BEVERAGES DURING PREGNANCY BECAUSE OF THE RISK OF BIRTH DEFECTS. (2) CONSUMPTION OF ALCOHOLIC BEVERAGES IMPAIRS YOUR ABILITY TO DRIVE A CAR OR OPERATE MACHINERY, AND MAY CAUSE HEALTH PROBLEMS.

(12)   (16)

VARIETAL DOC/DOCG WINE

(3) ST. PAULS

EXCLUSIV

(6) Pinot bianco
Plötzner (7)

(2b) Südtirol   Alto Adige (2c)

Pinot Bianco (6)
(9) Südtirol • Alto Adige
Denominazione di origine controllata
»Plötzner« (7)

(10) WHITE WINE
(1) 2004
PRODUCED AND BOTTLED IN ITALY BY:
CANTINA SOCIALE SAN PAOLO - 39050 SAN PAOLO
ITALY

(11) NET. CONT. 750 ML
(12) ALCOHOL ▮ BY VOL 14%
IMPORTED BY:
(14) TESORI WINES
SANTA FE SPRINGS, CA
www.tesoriwines.com

 ITALIA
(15)

GOVERNMENT WARNING: (1) ACCORDING TO THE SURGEON GENERAL, WOMEN SHOULD NOT DRINK ALCOHOLIC BEVERAGES DURING PREGNANCY BECAUSE OF THE RISK OF BIRTH DEFECTS. (2) CONSUMPTION OF ALCOHOLIC BEVERAGES IMPAIRS YOUR ABILITY TO DRIVE A CAR OR OPERATE MACHINERY, AND MAY CAUSE HEALTH PROBLEMS. (16)

(17) CONTAINS SULFITES

APPELLATION DOC/DOCG WINE

(10) IMBOTTIGLIATO DA VIETTI CASTIGLIONE FALLETTO ITALIA

Vietti (3)
(1) 1993
BAROLO (2)
(2a) DENOMINAZIONE DI ORIGINE CONTROLLATA E GARANTITA
BRUNATE
ALCOHOL 13.50% BY VOL · BOTTLED BY VIETTI · RED WINE · 750 ML
PRODOTTE 3.234 BOTTIGLIE, 47 MAGNUM
PRODUCT OF ITALY

13,5 vol   e 1,5 lt
(12)   (15)   (11)

Note: This is an example of a European label as sold in Europe without U.S. label requirements and statements.

---

1 Vintage	**8** Varietal and appellation
2 Appellation	**9** DOC appellation - Südtirol/Alto Adige
a DOC/DOCG	**10** Name and address of producer
b German name of DOC	**11** Net contents
c Italian name of DOC	**12** Alcohol content
d IGT	**13** Estate bottled
3 Name of producer	**14** Name and address of importer
a Estate name	**15** Country of origin
4 Proprietary name	**16** Health warning statement
5 Legendary name	**17** Sulfites declaration
6 Varietal/varietals	**18** Marketing information
7 Vineyard name	

VARIETAL AND APPELLATION DOC/DOCG WINE

PROPRIETARY DOC/ DOCG WINE

LEGENDARY DOC/DOCG WINE

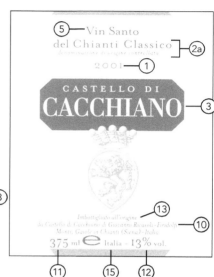

1   Vintage
2   Appellation
    a DOC/DOCG
    b German name of DOC
    c Italian name of DOC
    d IGT
3   Name of producer
    a Estate name
4   Proprietary name
5   Legendary name
6   Varietal/varietals
7   Vineyard name

8   Varietal and appellation
9   DOC appellation - Südtirol/Alto Adige
10 Name and address of producer
11 Net contents
12 Alcohol content
13 Estate bottled
14 Name and address of importer
15 Country of origin
16 Health warning statement
17 Sulfites declaration
18 Marketing information

Italy

SWITZERLAND

BELGIUM

AUSTRIA

AOSTA
VALLEY

Dolomites

TRENTINO-
ALTO ADIGE

LOMBARDY

Trento

Turin

Milan

FRIULI

PIEDMONT

VENETO

Po

Venice

LIGURIA

Genoa

EMILIA-ROMAGNA

Bologna

*Ligurian
Sea*

Arno

Florence

a

*Corsica
(FRANCE)*

TUSCANY

MARCHE

UMBRIA

ABRUZZO

Rome

SARDINIA

LATIUM/
LAZIO

MOLISE

Tyrrhenian Sea

CAMPANIA

Naples

APULIA

Cagliari

BASILICATA

*Mediterranean
Sea*

CALABRIA

Palermo

*Ionian
Sea*

SICILY

Adriatic Sea

N

	Aosta Valley
	Piedmont
	Liguria
	Lombardy
	Trentino-Alto Adige
	Veneto
	Friuli
	Emilia-Romagna
	Tuscany
	Umbria
	Marche
	Latium/Lazio
	Abruzzo
	Molise
	Campania
	Basilicata
	Apulia
	Calabria
	Sicily
	Sardinia

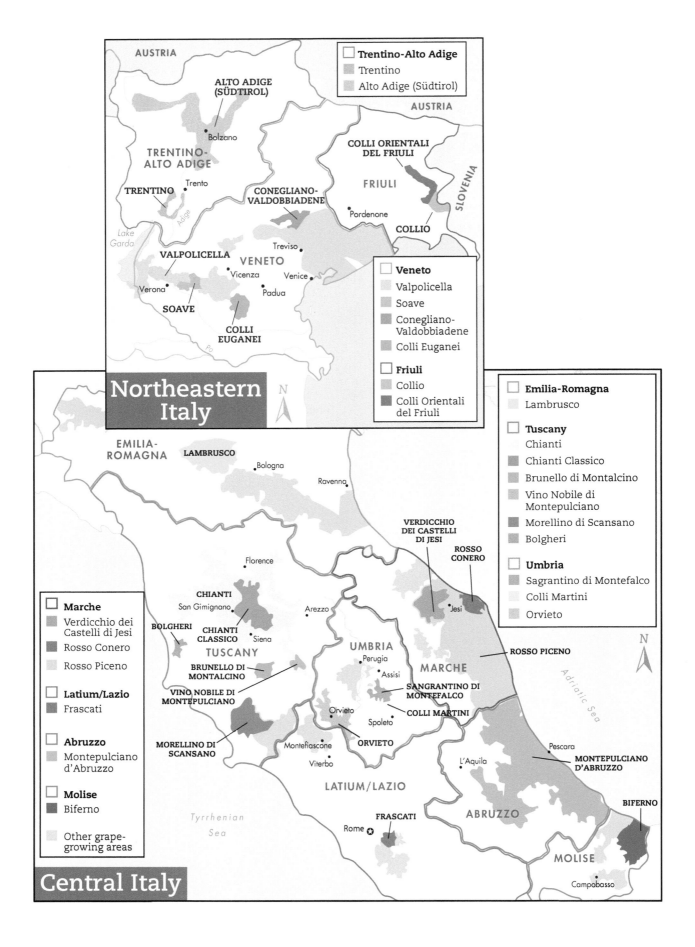

Northeastern Italy

- **Trentino-Alto Adige**
  - Trentino
  - Alto Adige (Südtirol)

- **Veneto**
  - Valpolicella
  - Soave
  - Conegliano-Valdobbiadene
  - Colli Euganei

- **Friuli**
  - Collio
  - Colli Orientali del Friuli

Central Italy

- **Emilia-Romagna**
  - Lambrusco

- **Tuscany**
  - Chianti
  - Chianti Classico
  - Brunello di Montalcino
  - Vino Nobile di Montepulciano
  - Morellino di Scansano
  - Bolgheri

- **Umbria**
  - Sagrantino di Montefalco
  - Colli Martini
  - Orvieto

- **Marche**
  - Verdicchio dei Castelli di Jesi
  - Rosso Conero
  - Rosso Piceno

- **Latium/Lazio**
  - Frascati

- **Abruzzo**
  - Montepulciano d'Abruzzo

- **Molise**
  - Biferno

  - Other grape-growing areas

Northwestern Italy

SWITZERLAND

AOSTA VALLEY

GATTINARA

GHEMME

FRANCIACORTA

LOMBARDY

FRANCE

Lake Orta

Lake Maggiore

Lake Como

Lake Iseo

Lake Garda

Brescia

Milan

PIEDMONT

Turin

ASTI

Asti

BARBARESCO

BAROLO

L I G U R I A

Genoa

FRANCE

- ☐ **Aosta Valley**
- ☐ **Piedmont**
  - Asti
  - Barolo
  - Barbaresco
  - Gattinara
  - Ghemme
- ☐ **Liguria**
- ☐ **Lombardy**
  - Franciacorta
- Other grape-growing areas

Southern Italy

Rome

Adriatic Sea

VERMENTINO DI GALLURA

GRECO DI TUFO

FIANO DI AVELLINO

TAURASI

APULIA

AGLIANICO DEL VULTURE

BASILICATA

CAMPANIA

SARDINIA

Tyrrhenian Sea

CANNONAU DI SARDEGNA

VERMENTINO DI SARDEGNA

SALICE SALENTINO

CALABRIA

Ionian Sea

CIRÒ

MARSALA

Palermo

ETNA

SICILY

Agrigento

Siracusa

PANTELLERIA

Mediterranean Sea

- ☐ **Sardinia**
  - Vermentino di Gallura
  - Cannonau di Sardegna
  - Vermentino di Sardegna
- ☐ **Sicily**
  - Marsala
  - Pantelleria
  - Etna
  - Other grape-growing areas

- ☐ **Campania**
  - Taurasi
  - Fiano di Avellino
  - Greco di Tufo
- ☐ **Basilicata**
  - Aglianico del Vulture
- ☐ **Apulia**
  - Salice Salentino
- ☐ **Calabria**
  - Cirò

SPAIN

Spanish Wine Classifications

Spain's Ley de la Viña y del Vino (Vineyard and Wine Act) specifies two general wine classifications: Quality Wines and Table Wines. Each classification includes several subclassifications. Each Denominación de Origen (DO) within the quality wine classification is governed by a Consejo Regulador, an independent control board composed of grape growers, wine producers, and enologists. The board defines the production area, authorizes varieties, establishes yields, and identifies approved grape-growing and winemaking techniques. The classifications from lowest to highest are VdM, VdlT, Indicación Geográfica, DO, DOCa, and VdP.

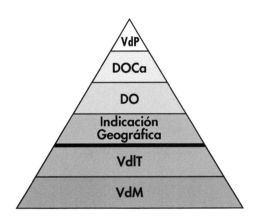

Spanish Wine Classifications

Vinos de Mesa

1 **Vinos de Mesa (VdM)**
This is the lowest category of wine. The grapes come from unclassified vineyards, or the wine is made by blending unclassified wines.

2 **Vinos de la Tierra (VdlT)**
Similar to France's Vin du Pays, VdlT regulations establish location, varieties, and types of wine, but are not as stringent as those for quality wines. The region is defined by vineyards that have an identifiable local character. At least 60 percent of the grapes must come from the region. The law defines the minimum alcohol content and describes sensory characteristics of the wine.

Vinos de Calidad Producido en Región Demarcada

1 **Vinos de Calidad con Indicación Geográfica**
Wine must be made in a specific region from grapes grown in that region. Its quality and characteristics are due to its environment and/or to the techniques practiced by growers or winemakers.

2 **Vinos de Denominación de Origen (DO)**
DOs must meet local standards set by their regional Consejo Regulador. To qualify the region must be recognized for at least five years as producing quality wines from a specified region.

3 **Vinos de Denominación de Origen Calificada (DOCa)**
To qualify for a DOCa designation, a region must produce superior DO wines for 10 years and meet stringent production requirements. Only three DOs have advanced to this level. The Rioja received the designation in 1991, followed by the Priorat in 2001 and Ribera del Duero in 2008.

4 **Vinos de Pago (VdP)**
VdP is the highest category, similar to a French Grand Cru. These wines are produced from grapes grown at a specific vineyard that is tied to the characteristics of its terroir. Wines must meet the requirements of a DOCa. In addition, they must be made and bottled at the estate where the vineyard is located. Only a handful of wines qualify for this classification.

Aging Classification System

In addition to the DO regulations, the law sets criteria for aging quality wines in wood and in bottle. There are minimum national standards that must be met by all DO producers, but many individual DOs and their Consejo Reguladors have established stricter standards. The minimum requirements are:

Joven These are generally young unaged wines, although some spend a small amount of time on oak.

Crianza Reds must age at least two years with at least six months on oak. Whites and rosados (rosés) must age at least one year with at least six months on oak.

Reserva Reds must age at least three years with at least one year on oak. Whites and rosados must age at least two years with at least six months on oak.

Gran Reserva These are made only in years with excellent vintages using the best grapes. Additionally, they must be aged at the bodega where they are produced. Reds must age at least five years with at least eighteen months on oak. Whites and rosados must age at least four years with at least six months on oak.

Spanish Wine Labels

The primary classes of Spanish wine sold in the international marketplace are DO and DOCa. VdM wines are beginning to show up internationally.

DO and DOCa

DO and DOCa wines follow the same labeling guidelines. The four quality levels, joven, crianza, reserva, and gran reserva, are printed on the label. Joven wines are rarely seen in the marketplace.

CRIANZA

RESERVA

GRAN RESERVA

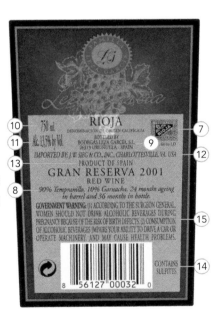

1   Vintage
2   Appellation/region
3   Wine name/brand
4   Producer
5   Classification
6   Quality Level and Aging Requirement
7   Official seal
8   Grape varieties
9   Name and address of producer
10   Net contents
11   Alcohol content
12   Name and address of importer
13   Country of origin
14   Sulfites declaration
15   Health warning statement

Spain

FRANCE

- San Sebastian
- Bílbao
- **BASQUE COUNTRY**
- Pamplona
- **NAVARRE**
- **RIOJA**
- Pyrenees
- Duero
- **ARAGON**
- **GALICIA**
- Pontevedra
- **CASTILE AND LEON**
- Valladolid
- Douro
- Duero
- Duero
- **CATALONIA**
- Barcelona
- **MADRID**
- Madrid
- Toledo
- *Balearic Sea*
- **CASTILE-LA MANCHA**
- **VALENCIA**
- Valencia
- **BALEARIC ISLANDS**
- **EXTREMADURA**
- Badajoz
- Ciudad Real
- **MURCIA**
- Alicante
- Cordoba
- Murcia
- Sevilla
- Granada
- **ANDALUSIA**
- Málaga
- Cádiz
- *Mediterranean Sea*
- PORTUGAL
- ATLANTIC OCEAN

CANARY ISLANDS

Legend:
- Galicia
- Castile and Leon
- Basque Country
- Navarre
- Rioja
- Aragon
- Catalonia
- Madrid
- Castile-La Mancha
- Valencia
- Extremadura
- Murcia
- Andalusia
- Balearic Islands
- Canary Islands

Northwestern Spain

- **GALICIA**
- **RÍAS BAIXAS**
- Pontevedra
- **BIERZO**
- **CASTILE AND LEON**
- **RIBERA DEL DUERO**
- **TORO**
- Valladolid
- **RUEDA**
- PORTUGAL
- Douro
- Duero
- ATLANTIC OCEAN

Legend:
- **Galicia**
  - Rías Baixas
- **Castile and Leon**
  - Bierzo
  - Toro
  - Rueda
  - Ribera del Duero

Northeastern Spain

Bilbao
San Sebastián
BASQUE COUNTRY
Pamplona
FRANCE
Pyrenees
N
NAVARRE
Lagroño
RIOJA
SOMANTANO
Huesca
ARAGON
CATALONIA
Zaragoza
Barcelona
PENDÈS
PRIORAT
BALEARIC SEA
BALEARIC ISLANDS
BINISSALEM MALLORCA
Binissalem Mallorca
MAJORCA
PLÀ I LEVANT

Basque Country
Rioja
Navarre
Aragon
Somantano
Catalonia
Penedès
Priorat
Balearic Islands
Binissalem Mallorca
Plà i Levant

Central & Southern Spain

Madrid
VINOS DE MADRID
PORTUGAL
UTIEL-REQUENA
Toledo
Valencia
LA MANCHA
Lisbon
Badajoz
VALENCIA
YECLA
EXTREMADURA
Ciudad Real
MURCIA
JUMILLA
VALDEPEÑAS
Alicante
MONTILLA-MORILÉS
Cordoba
Murcia
Sevilla
Granada
Huelva
Málaga
CONDADO DE HUELVA
JEREZ
Cádiz
MÁLAGA
SIERRA DE MÁLAGA
MEDITERRANEAN SEA
CANARY ISLANDS

Madrid
Vinos de Madrid
Extremadura
Castile-La Mancha
La Mancha
Valdepeñas
Valencia
Valencia
Utiel-Requena
Murcia
Jumilla
Murcia
Yecla
Andalusia
Condado de Huelva
Jerez
Málaga
Montilla-Morilés
Sierra de Málaga
Canary Islands
Tacoronte-Acontejo

GERMANY

German Wine Classifications

The German wine classification system is driven by the country's dominant grape variety, Riesling, although all German wines fall within the system. The quality classifications established by German wine law are based on the ripeness of the grapes when they are harvested. In addition, producers have created quality designations for dry wines from top-level vineyards.

Four Quality Classifications

As with other European countries, German wine law sets stringent requirements for growing grapes and producing wine, including identification of wine regions and subregions, grape varieties, and production methods.

The higher the ripeness of the grapes used in a wine, the higher up the classification pyramid the wine is placed. Grape ripeness does not indicate the sweetness level of the finished wine. The winemaker decides whether to ferment the wine dry or to leave residual sugar to create sweet wine.

The four legally defined quality classifications from lowest to highest are Deutscher Tafelwein, Deutscher Landwein, Qualitätswein bestimmter Anbaugebiete (QbA), and Prädikatswein.

German Quality Classifications

Deutscher Tafelwein

Tafelwein, comparable to jug wine in the United States or Vin de Table in France, is the lowest quality of German wine. The grapes are grown within broadly defined Tafelwein regions and are harvested when they obtain the normal ripeness level. Tafelwein must achieve 8.5 percent alcohol and chaptalization is allowed. It is rarely made from Riesling and is not exported to the United States.

Deutscher Landwein

Deutscher Landwein is a newer classification intended to mimic Vin de Pays of France. The grapes can come from one of 19 Landwein-designated regions. Chaptalization is allowed, and the wine can contain a slightly higher percentage of alcohol than the level established for Tafelwein. It can be dry (trocken) or off-dry (halbtrocken).

Qualitätswein bestimmter Anbaugebiete (QbA)

QbA wines are produced in one of 13 approved wine regions within Germany and often carry a village or vineyard name as well as the required region name. The grapes must obtain a specific level of ripeness to be used in quality wine. The wines are made from approved varieties, can be chaptalized, and must reach a minimum alcohol content of 7 percent. They can be trocken, halbtrocken, or sweet.

QbA trocken and halbtrocken wines can be divided into two classifications: Classic or Selection.

Classic Good-quality wines for daily drinking as a table wine. They must have a potential alcohol of 12 percent with no more than 12 grams per liter of residual sugar. The varieties used must be classic grapes of the region.

Selection The wines are higher in quality and, therefore, more expensive. They must be hand-harvested with a yield of no more than 60 hectoliters per hectare (less than four tons per acre) and must reach an alcohol level of 12.2 percent with a residual sugar level of no more than 12 grams per liter.

Wines under both classifications, Classic and Selection, are typically trocken; however, at the higher end of allowable sugar (12 grams per liter) they will be halbtrocken.

Prädikatswein (formerly Qualitätswein mit Prädikat, QmP)

The highest-quality levels of wines are labeled Prädikatswein. Prädikatswein wines must come from one of 39 subregions within the 13 designated quality appellations and, in addition, can have more specific village and/or vineyard appellation designations. These wines cannot be chaptalized.

Within the Prädikatswein category, there are six quality levels. The Prädikatswein quality level, which is clearly identified on the label, represents the increasing ripeness of the grapes when they were picked and a corresponding increase in price. The classifications, from lowest to highest quality, are:

Kabinett The lowest ripeness level of Prädikatswein wines. The grapes are picked when fully ripe during the main harvest. The wines are usually trocken or halbtrocken, but may be sweet and have alcohol levels as low as 7 to 8 percent.

Spätlese From late-harvest grapes. These grapes have been left on the vine longer and develop higher sugar levels with richer flavors and more body. Spätlese wines can be fermented in a trocken, halbtrocken, or sweet style.

Auslese From selected grape bunches. Only very ripe bunches are hand-selected and handpicked cluster by cluster for these wines. The grapes are left on the vine longer than the ones used for Spätlese and, therefore, are riper and higher in sugar. The wines are typically made in a sweet style although trocken and halbtrocken are permitted.

Beerenauslese (BA) Meaning selected berries. The overripe grapes are individually selected and picked. They make a very sweet and expensive dessert wine.

Eiswein In English, ice wine. The wine is made from grapes that are left on the vine to reach a very high sugar content and are not harvested until the grapes have frozen. These are very concentrated sweet dessert wines at the approximate sweetness level of a BA.

Trockenbeerenauslese (TBA) Meaning selected dry berries. Individual grapes that are overripe are allowed to dry on the vine. They are then hand-selected and -harvested. These grapes may be infected with *Botrytis cinerea* and produce the most unctuous sweet and expensive wines.

Grape Ripeness, Wine Sweetness

Grape ripeness is frequently confused with wine sweetness. Each Prädikatswein classification is based on grape ripeness at harvest, but it does not indicate how sweet the wine will be. As you ascend by classification, grapes are riper, with higher sugar and potential alcohol levels. It is fermentation that determines the sweetness of the wine. Complete fermentation results in a dry wine regardless of grape ripeness, whereas partial fermentation results in unfermented sugar and a sweeter wine. For example, it is possible to have a drier Spätlese and a sweeter Kabinett even though the Spätlese is made from riper grapes because more of the sugar was fermented.

Other Classifications

Producers were concerned that the legal classifications did not take into account quality from Germany's best vineyards. The Association of German Quality and Prädikat Wine Estates (Verband Deutscher Qualitäts- und Prädikatsweingüter) established criteria for dry wines produced at Germany's best vineyard sites that mimic France's Grand Cru designations. These designations established rigorous criteria for each category, including allowed varieties, yield, harvest, and release date. From lowest to highest, they are:

Gutswein To qualify for this designation, 80 percent of the vineyard must be planted in the traditional grapes of the region. These are high-quality wines, labeled with a proprietary name or the name of a village or region.

Klassifizierte Lage/Ortswein/Terroirwein These wines come from classified superior sites. The grapes are planted in traditional vineyards that have distinctive character. The wines are labeled with the vineyard name.

Erste Lage These are wines from top vineyards known for exceptional terroir. They are called Grosses Gewächs or Erstes Gewächs in the Rheingau. They are labeled with the vineyard name and a specially designed logo is printed on the label behind the name or is embossed on the bottle.

German Wine Labels

The classes of German wine sold in the international marketplace are QbA and Prädikatswein. All German wines are identified by grape variety.

QbA

QbA wines identify the grape variety and major region where the grapes were grown. They are labeled trocken, halbtrocken, or sweet. The term *feinherb* can be used in place of halbtrocken.

Prädikatswein

Prädikatswein wines, the top tier of German wines, are divided into six quality levels based on grape ripeness. Each is identified on the label.

PRÄDIKATSWEIN
(FORMERLY QmP)

QbA RHEINGAU

QbA MOSEL

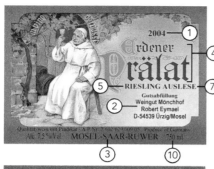

Weingut Mönchhof · Robert Eymael
D-54539 Ürzig / Mosel
www.moenchhof.de
2004 Erden Prälat Riesling Auslese
Gutsabfüllung A.P.Nr. 2602 029 009 05
Qualitätswein mit Prädikat
750 ml Mosel - Saar - Ruwer Alc.7.5%/ Vol

www.germanwine.net
IMPORTED BY CELLARS INTERNATIONAL INC.
SAN MARCOS CA 92078 USA
RUDI WIEST

GOVERNMENT WARNING:
(1) ACCORDING TO THE SURGEON GENERAL,
WOMEN SHOULD NOT DRINK ALCOHOLIC
BEVERAGES DURING PREGNANCY BECAUSE OF
THE RISK OF BIRTH DEFECTS. (2) CONSUMPTION
OF ALCOHOLIC BEVERAGES IMPAIRS YOUR ABILITY
TO DRIVE A CAR OR OPERATE MACHINERY, AND
MAY CAUSE HEALTH PROBLEMS.

CONTAINS SULFITES   PRODUCE OF GERMANY

2006 Rheingau Qualitätswein
Riesling

A.P.Nr. 40 060 026 07 - Gutsabfüllung

Weingut Franz Künstler D - 65239 Hochheim/Main

"Gourmet Magazine named Gunter Künstler as not only
one of the top wine making talents in all of Germany, but
in the world. Künstler's wines are rich and elegant examples
of Rheingau Riesling with a tantalizing acidity. Enjoy this
food flexible Riesling with your favorite dish or enjoy
as an aperitif."

750 ml e

7 67946 20200 3 Alc. 9.5% Vol.

PRODUCE OF GERMANY   CONTAINS SULFITES

GOVERNMENT WARNING: (1) ACCORDING TO THE SURGEON GENERAL,
WOMEN SHOULD NOT DRINK ALCOHOLIC BEVERAGES DURING PREGNANCY
BECAUSE OF THE RISK OF BIRTH DEFECTS. (2) CONSUMPTION OF ALCOHOLIC
BEVERAGES IMPAIRS YOUR ABILITY TO DRIVE A CAR OR OPERATE MACHINERY,
AND MAY CAUSE HEALTH PROBLEMS.

www.germanwine.net
IMPORTED BY CELLARS INTERNATIONAL INC.
SAN MARCOS CA 92078 USA

RUDI WIEST

EITELSBACHER KARTHÄUSERHOFBERG
2008 RIESLING MEDIUM-DRY
MOSEL-SAAR-RUWER
Gutsabfüllung-Karthäuserhof-54292 Trier
A.P.Nr. 3 561 303 0409
Situated in the Ruwer valley, the Karthäuserhof's wine making
tradition dates back to Roman times. Carthusian monks were
making wine at the estate from 1335-1803. Today the Tyrells,
who purchased the estate from Napoleon in 1811, are in charge.
In 1997, Christoph Tyrell, the Karthäuserhof's current proprietor
was selected Winemaker of the Year by the prestigious Fein-
schmecker magazine in Germany. In 2005, he was awarded as
Producer of the Year by the bestselling German Wineguide
Gault Millau.

ALC. 11.0 % BY VOL   750 ML.

7 67946 11130 5

CONTAINS SULFITES   PRODUCE OF GERMANY

GOVERNMENT WARNING: (1) ACCORDING TO THE
SURGEON GENERAL WOMEN SHOULD NOT DRINK
ALCOHOLIC BEVERAGES DURING PREGNANCY
BECAUSE OF THE RISK OF BIRTH DEFECTS. (2)
CONSUMPTION OF ALCOHOLIC BEVERAGES IM-
PAIRS YOUR ABILITY TO DRIVE A CAR OR OPERATE
MACHINERY, AND MAY CAUSE HEALTH PROBLEMS.

www.germanwine.net
IMPORTED BY CELLARS INTERNATIONAL INC.
SAN MARCOS CA 92078 USA
RUDI WIEST

1 Vintage	**8** Estate bottled
2 Producer	**9** Name and address of producer
3 Wine region/appellation	**10** Net contents
4 Estate vineyard	**11** Alcohol level
5 Varietal	**12** Government approval number
6 Quality Classification	**13** Name and address of importer
a Prädikatswein	**14** Health warning statement
b Qualitätswein	**15** Country of origin
7 Ripeness level	**16** Sulfites declaration

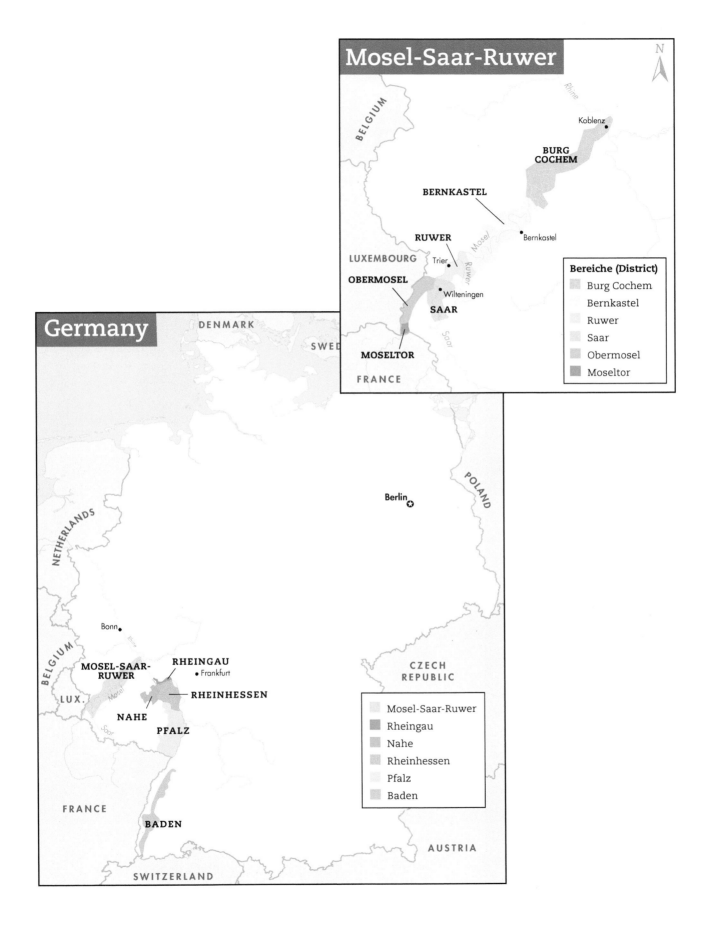

Mosel-Saar-Ruwer

N

BELGIUM

Rhine

Koblenz

BURG
COCHEM

BERNKASTEL

Bernkastel

RUWER

LUXEMBOURG

Mosel

Trier

Ruwer

OBERMOSEL

Wilteningen

SAAR

Saar

MOSELTOR

FRANCE

Bereiche (District)

Burg Cochem
Bernkastel
Ruwer
Saar
Obermosel
Moseltor

Germany

DENMARK

SWED

NETHERLANDS

Berlin

POLAND

Bonn

Rhine

BELGIUM

MOSEL-SAAR-
RUWER

RHEINGAU

Frankfurt

CZECH
REPUBLIC

LUX.

Mosel

RHEINHESSEN

NAHE

Saar

PFALZ

Mosel-Saar-Ruwer
Rheingau
Nahe
Rheinhessen
Pfalz
Baden

FRANCE

BADEN

AUSTRIA

SWITZERLAND

Distinguishing Characteristics of French Wines

Note: Grape varieties are listed according to most acreage planted.

REGION	APPELLATION, SUBAPPELLATION, OR VILLAGE	PRIMARY RED GRAPE VARIETIES	PRIMARY WHITE GRAPE VARIETIES	TYPE OF WINE— VARIETAL OR BLEND
Alsace	Haut-Rhin	Pinot Noir	Riesling Gewürztraminer Pinot Gris Auxerrois Sylvaner Pinot Blanc Muscat d'Alsace	Mostly whites Some sparkling (crémant) Some dessert Mostly varietal wines

Bordeaux	Médoc–Haut Médoc
	  Margaux
	  Pauillac
	  St. Estèphe
	  St. Julien
		Cabernet Sauvignon
		Merlot
		Cabernet Franc
		Malbec
		Petit Verdot
				Reds
				All blends

	Graves Pessac-Léognan	Cabernet Sauvignon Merlot Cabernet Franc Malbec	Sémillon Sauvignon Blanc Muscadelle	Reds, whites All blends

	Libournais
	  St. Émilion
	  Pomerol
	  LeLande de Pomerol
	  Fronsac
		Merlot
		Cabernet Franc
		Cabernet Sauvignon
				Reds
				All blends

	Sauternes Barsac		Sémillon Sauvignon Blanc	Sweet white dessert All blends
	Entre-Deux-Mers	Merlot Cabernet Sauvignon	Sauvignon Blanc Sémillon Muscadelle	Whites, reds, rosés Mostly blends Some white varietal wines

	The Côtes
	  Premiéres Côtes
	    de Bordeaux
	  Côtes de Castillon
	  Côtes de Francs
		Merlot
		Cabernet Sauvignon
		Cabernet Franc
			Sauvignon Blanc
			Sémillon
			Colombard
				Mostly reds
				Some whites, rosés
				All blends

Burgundy	Chablis		Chardonnay	100% varietal Chardonnay

	Côte de Nuits
	  Gevrey-Chambertin
	  Morey-St-Denis
	  Chambolle-Musigny
	  Vougeot
	  Vosne-Romanée
	  Nuits-St-Georges
		Pinot Noir		100% varietal Pinot Noir
				Few white wines

DISTINGUISHING CHARACTERISTICS IN VINEYARD	DISTINGUISHING CHARACTERISTICS IN WINEMAKING	DISTINGUISHING CHARACTERISTICS OF WINE
Sunny microclimate Limestone marl soil in best vineyards	Winemaking and aging in stainless steel or large, old-oak casks	Whites with rich, full flavors Mostly dry, some off-dry Some excellent dessert wines Pinot Noir likely to be light, lean, acidic
Close to Atlantic Ocean, maritime climate Coolest Bordeaux appellation Heavy rains, but good drainage on gravel soils Flat terrain	Traditional methods, with some producers following modern techniques Oak aged	Benchmark for Cabernet Sauvignon–dominant red wine: subtle, delicate, great finesse Most wines demonstrate moderate alcohol, acidity, good color, structure, and body Balanced Ageworthy in bottle
Warmest part of left bank Rolling hills Gravel soils	Barrel-fermented and -aged red and white wines	Cabernet Sauvignon–dominant red wine Most distinctive white wines: elegant, flavorful, nuanced, complex
Dry, warm climate for Bordeaux Flat, rolling hills More clay in soil	Many producers using modern techniques, creating ripe, juicy wines	Merlot-dominant red wine: fruitier, smoother, richer than left bank wines
Heavy morning fog, sunny afternoons High humidity Grapes subject to *Botrytis cinerea*	Aged in new and used oak barrels	Distinctive, world-class, botrytized dessert wines Honeyed, lush, exotic, opulent Age up to 30 years or more
Rolling hills Clay, limestone soils	Embracing some modern techniques Stainless steel tanks	White wine named Entre-Deux-Mers, but mostly red blends
Varied terrain: rolling hills to steeper sloped vineyards Limestone, clay soils	Traditional techniques	Less expensive than major AOCs Variable quality Good value
Cold climate Low yields Kimmeridgian limestone soils	Traditional techniques Mostly no oak aging	Varied quality High acid, steely, austere, flinty
Limestone, marl soils Best vineyards midway up east-facing slopes	Varied techniques Barrel aged Long aging produces best wines	Benchmark for Pinot Noir: rich, heady, aromatic Excellent bottle-aging potential

REGION	APPELLATION, SUBAPPELLATION, OR VILLAGE	PRIMARY RED GRAPE VARIETIES	PRIMARY WHITE GRAPE VARIETIES	TYPE OF WINE— VARIETAL OR BLEND

Burgundy, *continued*	Côte de Beaune
	  Aloxe-Corton
	  Puligny-Montrachet
	  Chassagne-Montrachet
	  Ladoix-Serrigny
		Pinot Noir	Chardonnay
			Aligoté
				100% varietal Pinot Noir and Chardonnay

	Côte Chalonnaise	Pinot Noir	Chardonnay Aligoté	Reds, whites Varietals

	Mâconnais
	  Pouilly-Fuissé
	  St. Véran
		Gamay
		Pinot Noir
			Chardonnay
			Aligoté
				100% varietal Chardonnay, Pinot Noir, and Gamay

	Beaujolais	Gamay		100% varietal Gamay
Champagne	Montagne de Reims Marne Valley Côte des Blancs Aube (Côte des Bar)	Pinot Noir Pinot Meunier	Chardonnay	Sparkling whites, rosés All blends
Corsica		Sangiovese Grenache Sciacarello Carignan Cabernet Sauvignon Mourvèdre	Vermentino (Rolle) Ugni Blanc Muscat Bianco Gentile Chardonnay	Reds, whites All blends
Jura/Savoy		Poulsard Pinot Noir Mondeuse (Refosco) Trousseau Gamay	Chardonnay Savagnin Jacquère Roussette (Altesse)	Reds, whites, rosés Sparkling Fortified and dessert
Languedoc-Roussillon		Carignan Grenache Mourvèdre Cinsaut Syrah Cabernet Sauvignon Merlot	Chardonnay Sauvignon Blanc Viognier Grenache Blanc Rolle (Vermentino) Roussanne Mauzac	Varietal VDP Blended AOC and VDP whites, reds, rosés Sparkling (crémant)

Loire	Upper Loire
	  Sancerre
	  Menetou-Salon
	  Quincy
	  Reuilly
	  Pouilly-Fumé
		Pinot Noir	Sauvignon Blanc	100% varietal Pinot Noir and Sauvignon Blanc

DISTINGUISHING CHARACTERISTICS IN VINEYARD	DISTINGUISHING CHARACTERISTICS IN WINEMAKING	DISTINGUISHING CHARACTERISTICS OF WINE
Limestone soils	Chardonnay aged in large oak casks, barrels	Benchmark for Chardonnay: rich, full-bodied, intense, complex aromas
Limestone, clay soils Higher yields than Côte d'Or	Traditional techniques	Less distinctive, less costly than Côte de Beaune and Côte de Nuits Good value
Limestone soils More sun, less rain than Côte d'Or Higher yields than Côte d'Or	Traditional techniques	Warmer temperatures than other areas of Burgundy create ripe wines Pouilly-Fuissé regarded as one of best Chardonnay wines
Granite soils	Use of oak aging for cru wines only Nouveau style for immediate release	Nouveau, Beaujolais, Beaujolais-Villages: bright, fresh, fruity Pinot Noir complexity in cru wines
Extremely cold temperatures Chalky soil	Basket pressing Méthode champenoise	High acid Benchmark for sparkling wines: crisp, tart
Mountainous island, sunny climate Best grapes on cool hillsides	Traditional techniques	Red wines can include Vermentino in blend Native varieties dominant
Cool climate Various soil types	Traditional techniques	Whites: bright, fresh, crisp Vin Jaune: like sherry Reds: light, mineral or dark, dense
Mediterranean climate Various soils	Mostly traditional techniques	VDP wines made in a modern style to appeal to the New World consumer AOC wines similar to Southern Rhône wines Rich, ample reds with herbal notes
Very cool climate Clay, limestone, gravel soils	Mostly no oak on whites	Wines relatively low in alcohol with high acidity Benchmark for Sauvignon Blanc: fragrant, crisp, tangy, mineral

Distinguishing Characteristics of French Wines - 3

REGION	APPELLATION, SUBAPPELLATION, OR VILLAGE	PRIMARY RED GRAPE VARIETIES	PRIMARY WHITE GRAPE VARIETIES	TYPE OF WINE— VARIETAL OR BLEND
Loire, *continued*	Central Loire Touraine - Chinon - Bourgueil - St. Nicolas de Bourgueil - Vouvray Anjou-Saumur - Anjou - Saumur - Coteaux du Layon	Cabernet Franc Cabernet Sauvignon Gamay Malbec (Côt) Pineau d'Aunis	Chenin Blanc Chardonnay Sauvignon Blanc	100% varietal Chenin Blanc Varietal and blended Cabernet Franc Sparkling Dessert
	Western Loire Pays Nantais		Melon de Bourgogne (Muscadet)	100% varietal Melon de Bourgogne
Provence	Côtes de Provence Bandol Cassis	Mourvèdre Grenache Cinsaut Carignan Syrah Cabernet Sauvignon	Bourboulenc Clairette Grenache Blanc Ugni Blanc Rolle (Vermentino) Marsanne	Mostly rosés Some reds Small amount of whites
Rhône	Northern Rhône Côte-Rôtie Hermitage Crozes-Hermitage St. Joseph Cornas Condrieu Château Grillet	Syrah	Viognier Marsanne Roussanne	100% varietal Syrah or blended with whites 100% varietal Viognier Blended whites
	Southern Rhône Côtes du Rhône Côtes du Rhône-Villages Châteauneuf-du-Pape Tavel Gigondas Vacqueyras Lirac Beaumes-de-Venise	Grenache Syrah Cinsaut Carignan Mourvèdre	Grenache Blanc Clairette Bourboulenc Roussanne Picpoul Muscat	Mostly reds Rosés, small amount of whites Dessert Virtually all blends
South-West	Bergerac Cahors Madiran Jurançon Irouléguy	Auxerrois (Malbec) Cabernet Franc Cabernet Sauvignon Merlot Négrette Tannat	Gros Manseng Petit Manseng Sémillon Sauvignon Blanc Muscadelle	Reds, whites, rosés Varietals, blends

DISTINGUISHING CHARACTERISTICS IN VINEYARD	DISTINGUISHING CHARACTERISTICS IN WINEMAKING	DISTINGUISHING CHARACTERISTICS OF WINE
Cool climate Chenin Blanc, tufa soils (chalky limestone)	White wines generally not oak aged Red wines may be aged in older oak barrels	Bright acids in both red, white varietals Wines range from light to medium body with fresh fruit and herbal flavors
Cool maritime climate	Sur lies aging	Neutral, high-acid wines
Mediterranean climate	Winemaking similar to that used in the Southern Rhône and Languedoc-Roussillon	Wide range of fruity rosés and red wines of varying quality Hot climate can cause some to be quite ripe and high in alcohol
Cool climate similar to Burgundy Côte-Rôtie: ultrasteep, terraced slopes	Traditional and modern techniques Mostly aged in older barrels, some new	Syrah ranges from light style with depth of fruit in Côte-Rôtie to luscious ripe fruit in Hermitage Viognier: very floral, expensive
Hot, sunny Mediterranean climate Broad range of soils Some vineyards covered with large, red stones	Traditional techniques	Mostly ripe, fruity reds, many with high alcohol Sturdy, earthy, bold reds Baumes-de-Venise: excellent dessert wines
Wide array of climates, soils	Traditional techniques	Mostly little known outside of France Unusual, native varieties used Many Bordeaux-style wines Great value

Distinguishing Characteristics of Italian Wines

Note: Star (*) indicates indigenous variety.

REGION	APPELLATION, SUBAPPELLATION, OR VILLAGE	PRIMARY RED GRAPE VARIETIES	PRIMARY WHITE GRAPE VARIETIES	TYPE OF WINE—VARIETAL OR BLEND
NORTHWEST				
Aosta Valley	Aosta Valley	Petit Rouge * Nebbiolo Fumin * Dolcetto	Moscato Bianco Müller-Thurgau Petite Arvine Chardonnay	Reds, whites Varietals, blends
Liguria		Dolcetto Rossese * Sangiovese	Pigato * Vermentino *	Whites Some reds Varietals
Lombardy	Franciacorta	Barbera Croatina Nebbiolo Cabernet Sauvignon Pinot Noir	Garganega Trebbiano Chardonnay Pinot Bianco Riesling	Reds, rosés, whites Sparkling Varietals, blends
Piedmont	Barolo Barbaresco Asti Gattinara Ghemme	Nebbiolo * Barbera * Brachetto * Dolcetto * Freisa *	Arneis * Cortese * Favorita Moscato Bianco Chardonnay	Mostly reds Few whites High volume of sparkling Varietals, blends
NORTHEAST				
Friuli	Collio Colli Orientali del Friuli	Refosco * Cabernet Franc Cabernet Sauvignon Merlot Pinot Noir Pignolo * Schioppettino *	Malvasia Istriana * Picolit * Ribolla Gialla * Tocai * Chardonnay Pinot Blanc Sauvignon Blanc Verduzzo	Mostly whites Reds Varietals
Trentino-Alto Adige	Trentino	Marzemino * Teroldego * Cabernet Sauvignon Merlot Lagrein	Pinot Grigio Chardonnay Pinot Bianco Müller-Thurgau	Whites, reds Sparkling Varietals
	Alto Adige (Südtirol)	Lagrein * Schiava (Vernatsch) * Cabernet Sauvignon Merlot	Gewürztraminer Müller-Thurgau Silvaner Pinot Bianco	Reds, whites Varietals, blends
Veneto	Valpolicella Soave Conegliano-Valdobbiadene Colli Euganei	Corvina * Molinara * Rondinella * Cabernet Franc Cabernet Sauvignon Merlot Pinot Noir	Garganega * Pinot Grigio Prosecco * Trebbiano di Soave * Verduzzo Trevigiano Chardonnay Pinot Blanc	Reds, whites Sparkling (Prosecco) Blends

DISTINGUISHING CHARACTERISTICS IN VINEYARD	DISTINGUISHING CHARACTERISTICS IN WINEMAKING	DISTINGUISHING CHARACTERISTICS OF WINE
Extremely cold climate Rugged mountains	Traditional techniques	Mostly light, fruity wines from local varieties
Maritime influence Rugged terrain Terraced vineyards	Traditional techniques	Wines are usually undistinguished Whites work with regional seafood
Mostly flat plains	Traditional techniques Traditional method used for sparkling wines	Leading Pinot Noir grower World-class sparkling wines
Cool, wet, foggy climate Limestone, marl, chalk soils	Traditional and modern techniques	Nebbiolo: tannic, long aging More DOCs than any other region
Plentiful rainfall Sandy, gravelly alluvial plains Limestone, marl hillsides	Merlot most planted variety Sophisticated winemaking	Simple wines from plains Friulian style: clean, fresh, fruity, delicate Recrafting reds for more power
Hot, humid valleys Gravelly soil	Large bulk-wine producers Some small high-quality producers Many cooperatives	Produces most Chardonnay of all regions
Mountains: cold climate Valleys: hot summers South-facing slopes	Many high-quality cooperatives Increased use of barriques for red wines	Wines from wide range of international and indigenous varieties
Largest winemaking region Mostly flat plains Some high-quality international varieties grown	High-yield grapes Traditional techniques	Mostly indigenous red and white blends Prosecco: sparkling wine

Distinguishing Characteristics of Italian Wines - 2

REGION	APPELLATION, SUBAPPELLATION, OR VILLAGE	PRIMARY RED GRAPE VARIETIES	PRIMARY WHITE GRAPE VARIETIES	TYPE OF WINE—VARIETAL OR BLEND
CENTRAL				
Abruzzo	Montepulciano d'Abruzzo	Montepulciano *	Trebbiano	Reds, rosés, whites Varietals
Emilia-Romagna	Lambrusco	Bonarda Barbera Croatina Lambrusco Malvasia Sangiovese	Malvasia Bianco Trebbiano Albana Angellotta Sauvignon Blanc	Reds Sparkling Fortified Varietals, blends
Latium/Lazio	Frascati	Sangiovese Cabernet Sauvignon Merlot	Malvasia Bianco Trebbiano	Mostly whites Varietals, blends
Marche	Verdicchio dei Castelli di Jesi Rosso Cònero Rosso Piceno	Montepulciano Sangiovese	Trebbiano Verdicchio	Whites, reds Varietals
Molise	Biferno	Aglianico Malvasia Montepulciano *	Falanghina Malvasia Bianco Trebbiano	Whites, reds, rosés Blends

Tuscany	Chianti
	Chianti Classico
	Brunello di Montalcino
	Vino Noble di Montapulciano
	Morellino di Scansano
	Maremma
	Bolgheri
		Sangiovese *
		  Brunello
		  Prugnolo Gentile
		  Morellino
		Canaiolo Nero
		Cabernet Sauvignon
		Merlot
		Syrah
			Malvasia Bianco
			Pinot Grigio
			Trebbiano
			Vermentino
			Vernaccia *
				Mostly reds
				Whites
				Dessert
				Varietals, blends

Umbria	Sagrantino di Montefalco Colli Martani Orvieto	Sagrantino * Sangiovese Cabernet Sauvignon	Grechetto * Trebbiano Chardonnay Verdello	Whites, reds Sparkling Blends
SOUTH				
Apulia	Salice Salentino	Montepulciano Negroamaro * Primitivo Sangiovese Uva di Troia *	Bombino Bianco Malvasia Bianco Trebbiano	Reds, rosés Dessert Varietals, blends
Basilicata	Aglianico del Vulture	Aglianico *		Reds only Varietals
Calabria	Cirò	Gaglioppo * Cabernet Sauvignon Magliocco *	Greco * Chardonnay Sauvignon Blanc	Mostly reds Whites Some dessert Varietals, blends

DISTINGUISHING CHARACTERISTICS IN VINEYARD	DISTINGUISHING CHARACTERISTICS IN WINEMAKING	DISTINGUISHING CHARACTERISTICS OF WINE
Mountainous with some coastal vineyards Generally large yields	Large cooperatives dominate	Montepulciano made in two styles: light, drinkable; rich, earthy, undistinguished
	Large cooperatives dominate	Known for passito, Lambrusco Sangiovese: fruity
Volcanic hills	Traditional techniques	Mostly undistinguished whites
Coastal Adriatic Sea: Mediterranean climate Coastal plains: rolling hills, limestone soil		Fresh, fruity whites with depth of flavor Montepulciano
Mountainous with coastal vineyards	Large cooperatives dominate Mostly traditional techniques with modern influence	Mostly undistinguished wines, with a few of good quality
Hilly Best vineyards on sloped sites	Tremendously variable by region, variety, style Sangiovese dominant in many styles, traditional and modern	Wide range of high-profile Sangiovese- based reds: Chianti, Brunello di Montalcino, Super Tuscans, Vino Noble di Montepulciano Vin Santo: dessert wine Vernaccia: notable whites
Vineyard characteristics similar to Tuscany	Mostly traditional techniques Some barrel-fermented whites	Orvieto dominant Sagrantino most highly regarded
Hot, dry plains Salento Peninsula, ideal climate: cool nights, hot days	Large cooperatives dominate Macerate Negroamaro for rosés	High-alcohol wines used for distilling Primitivo same as California Zinfandel
Arid, cold Mountainous, limited vineyard space Volcanic soil	Limited production	Aglianico del Vulture: deep color, dense, robust, powerful
Limited acreage Most on coast of Adriatic Sea	Heat-loving varieties best	Reds dominated by Gaglioppo Cabernet Sauvignon, Gaglioppo blends Greco: dessert wine

Distinguishing Characteristics of Italian Wines - 3

REGION	APPELLATION, SUBAPPELLATION, OR VILLAGE	PRIMARY RED GRAPE VARIETIES	PRIMARY WHITE GRAPE VARIETIES	TYPE OF WINE— VARIETAL OR BLEND
Campania	Taurasi Fiano di Avellino	Aglianico * Piedirosso *	Falanghina * Fiano * Greco di Tufo *	Reds, whites Varietals, blends
Sardinia	Vermentino di Gallura Vermentino di Sardegna Cannonau di Sardegna	Bovale Cannonau Carignano Syrah	Malvasia Bianco Vermentino	Reds, whites Varietals, blends
Sicily	Marsala Pantelleria Etna	Nerello Mascalese * Nero d'Avola * Cabernet Sauvignon Syrah	Catarratto * Greco * Grillo * Inzolia * Malvasia Bianco Zibibbo (Muscat) Chardonnay	Reds, whites Fortified Varietals, blends

Distinguishing Characteristics of Spanish Wines

Note: Star (*) indicates indigenous variety.

REGION	APPELLATION, SUBAPPELLATION, OR VILLAGE	PRIMARY RED GRAPE VARIETIES	PRIMARY WHITE GRAPE VARIETIES	TYPE OF WINE— VARIETAL OR BLEND
NORTHWEST				
Castile and Leon	Bierzo	Mencía * Garnacha	Doña Blanca Godello	Mostly reds Varietals, blends
	Ribera del Duero	Tinto Fino (Tempranillo) * Cabernet Sauvignon Garnacha Merlot		Mostly reds Varietals, blends
	Rueda	Tinta de Toro (Tempranillo)	Verdejo * Sauvignon Blanc Viura (Macabeo)	Primarily whites Varietals, blends
	Toro	Garnacha Tinta Tinta de Toro (Tempranil- lo) *	Malvasía Verdejo	Mostly reds Varietals, blends
Galicia	Rías Baixas		Albariño *	Mostly whites 100% varietal Albariño Blends

DISTINGUISHING CHARACTERISTICS IN VINEYARD	DISTINGUISHING CHARACTERISTICS IN WINEMAKING	DISTINGUISHING CHARACTERISTICS OF WINE
Coastal Mediterranean climate Diverse geography, soils Volcanic soil in some areas	Traditional and modern techniques	All indigenous varieties Diversity of reds and whites
Hot, dry climate Granite, volcanic soils		Rich, alcoholic wines
Very hot, arid climate Irrigation permitted Volcanic soil in Mt. Etna region	High-color, acidic varieties best	Indigenous reds most distinctive International whites make strong statement

DISTINGUISHING CHARACTERISTICS IN VINEYARD	DISTINGUISHING CHARACTERISTICS IN WINEMAKING	DISTINGUISHING CHARACTERISTICS OF WINE
Wet climate Hilly terrain Slate, granite soils; good drainage	Mencía thought to be Cabernet Franc Low yields Modern techniques	Traditional: fruit, acid, lower alcohol Modern: dark, heavy, higher alcohol
Continental climate: bitterly cold winters, long hot summers Arid, rolling plain	Tempranillo dominant Modern techniques	Powerful, rich, fruity, lightly oaked reds Best: Reservas, Gran Reservas

Continental climate: bitterly cold winters,
   long hot summers
Arid, rolling plain
	Fermentation mostly stainless steel
	   or concrete
		Powerful, flavorful whites with moderate
		   alcohol
		100% Verdejo or blended with Sauvignon
		   Blanc

Continental climate: bitterly cold winters,
   long hot summers
Arid, rolling plain
	Oak aging	Best: 100% Tempranillo, minimum 75%
		   Tempranillo blended with Garnacha

Coolest, wettest, greenest region
Atlantic Ocean influence, but sunny days
Pergola trellising
	Fermentation mostly stainless steel or
	   concrete
	Some with oak
		Best Albariño: fresh, fruity, floral, oily,
		   long, finish, low alcohol (below 13%)
		Some barrel aged

Distinguishing Characteristics of Spanish Wines - 2

REGION	APPELLATION, SUBAPPELLATION, OR VILLAGE	PRIMARY RED GRAPE VARIETIES	PRIMARY WHITE GRAPE VARIETIES	TYPE OF WINE— VARIETAL OR BLEND
NORTHEAST				
Aragon	Somontano	Parraleta * Moristel * Cabernet Sauvignon Garnacha Merlot Tempranillo	Alcañon * Chardonnay Garnacha Blanca Gewürztraminer Macabeo (Viura)	Reds, rosés, whites Varietals, blends
Balearic Islands	Binissalem Mallorca Plà i Llevant	Callet * Manto Negro * Listan Negro Cabernet Sauvignon	Moll * Moscatel * Chardonnay Parellada Listan Blanco	Mosty reds Some whites Varietals, blends
Basque Country	Chacolí		Hondarribi Zuri *	Varietal whites
Catalonia	Penedès	Ull de Llebre (Tempranillo) Garnacha Monastrell Cariñena (Mazuelo) Merlot Cabernet Sauvignon	Parellada * Macabeo (Viura) * Xarel-lo * Chardonnay	Reds, whites Sparkling (Cava) Varietals, blends
	Priorat	Garnacha * Cabernet Sauvignon Cariñena (Mazuelo)		Mostly reds Rosés Varietals, blends
Navarre	Navarre	Garnacha * Tempranillo * Cabernet Sauvignon Graciano Mazuelo	Viura (Macabeo) * Chardonnay Garnacha Blanca Malvasía	Rosés, reds, whites Varietals, blends
Rioja	Rioja	Tempranillo * Garnacha Mazuelo (Cariñena) Graciano	Viura (Macabeo) * Garnacha Blanca Malvasía	Mostly reds Some whites, rosés Mostly blends
CENTRAL				
Castile-La Mancha	La Mancha	Cencibel (Tempranillo) * Cabernet Sauvignon Garnacha Tinta Moravia	Airén * Chardonnay Sauvignon Blanc Viura (Macabeo)	Reds, whites, rosés Fortified and dessert Varietals, blends
	Valdepeñas	Cencibel (Tempranillo) * Cabernet Sauvignon Garnacha Tinta	Airén * Viura (Macabeo)	Reds, whites, rosés Fortified and dessert
Extremadura	Ribera del Guadiana 6 subregions	Tempranillo * Cabernet Sauvignon Garnacha *	Alarije Pardina * Borba Cayetana Blanca *	Reds, rosés, whites Varietals, blends

DISTINGUISHING CHARACTERISTICS IN VINEYARD	DISTINGUISHING CHARACTERISTICS IN WINEMAKING	DISTINGUISHING CHARACTERISTICS OF WINE
Abundant rainfall	Traditional and modern techniques	Blend indigenous varieties with international varieties Widely varied wines
Hot summers Rolling plateau Very ripe grapes	Traditional techniques	Little exported
Cool, damp climate Varied soils	Traditional winemaking No oak	Chacolí: young, fresh white wine with spritz Joven only
Varied terrain, microclimates	Innovative techniques Traditional method used for sparkling wines	Good, high-quality wines
Mountainous, terraced vineyards Volcanic soil	Ripe grapes Oak aging	Reds: Joven to Gran Reserva High alcohol
Several microclimates		Fresh, fruity, robust rosés Great diversity of rosés, reds, whites Emphasis on Crianza, Reserva, Gran Reserva
Alta and Alevesa: higher, cooler Baja: lower, hotter, more Mediterranean climate	Traditional and modern techniques	Many distinct Rioja styles Bottle age reds 30 years or more Barrel-fermented whites Best from Alta and Alevesa
Continental climate: bitterly cold winters, long hot summers Extremely arid, drought constant problem Covers enormous central plain	Traditional techniques	Largest-volume wine producer in world High-alcohol wines
Continental climate: bitterly cold winters, long hot summers Extremely arid, drought constant problem	Traditional techniques	Modest quality, good value wines Cencibel: Reservas, Gran Reservas
More temperate than other regions in area Diverse terrain: mountainous, valleys, plains	Many producers still use traditional techniques	Regional native varieties used in wine Moderate-alcohol wines Substantial whites, fruity rosés, powerful, rich, ripe reds

Distinguishing Characteristics of Spanish Wines - 3

REGION	APPELLATION, SUBAPPELLATION, OR VILLAGE	PRIMARY RED GRAPE VARIETIES	PRIMARY WHITE GRAPE VARIETIES	TYPE OF WINE— VARIETAL OR BLEND
Madrid	Vinos de Madrid	Tinto Fino (Tempranillo) * Garnacha * Merlot Syrah Cabernet Sauvignon	Airén * Albillo * Malvar * Moscatel Parellada	Reds, whites, rosés Varietals, blends
Murcia	Jumilla	Monastrell *	Airén *	Mostly reds Mostly varietals
	Yecla	Monastrell *	Merseguera Viura (Macabeo)	Mostly reds Mostly varietals
Valencia	Utiel-Requena	Bobal * Tempranillo Garnacha	Chardonnay Macabeo (Viura) Merseguera	Reds, rosés Few whites Varietals, blends
SOUTH				
Andalusia	Condado de Huelva		Zalema * Listán (Palomino)	Whites Varietals, blends

	Jerez
	  Jerez de la Frontera
	  El Puerto de Santa Maria
	  Sanlúcar de Barrameda
			Moscatel
			 Palomino Fino (Listán) *
			 Pedro Ximénez
				Mostly fortified and dessert
				 Reds, whites
				 Blends

	Málaga		Moscatel Pedro Ximénez *	Mostly fortified and dessert Whites only, blends
	Montilla-Moriles		Pedro Ximénez * Moscatel	Whites Fortified Varietals, blends
	Sierra de Málaga	Romé Cabernet Sauvignon Merlot Syrah Tempranillo	Chardonnay Colombard Moscatel Pedro Ximénez Sauvignon Blanc Viura (Macabeo)	Reds, rosés, whites Varietals, blends
Canary Islands	Tacoronte-Acentejo	Listán Negro * Negramoll *	Listán Blanca (Palomino) * Malvasía	Reds, whites, rosés Fortified and dessert Varietals, blends

DISTINGUISHING CHARACTERISTICS IN VINEYARD	DISTINGUISHING CHARACTERISTICS IN WINEMAKING	DISTINGUISHING CHARACTERISTICS OF WINE
Continental climate: bitterly cold winters, long hot summers	Use indigenous, international varieties Traditional techniques	Mostly young wines meant for early consumption
Continental climate: bitterly cold winters, long hot summers Rainfall varies between subregions Extremely arid, drought constant problem	Traditional winemaking	Monastrell dominant: ripe, black fruit, powerful
Southern climate with hot summers Little rain	Traditional techniques	Monastrell dominant: ripe, rich, dark
Continental climate moderated by Mediterranean influence Variable soils	Traditional and modern techniques	Bobal: used for rosés Garnacha, Macabeo: for modern-style wines Very dark, alcoholic reds used for blending
Mediterranean and Atlantic climate Albariza soil	Traditional techniques for young whites	Zalema: light, floral white wine
Coastal: moderate temperatures, humid Albariza soil	Solera system for aging	World's best, most traditional sherry
Coastal: moderate temperatures, humid	Traditional techniques Solera system for aging	Traditional fortified and dessert wines
Continental climate Mostly head pruned No irrigation Albariza soil		
Hot, dry summers with cold winters Mountainous	Mostly modern techniques Experimental winemaking	New region producing only table wines Ripe red wines
Many microclimates Volcanic soils	Traditional techniques	Many indigenous varieties not found elsewhere

Distinguishing Characteristics of German Wines

REGION	APPELLATION, SUBAPPELLATION, OR VILLAGE	PRIMARY RED GRAPE VARIETIES	PRIMARY WHITE GRAPE VARIETIES	TYPE OF WINE— VARIETAL OR BLEND
Baden	9 Bereiche (districts)	Spätburgunder (Pinot Noir)	Müller-Thurgau Grauburgunder (Pinot Gris)	Reds, whites Varietals
Mosel-Saar Ruwer	3 separate areas along the Mosel River and its tributaries, the Saar and Ruwer 6 districts: most famous Bernkastel		Riesling Müller-Thurgau	Whites Dry to very sweet Sparkling (Sekt) Mostly varietals
Nahe	3 major areas between Monzinger and Bad Münster-am-Stein		Riesling Müller-Thurgau Kerner Silvaner	Mostly whites Varietals, blends
Pfalz	2 districts Mittelhaardt/Deutsche Weinstrasse Südliche Weinstrasse	Dornfelder Portugieser	Rielsling Müller-Thurgau Kerner Grauburgunder	Mostly whites Varietals, blends
Rheingau		Spätburgunder	Riesling Müller-Thurgau	Whites Dry to very sweet Varietals
Rheinhessen	Many subregions	Dornfelder Portugieser Spätburgunder	Müller-Thurgau Riesling Silvaner Kerner Scheurebe	Whites Dry to very sweet Sparkling Mostly varietals, some blends

DISTINGUISHING CHARACTERISTICS IN VINEYARD	DISTINGUISHING CHARACTERISTICS IN WINEMAKING	DISTINGUISHING CHARACTERISTICS OF WINE
Warmer than other parts of Germany Low yields	Many cooperatives	Diverse wines Higher in alcohol than other German wines Similar to wines of Northern France, light
Cold climate Late ripening Ultrasteep vineyards Blue and red slate soil	No barrel fermentation or aging	Benchmark for Riesling Some of finest dessert wines
Temperate, sunny	Few cooperatives Traditional and modern techniques	Variable quality Some excellent values
Sunniest, driest region Protected by mountains Diverse soils	Cooperatives and important estates	Broad range of varietal wines and blends Making smaller amounts of Liebfraumilch Increase in great diversity of higher-quality wines
Flat, rolling to steeply sloped vineyards Varied soils Low yields	No barrel fermentation or aging	Focus on dry Riesling Light-style Pinot Noir
Much of region is flat Some steep vineyards Sandstone, red slate soils	High-yield vineyards in flat areas Many cooperatives	High-quality, low-yield varietal wines Mostly bulk blends Drier varietal wines increasing

Appendix

3

New World Wine Regions

The following New World wine regions were selected because they are the dominant producers of wines outside of Old World countries. They produce the most wine, which is of excellent quality and widely distributed in the international marketplace.

New World wine laws are less restrictive than those of Old World countries. Some establish geographic areas and most have enacted labeling laws that specify requirements for varietal, appellation, and vintage.

NORTH AMERICA

U.S. Wine Laws and Labels

Wines are controlled by the Alcohol and Tobacco Tax and Trade Bureau (TTB). The TTB determines specific American Viticultural Areas (AVAs) and grape-growing and winemaking practices through its labeling guidelines. For all TTB alcoholic beverage requirements, including those for wine, see Appendix 1. The TTB does not establish any additional requirements for grape growing or winemaking that are not part of the labeling requirements.

WASHINGTON STATE WINE

NEW YORK-FINGER LAKES WINE WITH IRF SWEETNESS SCALE

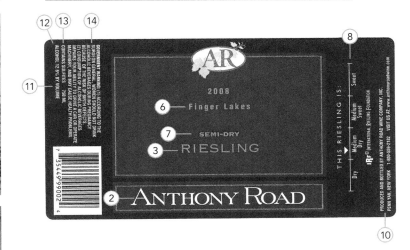

1 Vintage
2 Producer
3 Varietal/varieties
4 Indication of quality (no standard in the United States)
5 Same as grown, produced, and bottled by on back label
6 AVA
7 Sweetness indication
8 IRF Sweetness Scale
9 Same as estate bottled on front label
10 Name and address of winery
11 Net contents (not shown-embedded on bottle)
12 Alcohol content
13 Sulfites declaration
14 Health warning statement

North America:
U.S. and Canada

Legend:
- Main California wine areas
- Other U.S. wine areas
- Canadian wine areas

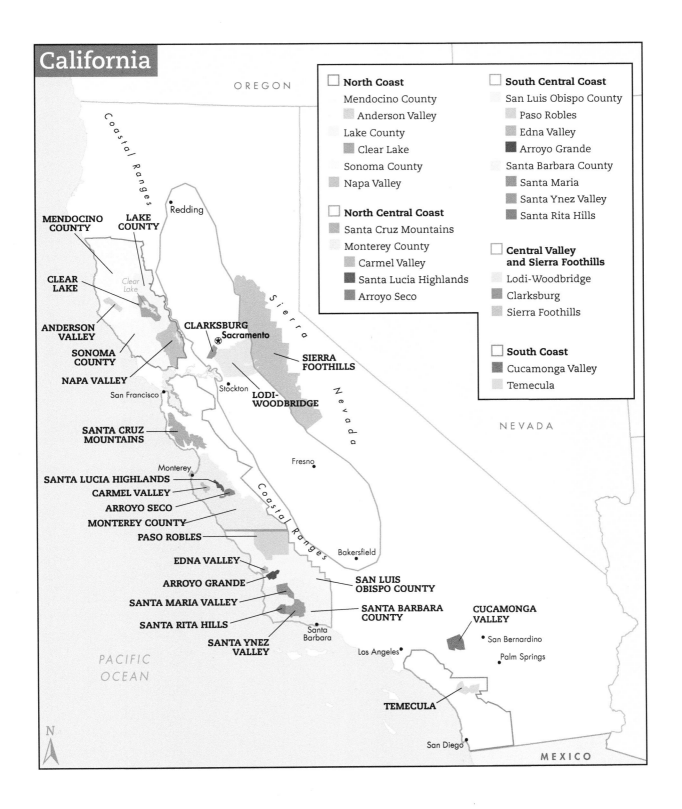

California

North Coast
Mendocino County
Anderson Valley
Lake County
Clear Lake
Sonoma County
Napa Valley

North Central Coast
Santa Cruz Mountains
Monterey County
Carmel Valley
Santa Lucia Highlands
Arroyo Seco

South Central Coast
San Luis Obispo County
Paso Robles
Edna Valley
Arroyo Grande
Santa Barbara County
Santa Maria
Santa Ynez Valley
Santa Rita Hills

**Central Valley
and Sierra Foothills**
Lodi-Woodbridge
Clarksburg
Sierra Foothills

South Coast
Cucamonga Valley
Temecula

OREGON

Coastal Ranges

Redding

MENDOCINO
COUNTY

LAKE
COUNTY

Clear
Lake

CLEAR
LAKE

ANDERSON
VALLEY

SONOMA
COUNTY

NAPA VALLEY

CLARKSBURG

Sacramento

Sierra

SIERRA
FOOTHILLS

Stockton

LODI-
WOODBRIDGE

Nevada

San Francisco

SANTA CRUZ
MOUNTAINS

Monterey

SANTA LUCIA HIGHLANDS

CARMEL VALLEY

ARROYO SECO

MONTEREY COUNTY

PASO ROBLES

Fresno

Coastal Ranges

Bakersfield

EDNA VALLEY

ARROYO GRANDE

SANTA MARIA VALLEY

SANTA RITA HILLS

SAN LUIS
OBISPO COUNTY

SANTA BARBARA
COUNTY

NEVADA

CUCAMONGA
VALLEY

San Bernardino

SANTA YNEZ
VALLEY

Santa
Barbara

Los Angeles

Palm Springs

PACIFIC
OCEAN

TEMECULA

N

San Diego

MEXICO

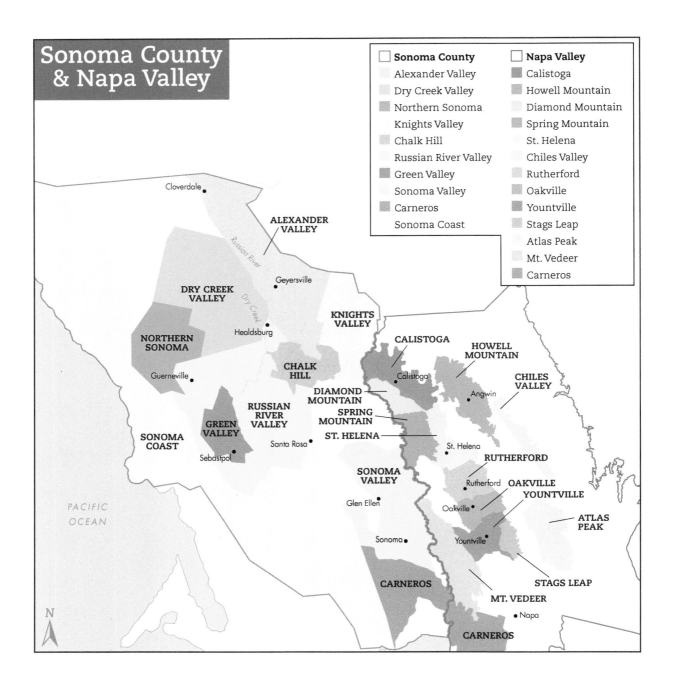

Sonoma County & Napa Valley

Sonoma County
- Alexander Valley
- Dry Creek Valley
- Northern Sonoma
- Knights Valley
- Chalk Hill
- Russian River Valley
- Green Valley
- Sonoma Valley
- Carneros
- Sonoma Coast

Napa Valley
- Calistoga
- Howell Mountain
- Diamond Mountain
- Spring Mountain
- St. Helena
- Chiles Valley
- Rutherford
- Oakville
- Yountville
- Stags Leap
- Atlas Peak
- Mt. Vedeer
- Carneros

Cloverdale

ALEXANDER VALLEY

Russian River

Geyersville

DRY CREEK VALLEY

Dry Creek

Healdsburg

NORTHERN SONOMA

KNIGHTS VALLEY

CALISTOGA

Guerneville

CHALK HILL

Calistoga

HOWELL MOUNTAIN

Angwin

CHILES VALLEY

DIAMOND MOUNTAIN

SPRING MOUNTAIN

RUSSIAN RIVER VALLEY

ST. HELENA

St. Helena

GREEN VALLEY

Santa Rosa

RUTHERFORD

SONOMA COAST

Sebastpol

Rutherford

OAKVILLE

YOUNTVILLE

SONOMA VALLEY

Oakville

Glen Ellen

ATLAS PEAK

Yountville

PACIFIC OCEAN

Sonoma

STAGS LEAP

MT. VEDEER

Napa

CARNEROS

N

CARNEROS

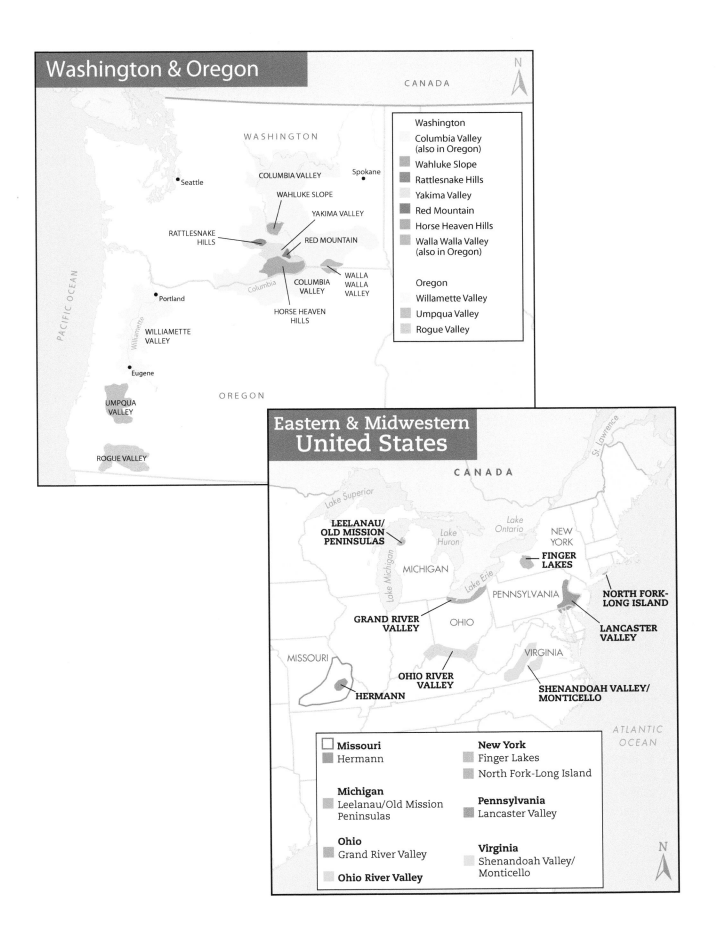

Washington & Oregon

CANADA

N

WASHINGTON

Seattle

COLUMBIA VALLEY

Spokane

WAHLUKE SLOPE

YAKIMA VALLEY

RATTLESNAKE
HILLS

RED MOUNTAIN

Columbia

WALLA
WALLA
VALLEY

COLUMBIA
VALLEY

PACIFIC OCEAN

Portland

HORSE HEAVEN
HILLS

Willamette

WILLIAMETTE
VALLEY

Eugene

OREGON

UMPQUA
VALLEY

ROGUE VALLEY

Washington

Columbia Valley
(also in Oregon)

Wahluke Slope

Rattlesnake Hills

Yakima Valley

Red Mountain

Horse Heaven Hills

Walla Walla Valley
(also in Oregon)

Oregon

Willamette Valley

Umpqua Valley

Rogue Valley

Eastern & Midwestern
United States

St. Lawrence

CANADA

Lake Superior

Lake
Ontario

NEW
YORK

LEELANAU/
OLD MISSION
PENINSULAS

Lake
Huron

Lake Michigan

MICHIGAN

FINGER
LAKES

Lake Erie

PENNSYLVANIA

NORTH FORK-
LONG ISLAND

GRAND RIVER
VALLEY

OHIO

LANCASTER
VALLEY

MISSOURI

VIRGINIA

OHIO RIVER
VALLEY

SHENANDOAH VALLEY/
MONTICELLO

HERMANN

ATLANTIC
OCEAN

N

Missouri
Hermann

Michigan
Leelanau/Old Mission
Peninsulas

Ohio
Grand River Valley

Ohio River Valley

New York
Finger Lakes

North Fork-Long Island

Pennsylvania
Lancaster Valley

Virginia
Shenandoah Valley/
Monticello

Canadian Wine Laws and Labels

Because Canada's wine laws are very unrestrictive, Canadian producers formed the Vintner's Quality Alliance (VQA). This group established regulatory criteria similar to those of European countries for important qualifications, including appellations and grape variety, as well as vineyard designation and winemaking practices. Membership in the Alliance is strictly voluntary, but members who follow the guidelines are allowed to use the VQA seal on their winery labels. The strict guidelines cover quality standards, geographic region, vineyard designation, estate bottling, wine category, and labeling guidelines. The key regulations include:

Quality Standards All wines must be made from grapes. The addition of water is prohibited.

Geographic Origin Grapes must come from the stated region of origin.

Vineyard Designation Wines must come from the specific vineyard identified on the label.

Tasting Panels A tasting panel of experts must certify the wines.

Canadian wines are currently produced in four regions: British Columbia, Ontario, Nova Scotia, and Quebec. Currently only two, British Columbia and Ontario, meet the stringent requirements of the VQA.

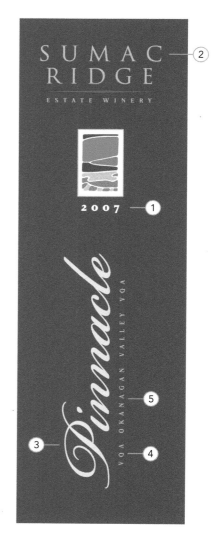

1   Vintage
2   Producer
3   Brand/proprietary blend
4   Certified Vinter's Quality Alliance regulation
5   Appellation
6   Varietals
7   Name and address of producer
8   Net contents
9   Alcohol content
10 Sulfites declaration

British Columbia

N

Legend:
- Vancouver Island
- Gulf Islands
- Fraser Valley
- Similkameen Valley
- Okanagan Valley

Kamloops

OKANAGAN VALLEY

Okanagan Lake

Kelowna

FRASER VALLEY

SIMILKAMEEN VALLEY

Okanagan

VANCOUVER ISLAND

Vancouver

GULF ISLANDS

WASHINGTON

Victoria

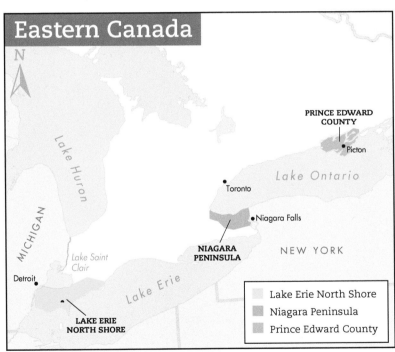

Eastern Canada

N

PRINCE EDWARD COUNTY

Picton

Lake Huron

Lake Ontario

Toronto

MICHIGAN

•Niagara Falls

Lake Saint Clair

NIAGARA PENINSULA

NEW YORK

Detroit

Lake Erie

LAKE ERIE NORTH SHORE

Legend:
- Lake Erie North Shore
- Niagara Peninsula
- Prince Edward County

SOUTH AMERICA

Chilean Wine Laws and Labels

Like laws of other New World wine-producing countries, Chile's laws are not restrictive. The laws enacted in 1995 were a joint effort by the Servicio Agricola Ganadero, the Ministerio de Agricultura, and the wineries. They define appellations, subregions, and zones within the regions. Chilean labeling laws stipulate that:

- A wine named by grape variety must have at least 75 percent of the named variety in the wine. Most Chilean producers use at least 85 percent to meet international requirements.
- Blended wines with up to three varieties must list them on the label in descending order by volume.
- A wine identified from a specific appellation, subregion, or region must have at least 75 percent of the wine from that specified place. Most Chilean producers use at least 85 percent to meet international requirements.
- A wine with a specified vintage must have at least 75 percent of the wine from that vintage. Most Chilean producers use at least 85 percent to meet international requirements.
- Special designation may be awarded to reserve wines:
  Reserva minimum 12 percent alcohol, oak optional
  Reserva Especial minimum 12 percent alcohol, oak required
  Gran Reserva 12.5 percent alcohol, oak required

VARIETAL WINE

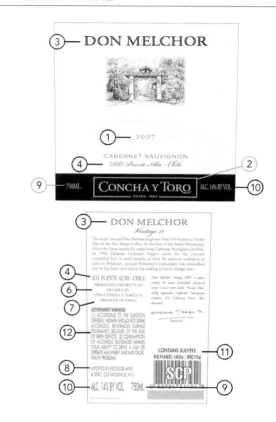

1 *Vintage*
2 *Producer*
3 *Brand/vineyard designation*
4 *Appellation*
5 *Statement of farming methods*
  a Biodynamic
  b Organic
6 *Name and address of producer*
7 *Country of origin*
8 *Name and address of importer*
9 *Net contents*
10 *Alcohol content*
11 *Sulfites declaration*
12 *Health warning statement*

ORGANIC/BIODYNAMIC WINE

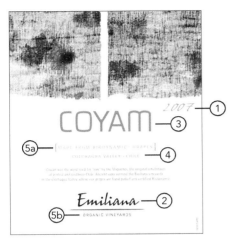

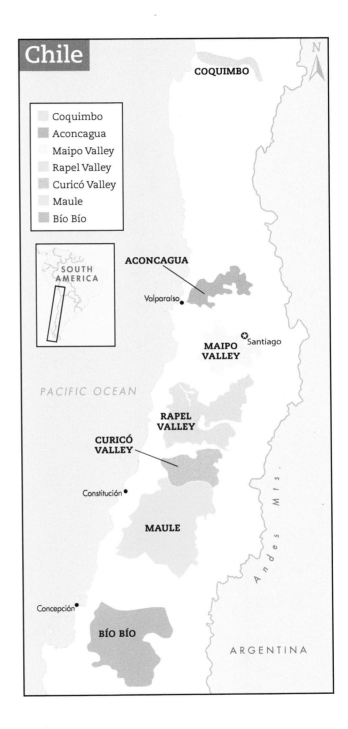

Argentinean Wine Laws and Labels

The only Argentinean law and label requirement is that a wine that is identified by grape variety must be made from at least 80 percent of that variety. Wines for the international marketplace follow more stringent international guidelines.

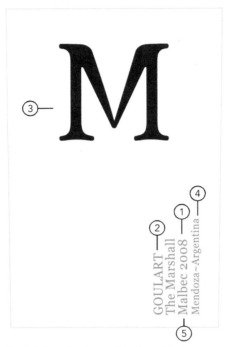

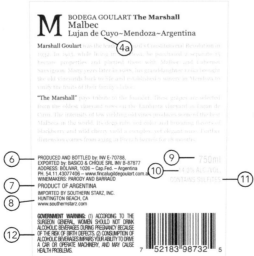

1 Vintage
2 Producer
3 Brand/proprietary name
4 Appellation
   a Appellation and subregion
5 Varietal/varieties
6 Name and address of producer or exporter
7 Country of origin
8 Name and address of importer
9 Net contents
10 Alcohol conent
11 Sulfites declaration
12 Health warning statement

OTHER NEW WORLD WINE REGIONS

Australian Wine Laws and Labels

The Australian Wine and Brandy Corporation establishes regulations that define wine areas and labeling requirements. Unlike those of the regulatory environment in France, Australia's wine laws are not very restrictive and do not regulate grape growing and wine production. The country is divided into five wine regions, South Australia, New South Wales, Victoria, Tasmania, and Western Australia, with specific wine districts, known as Geographic Indications (GIs), identified within the regions. Grapes do not have to come from a single vineyard or appellation. They can come from anywhere within the broad wine region although the premium wines generally are made from grapes that are grown in smaller, well-known wine districts.

Australian labeling laws stipulate that:

- When a GI or appellation is indicated on the label, 85 percent of the wine must come from that GI.
- If a variety is named on the label, 85 percent of the wine must be made from the named grape.
- If two varieties are used in a blend and neither represents 85 percent of the wine, both grapes must be named on the label. If there are more than two grapes, each grape consisting of 20 percent or more of the blend must be named.
- The tolerance for listed alcohol content is 1.5 percent for wine and sparkling wine and 0.5 percent for fortified wine.

AUSTRALIAN VARIETAL WINE

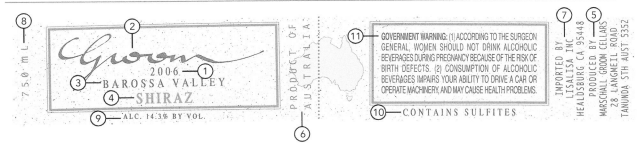

AUSTRALIAN BLENDED WINE

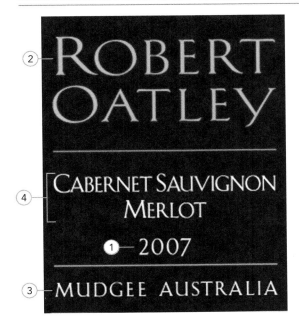

1   *Vintage*
2   *Brand name*
3   *Geographic indication*
4   *Varietal/varieties*
5   *Name and address of producer*
6   *Country of origin*
7   *Name and address of importer*
8   *Net contents*
9   *Alcohol content*
10   *Sulfites declaration*
11   *Health warning statement*

Australia
including Western Australia

N

INDIAN
OCEAN

NORTHERN
TERRITORY

QUEENSLAND

Alice Springs

WESTERN
AUSTRALIA

SOUTH
AUSTRALIA

Brisbane

NEW SOUTH
WALES

Perth Hills

Perth

Sydney

Adelaide

Canberra

Margaret
River

Great
Southern

Pemberton

VICTORIA

Melbourne

Tasman Sea

SOUTHERN
OCEAN

Western Australia

Perth Hills

Margaret River

Pemberton

Great Southern

TASMANIA

Hobart

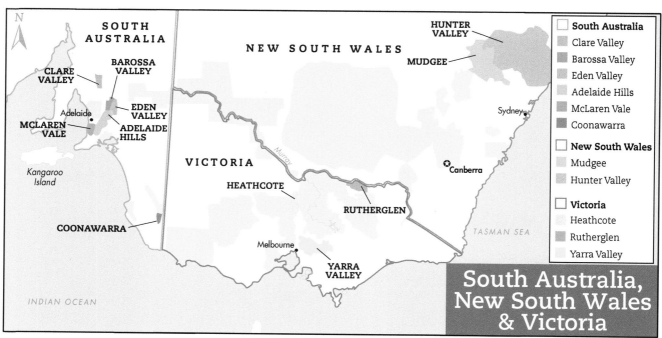

N

SOUTH
AUSTRALIA

HUNTER
VALLEY

NEW SOUTH WALES

MUDGEE

BAROSSA
VALLEY

CLARE
VALLEY

EDEN
VALLEY

Adelaide

Sydney

MCLAREN
VALE

ADELAIDE
HILLS

Canberra

Kangaroo
Island

VICTORIA

HEATHCOTE

TASMAN SEA

RUTHERGLEN

COONAWARRA

Murray

Melbourne

INDIAN OCEAN

YARRA
VALLEY

South Australia

Clare Valley

Barossa Valley

Eden Valley

Adelaide Hills

McLaren Vale

Coonawarra

New South Wales

Mudgee

Hunter Valley

Victoria

Heathcote

Rutherglen

Yarra Valley

South Australia,
New South Wales
& Victoria

New Zealand Wine Laws and Labels

In New Zealand wine laws are very similar to those of Australia and deal primarily with labeling. There are no requirements for grape growing and wine production. As of 2007 New Zealand labeling laws include the following:

– For a wine identified from a specific appellation, sub-region, or region at least 75 percent of the wine must come from that specified place. For export to the United States or the European Union, at least 85 percent of the wine must be from the stated appellation.

– For a wine designated with a vintage at least 85 percent of the wine must be from that vintage.

– If a wine label states a grape variety, the wine must include at least 85 percent of the stated variety.

– When two or more grapes are identified on the label, each variety in the wine must be listed in descending order by volume.

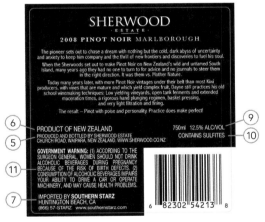

1 Vintage
2 Producer
3 Appellation
4 Varietal
5 Name and address of producer
6 Country of origin
7 Name and address of importer
8 Net contents
9 Alcohol conent
10 Sulfites declaration
11 Health warning statement

South African Wine Laws and Labels

The Wine and Spirits Board is the primary regulatory organization governing South African wines. The South African Wine Industry and System (SAWIS) sets the guidelines, which are strict and measure up to or exceed standards in other wine-growing countries. A primary focus is on identifying a Wine of Origin Scheme (WO). To be labeled a wine of origin, 100 percent of the grapes must come from the named appellation. The requirements for labeling South African wines were upgraded effective January 2006. South African labeling laws include the following:

– If identified by grape variety, 85 percent of the wine must be from the identified grape.
– If identified by vintage, 85 percent of the wine must be from the stated vintage and be certified by the Wine and Spirits Board.
– If identified by a specific appellation, 100 percent must be from that appellation.
– The grape varieties in a wine must be listed in order of prominence on the label.
– If a wine states a grape variety, vintage, and appellation, it must be certified by the Wine and Spirits Board to ensure its appropriate clarity, color, flavor, and taste.

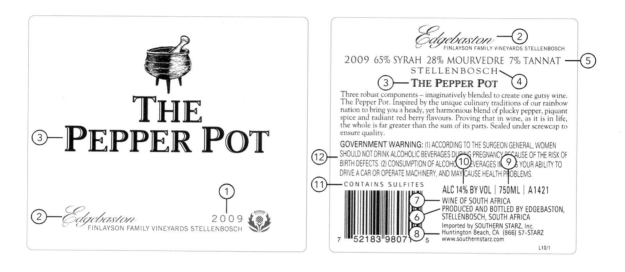

1   Vintage
2   Producer
3   Brand/proprietary name
4   Region/appellation
5   Varietal/varieties
6   Name and address of producer
7   Country of origin
8   Name and address of importer
9   Net contents
10 Alcohol conent
11 Sulfites declaration
12 Health warning statement

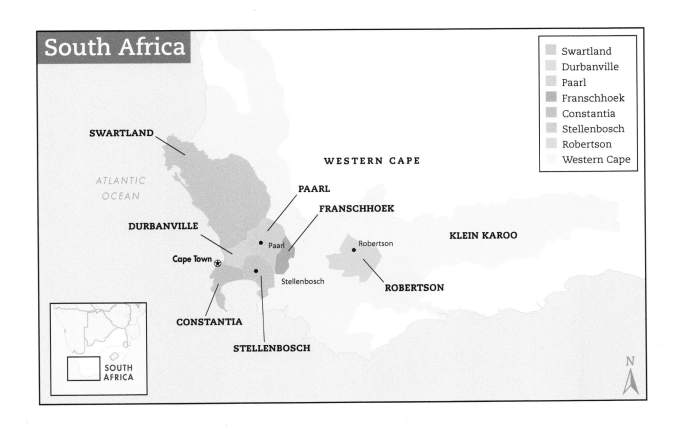

South Africa

Legend:
- Swartland
- Durbanville
- Paarl
- Franschhoek
- Constantia
- Stellenbosch
- Robertson
- Western Cape

SWARTLAND

ATLANTIC OCEAN

WESTERN CAPE

PAARL

FRANSCHHOEK

DURBANVILLE

KLEIN KAROO

• Robertson

• Paarl

Cape Town ✪

Stellenbosch

ROBERTSON

CONSTANTIA

STELLENBOSCH

SOUTH AFRICA

N

Distinguishing Characteristics of North American Wines

REGION	APPELLATION, SUBAPPELLATION, OR VILLAGE		PRIMARY RED GRAPE VARIETIES	PRIMARY WHITE GRAPE VARIETIES	TYPE OF WINE— VARIETAL OR BLEND
UNITED STATES					
California	Mendocino County Anderson Valley Inland Valleys		Pinot Noir Zinfandel Cabernet Sauvignon Petite Sirah	Riesling Gewürztraminer Chardonnay Sauvignon Blanc	Reds, whites Sparkling Mostly varietals
	Lake County		Cabernet Sauvignon	Sauvignon Blanc	Whites, reds Mostly varietals
	Napa Valley	Valley Floor Calistoga Oakville Rutherford St. Helena Yountville	Cabernet Sauvignon Merlot Zinfandel Syrah	Chardonnay Sauvignon Blanc	Whites, reds Sparkling Varietals, blends
		Elevated Diamond Creek Howell Mountain Mt. Veeder Spring Mountain Stags Leap District	Cabernet Sauvignon Merlot Zinfandel Syrah	Chardonnay Sauvignon Blanc	Whites, reds Varietals, blends
		Other Carneros	Pinot Noir Syrah	Chardonnay	Reds, whites Sparkling Mostly varietals
	Sonoma County	Elevated Dry Creek Valley Russian River Valley Sonoma Coast	Pinot Noir Zinfandel Cabernet Sauvignon Merlot Syrah	Sauvignon Blanc Chardonnay	Reds, whites Sparkling Varietals, blends
		Valley Floor Alexander Valley Knights Valley Sonoma Valley	Cabernet Sauvignon Merlot Zinfandel	Chardonnay Sauvignon Blanc	Reds, whites Varietals, blends
	Santa Cruz Mountains		Cabernet Sauvignon Pinot Noir	Chardonnay	Whites, reds Mostly varietals, some blends
	Monterey County		Pinot Noir Cabernet Sauvignon Merlot	Chardonnay Riesling Chenin Blanc Gewürztraminer	Reds, whites Varietals, blends
	Sierra Foothills		Zinfandel Syrah Petite Sirah		Mostly reds Mostly varietals

DISTINGUISHING CHARACTERISTICS IN VINEYARD	DISTINGUISHING CHARACTERISTICS IN WINEMAKING	DISTINGUISHING CHARACTERISTICS OF WINE
Climate varies significantly: Anderson Valley cool, inland valleys warm Elevation varies significantly	Mix of modern and traditional techniques	Pinot Noir, floral whites, and sparkling wines some of best in California
Warm to hot continental climate Cooler at elevation		Distinctive Sauvignon Blanc
Warmer valley floors Various soils	Modern techniques	Benchmark for warm-climate Cabernet Sauvignon
Cooler, elevated, sloped vineyards Various soils	Modern techniques	Benchmark for cool-climate California Cabernet Sauvignon
Cool climate with fog, Pacific Ocean breezes	Traditional method used for sparkling wines	California cool-climate examples of Pinot Noir and Chardonnay
Diverse climate Diverse soils: volcanic to sandy loam	Diverse techniques	Benchmark for California Sauvignon Blanc and Pinot Noir
Diverse climates Diverse soils	Modern techniques	Ripe, juicy wines
Cool climate Mountainous, steeply sloped vineyards Various soils		Long-lived Cabernet Sauvignon Delicious Pinot Noir
Little rain, irrigation required More arid inland Best Pinot Noir vineyards located close to Pacific Ocean or inland at high elevation (Mt. Harlan and Calara) Cabernet Sauvignon and Merlot grown primarily on alluvial soil	Mostly modern techniques Traditional techniques for Pinot Noir	Old World and New World Pinot Noir styles Other reds in New World style Chardonnay: rich, bold, powerful, heavily oaked
Continental climate: hot summers, cold winters Various elevations, soils		Old-vine Zinfandel

REGION	APPELLATION, SUBAPPELLATION, OR VILLAGE		PRIMARY RED GRAPE VARIETIES	PRIMARY WHITE GRAPE VARIETIES	TYPE OF WINE— VARIETAL OR BLEND
California, continued	Central Valley	San Joaquin Valley	Cabernet Sauvignon Merlot Various	Chardonnay Various	Reds, whites Varietals, blends
		Clarksburg		Chenin Blanc	Only 100% varietal Chenin Blanc
		Lodi/Woodbridge	Zinfandel Ruby Cabernet	Chardonnay Viognier	Whites, reds Mostly varietals
	Paso Robles		Cabernet Sauvignon Syrah Zinfandel	Chardonnay Viognier	Reds, whites Mostly varietals
	Santa Barbara County Santa Rita Hills Santa Maria Santa Ynez Valley		Pinot Noir Syrah Rhône varieties	Chardonnay Sauvignon Blanc Rhône varieties	Reds, whites Varietals
	Southern California Temecula Cucamonga Valley		Zinfandel Mission Merlot Grenache	Chardonnay Sauvignon Blanc Palomino Muscat	Reds, whites Sparkling Fortified Varietals, blends
Michigan	Leelanau and Old Mission Peninsulas		Pinot Noir Cabernet Franc	Aromatic whites particularly Riesling	Whites, reds Sparkling Mostly varietals
Missouri			Norton Chambourcin	Seyval Blanc Vidal Blanc Vignoles	Whites, reds Varietals, blends
New York	Long Island		Cabernet Franc Merlot Cabernet Sauvignon	Sauvignon Blanc Chardonnay	Whites, reds Bordeaux-style blends
	Finger Lakes		Pinot Noir Concord	Riesling Chardonnay	Whites, reds Mostly varietals
Oregon	Willamette Valley		Pinot Noir	Pinot Gris Chardonnay Riesling	Reds, whites Mostly varietals
Virginia			Cabernet Franc Merlot Cabernet Sauvignon Norton	Chardonnay Viognier	Whites, reds Varietals, blends

DISTINGUISHING CHARACTERISTICS IN VINEYARD	DISTINGUISHING CHARACTERISTICS IN WINEMAKING	DISTINGUISHING CHARACTERISTICS OF WINE
Vast valley of vineyards Heavily irrigated	Huge bulk-wine processing	Jug wines
Fog Deep, sandy soil		Only noted Chenin Blanc appellation in California
Hot plains		Zinfandel: distinctive, ripe, plummy
East: rolling plains West: high hills	Modern techniques	Juicy, ripe, high-alcohol wines
West side: very cool climate with Pacific Ocean influence East side: warmer	Almost all small producers	Superlative Pinot Noirs
Mediterranean climate to hot desert	Traditional and modern techniques	Temecula: excellent wines from little-known area Cucamonga Valley: historic area, classic fortified wines
Lake effect protects vines from freezing temperatures	Very small producers	World-class Riesling
Bitter winters	Hybrid and native American varieties	Regularly top winners at wine competitions Only available in Midwest and East Norton: best wine for barbecue
Moderating influence of Atlantic Ocean Lake effect protects vines from freezing temperatures Sandy loam soil with good drainage	Small, independent producers	Good-quality Bordeaux blends
Lake effect protects vines from freezing temperatures Calcareous soils	Small, independent producers	World-class Riesling
Cool, damp climate affected by the Pacific Ocean Moderately sloped vineyards Volcanic soils, marine sediment	Mostly small-batch winemaking	Best Pinot Noir appellation in the United States Burgundy-style wines
Difficult warm, muggy climate	Small producers Handcrafted wines	Exceptional Viognier

Distinguishing Characteristics of North American Wines - 3

REGION	APPELLATION, SUBAPPELLATION, OR VILLAGE		PRIMARY RED GRAPE VARIETIES	PRIMARY WHITE GRAPE VARIETIES	TYPE OF WINE— VARIETAL OR BLEND
Washington	Columbia Valley	Yakima Valley Horse Heaven Hills Red Mountain Wahluke Slope	Cabernet Sauvignon Merlot Cabernet Franc Syrah	Riesling Chardonnay Sémillon	Reds, whites Sparkling Mostly varietals
	Walla Walla Valley		Merlot Cabernet Sauvignon Syrah	Chardonnay	Mostly reds, some whites Varietals, blends
CANADA					
British Columbia	Okanagan Valley		Merlot Pinot Noir Cabernet Sauvignon Cabernet Franc Syrah	Pinot Gris Pinot Blanc Riesling Chardonnay Gewürztraminer	Whites, reds Late-harvest, ice wines Varietals, blends
Ontario			Pinot Noir	Riesling Gewürztraminer	Whites, reds Ice wines Mostly varietals

Distinguishing Characteristics of South American Wines

REGION	APPELLATION, SUBAPPELLATION, OR VILLAGE		PRIMARY RED GRAPE VARIETIES	PRIMARY WHITE GRAPE VARIETIES	TYPE OF WINE— VARIETAL OR BLEND
CHILE					
Coquimbo	Elqui		Cabernet Sauvignon	Chardonnay Sauvignon Blanc	Whites, reds Varietals, blends
Aconcagua	Aconcagua Casablanca		Cabernet Sauvignon Syrah Pinot Noir Merlot Carmenère	Chardonnay Sauvignon Blanc	Whites, reds Mostly varietals
Central Valley	Maipo Valley Maule Valley Rapel Valley Cachapoal Colchagua Curicó Valley		Cabernet Sauvignon Merlot Carmenère Syrah	Chardonnay Sauvignon Blanc	Whites, reds Varietals, blends
Southern Regions	Bío Bío		Cabernet Sauvignon Pinot Noir	Chardonnay Sauvignon Blanc Gewürztraminer	Whites, reds Mostly varietals
ARGENTINA					
	Mendoza		Malbec Cabernet Sauvignon Bonarda	Torrontés Chardonnay	Reds, whites Large amount of rosés regionally Varietals, blends

DISTINGUISHING CHARACTERISTICS IN VINEYARD	DISTINGUISHING CHARACTERISTICS IN WINEMAKING	DISTINGUISHING CHARACTERISTICS OF WINE
Continental climate: hot summers, cold winters Dry, irrigation required Large day-night temperature variation	Traditional method used for sparkling wines Modern techniques	Largest producer of Riesling in the United States Second-largest wine producer after California
Soils: volcanic, minerals, granite Bitterly cold winters Irrigation required	Modern techniques	Quality Merlot, Cabernet Sauvignon, and Syrah
Lake effect protects vines from freezing temperatures		Excellent cool-climate wines, mostly whites Excellent ice wines
Lake effect protects vines from freezing temperatures		Excellent ice wines

DISTINGUISHING CHARACTERISTICS IN VINEYARD	DISTINGUISHING CHARACTERISTICS IN WINEMAKING	DISTINGUISHING CHARACTERISTICS OF WINE
Andean and coastal influences Extreme heat	Newer region, more modern techniques	Lighter, varietal wines
Andean and coastal influences Cool climate Irrigation common No phylloxera, own-rooted vines	Transitioning from traditional to modern techniques	Bordeaux varieties and blends dominate Pinot Noir showing cool-climate character of Casablanca region
Andean and coastal influences Irrigation common No phylloxera, own-rooted vines Diverse soils	Transitioning from traditional to modern techniques	Bordeaux varieties and blends Lean, clean, fruity wines at lower price points Well-structured, barrel-aged wines at higher price points
Andean and coastal influences Heavy rain Irrigation common No phylloxera, own-rooted vines Stony soils	Newer region using more modern techniques	Mostly white cool-climate wines Pinot Noir increasing
Continental climate Little rainfall, irrigation common No phylloxera, mostly own-rooted vines Alluvial soils	Changing from traditional bulk winemaking techniques to modern techniques	Malbec varietal wines dominant internationally Bonarda dominant locally Spanish variety Torrontés: most distinguished white wines

Distinguishing Characteristics of Other New World Wines

REGION	APPELLATION, SUBAPPELLATION, OR VILLAGE	PRIMARY RED GRAPE VARIETIES	PRIMARY WHITE GRAPE VARIETIES	TYPE OF WINE—VARIETAL OR BLEND
AUSTRALIA				
New South Wales	Hunter Valley	Shiraz Cabernet Sauvignon	Sémillon Chardonnay	Reds, whites Varietals
	Mudgee	Cabernet Sauvignon Shiraz	Chardonnay Sémillon	Reds, whites Fortified Varietals
South Australia	Adelaide Hills	Shiraz Pinot Noir	Sauvignon Blanc Chardonnay Viognier	Whites, reds Sparkling Varietals, blends
	Barossa Valley	Shiraz Cabernet Sauvignon	Sémillon Chardonnay Riesling	Whites, reds Varietals, blends
	Eden Valley	Shiraz Cabernet Sauvignon	Riesling Chardonnay	Reds, whites Varietals
	McLaren Vale	Cabernet Sauvignon Grenache Shiraz	Chardonnay Sauvignon Blanc	Reds, whites Varietals, blends
	Coonawarra	Cabernet Sauvignon Shiraz	Chardonnay Riesling	Reds, whites Mostly blends, usually Cabernet Sauvignon with Shiraz
	Clare Valley	Cabernet Sauvignon Shiraz	Riesling	Reds, whites Varietals, some red blends
Tasmania		Pinot Noir	Riesling Chardonnay Sauvignon Blanc	Whites, reds Sparkling Mostly varietals
Victoria	Yarra Valley	Cabernet Sauvignon Merlot Pinot Noir Shiraz	Chardonnay Riesling	Reds, whites Varietals, blends
	Heathcote	Shiraz Cabernet Sauvignon		Mostly red varietals
Western Australia	Margaret River	Cabernet Sauvignon Merlot	Chardonnay Sémillon Riesling Sauvignon Blanc Chenin Blanc	Whites, reds Varietals, blends

DISTINGUISHING CHARACTERISTICS IN VINEYARD	DISTINGUISHING CHARACTERISTICS IN WINEMAKING	DISTINGUISHING CHARACTERISTICS OF WINE
Hot, humid Alluvial plains, rolling hills		Long-aging Sémillon Long-aging Shiraz
Little rain, lots of sun Irrigation required Well-drained soils	Barrel-fermented Chardonnay	
Cool climate, winter rains Moderately high vineyards Variable soils		Benchmark for premium Sauvignon Blanc Northern Rhône–style reds (Syrah with Viognier)
Vineyards located in cool microclimates Low-fertility soils: sandy loam, clay, loam Old vines prevalent		Most famous Australian wine region Old-vine Shiraz Benchmark for Sémillon GSM blends
Cool microclimates Rolling hills, open and windy Moderate elevation Soils: sandy loam, clay, loam		Ageworthy Riesling Rich New World Chardonnay Ripe Shiraz
Ocean influence Irrigation required Various soils		
Terra rossa limestone soil yields premium Cabernet Sauvignons		Rich Chardonnay Ageworthy Riesling Ripe Shiraz
Variable climate, mostly continental Irrigation required		Benchmark for dry Riesling Full-bodied Cabernet Sauvignon
Island maritime climate Soils: volcanic rock, sandstone	Traditional method used for sparkling wines	Very small production of excellent wines

Variable climate
Many microclimates fit for
  different varieties
Flat valley vineyards to steeply
  sloped vineyards
	Various winemaking techniques	Lighter-style Bordeaux blends
		Most Pinot Noir in Australia

Sloped vineyards Red calcareous clay soil		Ripe, potentially long-aging Shiraz
Maritime influence Rolling hills along Indian Ocean Soils: gravel, sandy loam	Use all available winemaking techniques with Chardonnay	

Distinguishing Characteristics of Other New World Wines - 2

REGION	APPELLATION, SUBAPPELLATION, OR VILLAGE	PRIMARY RED GRAPE VARIETIES	PRIMARY WHITE GRAPE VARIETIES	TYPE OF WINE— VARIETAL OR BLEND
NEW ZEALAND				
North Island	Aukland	Cabernet Sauvignon Merlot	Chardonnay	Whites, reds Varietals
	Hawkes Bay Gimblett Gravels	Merlot Cabernet Sauvignon Pinot Noir Shiraz	Chardonnay Sauvignon Blanc	Reds, whites Bordeaux-style blends
	Martinborough	Pinot Noir	Sauvignon Blanc Chardonnay Riesling Pinot Gris Gewürztraminer	Whites, reds Mostly varietals
South Island	Central Otago	Pinot Noir	Chardonnay Riesling	Reds, whites Varietals
	Marlborough	Pinot Noir	Sauvignon Blanc Chardonnay Riesling	Whites, reds Varietals
	Nelson	Pinot Noir	Sauvignon Blanc Riesling Chardonnay	Whites, reds Varietals
SOUTH AFRICA				
	Stellenbosch Paarl Constantia	Cabernet Sauvignon Shiraz Merlot Pinotage	Chenin Blanc Colombard Chardonnay Sauvignon Blanc	Reds, whites Varietals, blends

DISTINGUISHING CHARACTERISTICS IN VINEYARD	DISTINGUISHING CHARACTERISTICS IN WINEMAKING	DISTINGUISHING CHARACTERISTICS OF WINE
Heavy clay soil	Modern techniques	Chardonnay best
Warmer temperatures create riper, richer wines Alluvial soil	Sets standard for high-tech winemaking	Best Merlot planted on Gimblett Gravels
Coolest, driest part of North Island	Modern techniques	Various wines from small producers
Coolest wine region below equator	Modern techniques	Burgundy-style cool-climate wines
Flat, undulating vineyards	Modern winemaking Sauvignon Blanc not barrel fermented	Benchmark for Sauvignon Blanc: clean, grassy, citrus, unoaked Pinot Noir: burgundy-like
Coolest, wettest South Island climate	Modern techniques	Excellent white and red wines
Mediterranean climate Terroir key factor to growers	Ripe grapes require acidification Barrel aging popular in best wines	Bordeaux blends Pinotage unique

Glossary

ACID/ACIDITY A prominent taste component of wine that gives wine a crisp, tart, sour taste. In grapes it is most commonly tartaric acid, malic acid, plus a small amount of citric acid that constitute the primary acids.

ACIDIFICATION The artificial process of adding acid back into a fermented wine to create balance by recapturing the acidity that was lost because of extended grape ripening.

ACRE The standard measurement of area used to describe vineyard size in the United States. Equal to 4,840 square yards or 4,047 square meters.

AFTERTASTE The flavor and mouthfeel of a wine after it has been swallowed. The same as finish.

ALCOHOL BY VOLUME (ABV) The percentage of alcohol in relation to water in a beer or distilled spirit.

ALE The common name for any beer produced by top fermentation.

AMERICAN VITICULTURAL AREA (AVA) A grape-growing area with a defined boundary established and approved by the TTB.

ANTHOCYANINS The color agents found in red grapes and retained in red wine that can range from shades of red to shades of blue.

ANTIOXIDANTS Phenolic compounds found in all plant-derived foods, but in a much higher proportion in red grapes, that are believed to help prevent heart disease.

APPELLATION A specific geographic location with defined boundaries that produces specified wines.

APPELLATION D'ORIGINE CONTRÔLÉE (AOC) The highest quality category of wine in France determined by the Institut National des Appellations d'Origine, which governs controlled appellations for a wide range of food products, including wine.

ASTRINGENCY/ASTRINGENT A quality that describes a component of mouthfeel that draws moisture from the inside of your cheeks, leaving a drying sensation.

AUTOLYSIS The breakdown of yeast cells by their own enzymes during the aging period of wine. Most typical of sparkling wine but can occur with any wine that is sur lies aged.

BALANCE A description indicating that no one component of a wine overpowers another. The normal components considered to assess balance are fruit, alcohol, acidity, sweetness, and bitterness.

BARREL The generic term used to describe a wooden vessel of unspecified size that is commonly made of oak and used to ferment and age wine, and also to age some spirits.

BARRIQUE The French term used to identify a French oak barrel, typically the Bordeaux style that is 225 liters/59 gallons in capacity.

BITTERNESS One of the three distinctive taste components of wine (with acidity and sweetness) that is normally distinguishable on the back of the tongue. It is frequently misreferenced as astringency, a mouthfeel.

BODY The weight or viscosity of a wine, beer, or spirit, which normally increases as alcohol level increases.

BOTRYTIS CINEREA A beneficial form of bunch rot, commonly identified as noble rot, that attacks grapes under certain humid and damp conditions, causing reduced moisture and increased sugar in grapes, and adding a honey character to wine.

BOTTE (PL. BOTTI) A large wooden barrel or cask frequently made of Slovenian oak used for aging wines in Italy.

BOTTOM FERMENTATION One of two basic beer fermentation methods. Yeast tends to settle to the bottom of the fermenting beer (see Top Fermentation).

BOUQUET The combination of individual aromas found in wine, developed from the grapes in winemaking and during aging.

BRETTANOMYCES (BRETT) A genus of yeast that can infect wine and give it an earthy, mushroom character.

BRIX The unit of measure given in degrees of sugar concentration in grapes that is equivalent to the percentage of sugar in grapes. A reading of 24° Brix approximates 24 percent sugar in the grapes.

CANOPY The leaf structure of a vine that covers and shades the grapes.

CANOPY MANAGEMENT The process of controlling the amount of leaf coverage over grapes by determining which leaves to leave on the vine during ripening and which to remove.

CARBON DIOXIDE (CO$_2$) A naturally occurring by-product of fermentation that is usually allowed to dissipate into the atmosphere but is trapped during sparkling-wine production to give the wine its effervescence.

CASK A wooden barrel of any size.

CHAPTALIZATION The process of adding sugar to must to create more alcohol, not additional sweetness. This normally occurs in Old World wine regions where grapes do not ripen fully.

COLD SOAK Allowing the solids of the must to remain in contact with the juice at cold temperatures prior to fermentation to extract color and flavor into the juice.

COLUMN STILL (CONTINUOUS STILL) Two or more tall, narrow, metal columns connected by pipes that are used to distill beverages in a continuous process.

COMPETITIVE STATES States that license private businesses to distribute and sell alcoholic beverages.

CONGENERS The substances that give distilled spirits their characteristic tastes, aromas, and body.

CONTROL STATES Currently nineteen states that directly regulate the distribution and sale of at least some alcoholic beverages.

CORKED WINE A term that describes a wine that has been infected with the compound trichloroanisole (TCA) from the cork.

CROSS A grape variety derived from two varieties of the same species.

CRU The French term for growth used to identify specific wines that come from specific vineyards that are typically considered to be of excellent quality.

CRYSTAL MALT A specialty grain produced by heating sugar-rich malted barley to a point where the sugar crystallizes. Used to impart sweetness and copper color to beer.

CUVE The French term for a vat, tank, or cask.

CUVÉE The French term for blend. A cuvée can be a wine made from a blend of grape varieties, vintages, different vineyards, or winemaking methods.

DENOMINACIÓN DE ORIGEN (DO) Spanish for controlled appellation and the core of Spain's quality wine classifications.

DENOMINAZIONE DI ORIGINE CONTROLLATA (DOC) Italian for controlled appellation and the core of Italy's quality wine classifications.

DENOMINAZIONE DI ORIGINE CONTROLLATA E GARANTITA (DOCG) Italian for controlled and guaranteed appellation of origin. A higher and more stringent classification than DOC.

DISTILLATE The alcoholic liquid collected after distillation.

DISTILLATION The process of extracting alcohol from a fermented liquid made from grains, fruits, vegetables, or other plant materials.

DISTILLED SPIRITS Alcoholic beverages made by distillation.

DRY The opposite of sweet. The absence of sugar in wine that occurs when fermentation converts all of the sugar in grapes into alcohol.

ETHANOL Most common type of alcohol found in alcoholic beverages.

FERMENTATION The process of converting the sugar in any plant material into alcohol.

FILTRATION The use of very fine filter pads to strain out every possible particle left in a wine.

FINING A clarification process used to remove microscopic particles in a wine.

FINISH How a wine tastes and feels after it has been swallowed (see Aftertaste).

FLOR A special type of yeast native to Jerez, Spain, that coats sherry wines in the barrel, moderating oxidation and adding a distinctive flavor component to the wine.

FREE-RUN JUICE The juice that naturally flows from grapes before they are pressed. Considered to be the highest-quality juice.

FUSEL OILS Heavy alcohols that can add character to some types of distilled spirits, but can be harsh and unpleasant if overused.

GRAFTING Combining a cutting from one grapevine with a rootstock of a different grapevine to produce a new vine that grows grapes of the same variety as that of the cutting.

GRAIN WHISKEY (WHISKY) Whiskey made from a combination of raw grains. Called grain whisky in reference to Scotch.

GRIST Ground mix of malted grain and specialty grains made ready for mashing.

HEAT SUMMATION SCALE The measure of the amount of heat units for a particular appellation. Used to classify the climate from cool to hot.

HECTARE The standard measure of area used to describe vineyard size in countries where the metric system is standard, including Europe and most other countries. One hectare is 10,000 square meters, or the equivalent of 2.47 acres.

HECTOLITER The standard metric measure of volume used for large quantities of liquid. One hectoliter is 100 liters, or the equivalent of 26.4 gallons.

HOPS The generic name for the soft, pinecone-like female flower of a perennial climbing vine, also called hops. Used to impart flavor to beer.

HOT The term used to describe a wine that is high in alcohol that leaves a burning sensation on the tongue.

HYBRID A grapevine that is derived from two varieties of different species (e.g., a native American grape variety and a *vinifera* variety).

LAGER The common name for any beer produced by bottom fermentation.

LEES The solids left after fermentation that includes dead yeast cells and remaining particles of grape solids.

LEGS The rivulets of wine that form on the inside of a wineglass caused by high alcohol levels. It is a sign of the wine's alcohol content but is not an indicator of quality. Also known as tears.

MACERATION The process of allowing grape solids to remain in contact with must before fermentation (cold soak) or with wine after fermentation to extract color and flavor from the solids.

MALOLACTIC FERMENTATION (MLF) A secondary fermentation that occurs with the introduction of lactic acid bacteria that converts tart malic acid into softer lactic acid.

MALT WHISKEY A whiskey, in particular, Scotch whisky, that is made from malted barley.

MALTING A two-step process that breaks down the starch in grain into the sugar needed for fermentation. Grain is soaked until the kernels begin to germinate, then it is kiln-dried to stop growth. Used in beer and whiskey production.

MASH The product obtained when grains are cooked with water to dissolve starches and make sugar available for fermentation. Used in beer and whiskey production.

MERITAGE An American term created to describe a wine made outside Bordeaux (most commonly in the United States) using only the approved Bordeaux grape varieties.

MÉTHODE CHAMPENOISE The process of bottle fermentation created in the Champagne region of France to produce sparkling wine.

MOUTHFEEL The tactile sensation of wine or other beverages in your mouth.

MUST The slurry of skins, seeds, and juice from grapes after crushing before the juice is pressed or fermented.

NEGOTIANT In France, a person or organization that buys grapes or wine, makes a finished wine, and then markets it under the negotiant's label.

NEUTRAL SPIRITS Colorless, odorless, and flavorless alcoholic product of distillation that is almost 100 percent pure alcohol. It is used as a base for many types of spirits and as the fortifying alcohol for fortified wines.

NEW WORLD A reference to countries with a relatively new history of winemaking commonly found outside of Europe. New World countries generally have more sunshine, creating riper grapes and a riper, higher-alcohol wine than Old World wines.

OFF-PREMISE Refers to a place of business that sells alcohol in closed containers, typically bottles of wine, beer, or spirits that are not consumed on-site.

OLD WORLD A reference to European countries that have been dominant in winemaking for several centuries. It also refers to a wine style that is typically made from grapes receiving less sun, resulting in lower-alcohol wines.

ON-PREMISE Refers to a place of business that sells alcoholic beverages to be consumed on-site, such as a restaurant or brewpub.

OWN-ROOTED A grapevine that is grown on its own natural roots as opposed to being grafted onto rootstock.

OXIDATION The excessive addition of oxygen to a wine causing it to turn brown and take on a nutty character or become sherrylike.

OXIDIZED Describes a wine that is flawed from too much contact with oxygen. Only in fortified wines is it considered an attribute.

pH The strength of acidity in a wine. The lower the pH, the stronger the acidity.

PHENOLICS OR POLYPHENOLICS The chemical compounds found in wine that produce color, flavor, and mouthfeel characteristics.

POT STILL (BATCH STILL) A large, round kettlelike vessel with a narrow neck and tubes for collecting alcohol. The distillate must run through the pot still two or three times to condense the alcohol and extract flavors.

PROHIBITION The period of history in the United States from 1919 to 1933 when it was illegal under most circumstances to make, transport, or sell alcoholic beverages.

PROOF A term used only in the United States. Proof is double the percentage of alcohol in a distilled beverage.

PROPRIETARY NAME The name of a wine created by the proprietor of a winery. It can only be used for a specific wine produced by the proprietor.

PUMPING OVER The process of drawing wine out of the bottom of a tank and pouring it over the top of the wine remaining in the tank during fermentation. This allows the juice to filter through the solids to increase color, flavors, and tannins.

PUNCHING DOWN The process of reincorporating wine solids that collect at the top of the tank during fermentation. Solids are forced back down into the juice manually or mechanically to increase color, flavors, and tannins.

RACKED/RACKING The process of drawing off wine from one tank or barrel into another tank or barrel, leaving the wine solids in the original vessel. Racking is one step in the clarification process and can also add oxygen to the wine at an appropriate time.

REDUCTION The opposite of oxidation. No air comes in contact with the wine.

RESIDUAL SUGAR The amount of sugar remaining in wine after fermentation. It can be measured as grams per liter or as a percentage of volume. 25 grams per liter of sugar is the equivalent of 2.5 percent residual sugar.

ROASTED GRAIN A specialty grain produced by roasting unmalted or malted grain to a darker color. Used to impart roasted flavor and color ranging from light chocolate to black, to beer.

ROTARY FERMENTER A specific type of fermenter that lies horizontally instead of vertically. It mechanically reincorporates solids with the juice eliminating pumping over and punching down.

SETTLE/SETTLING Allowing must or wine to sit in a vessel for a period of time. The solids naturally fall out of solution and sink to the bottom of the vessel.

SOUR MASH A process used in American whiskeys. Some of the residue from one batch of mash is used to start fermenting a new batch of mash.

SPRITZ A term used to describe a light fizziness in a table wine that has a touch of carbon dioxide.

STABILIZATION The process of ensuring that a wine remains clear and does not re-ferment after bottling through cold stabilization, fining, and filtration.

STUCK FERMENTATION A fermentation that stops unintentionally before all the sugar has been converted to alcohol.

SUBMERGED CAP A technique that artificially causes the solids in a wine to be constantly held below the liquid line that incorporates color and flavors from the solids into the wine.

SULFUR DIOXIDE (SO$_2$) A common chemical compound naturally found in wine but also added before, during, and after fermentation to retard spoilage.

SWEET/SWEETNESS The presence of sugar in wine. Sweet also describes one of the five basic tastes.

TANNINS Chemical compounds found in wine (mostly red) that produce an astringent character. Tannins are part of mouthfeel, not taste.

TEMPERANCE The religious and lifestyle belief that the consumption of alcohol is socially, morally, and physically harmful.

TERROIR A term used by Europeans to describe the distinguishing characteristics of a particular place where grapes are grown. It includes climate, soil, topography, sunlight, and water access.

TOAST A term used to describe the treatment of the inside of a barrel that has been charred over fire in order to transfer toast aromas to the wine, or a quality in a wine that has developed toast aromas from aging in a toasted barrel.

TOP FERMENTATION One of two basic fermentation methods using yeast that tends to rise to the top of the fermenting beer (see Bottom Fermentation).

TRADITIONAL METHOD (MÉTHODE TRADITIONNELLE) The term used outside Champagne to describe making wine by the méthode champenoise.

TRELLIS A metal, wood, or plastic structure that supports grapevines.

ULLAGE The airspace in a barrel between the wine and the stopper or in a bottle between the wine and the cork.

VARIETAL A wine made from a specific grape variety that uses the variety as its descriptor.

VARIETY The grape subspecies used to make a wine.

VIN DE PAYS (VDP) The second highest classification of French wines that are identified by grape variety.

VINE TRAINING Forcing a grapevine to grow in a certain way to produce the best quality and quantity of grapes under the particular conditions of a specific appellation or vineyard.

VINICULTURE The broad term used to describe making wine that can include growing grapes.

VINIFERA The European species of grapes that is most frequently made into wine.

VINIFICATION The process of actually making wine. The fermentation process up to the aging of wines.

VITICULTURE The process of selecting grape varieties and vineyard sites, and growing grapes.

VITIS The genus of vine plants that includes grapes.

WINE STYLE A wine that has been created with certain characteristics. A blend of grape varieties, a level of acidity or ripeness, or a defined period of aging can exemplify wine style.

WORT The bittersweet liquid produced by boiling a mixture of mash and hot water, with hops. The cooled wort contains the sugar to be fermented by yeast.

Notes

CHAPTER 2, ALCOHOL AND THE LAW

1 "Persons Killed, by Highest Blood Alcohol Concentration (BAC) in the Crashes, 1994–2006–State: USA," National Highway Transportation Safety Administration, U.S. Department of Transportation, www.nhtsa.dot.gov.

2 ".08 BAC Illegal per se Level," *Traffic Safety Facts: Laws,* Vol. 2:1, March 2004, National Highway Transportation Safety Administration, U.S. Department of Transportation, www.nhtsa.dot.gov.

3 "Alcohol and Your Body," Brown University, www.brown.edu; "The ABCs of BAC," www.stopimpaired-driving.org; "If You Drink, Don't Drive—Period," *Westways,* Jan./Feb. 2008, p.13.

4 "Blood Alcohol Concentration Test Refusal Laws," *Traffic Safety Facts: Laws,* Jan. 2006, National Highway Transportation Safety Administration, U.S. Department of Transportation, www.nhtsa.dot.gov.

5 "DUI/DWI Laws," Insurance Institute for Highway Safety, Jan. 2008, www.iihs.org.

6 "Statistics on Underage Drinking," Underage Drinking Research Initiative, National Institute on Alcohol Abuse and Alcoholism, www.niaaa.nih.gov.

7 "Highlight on Underage Drinking," Alcohol Policy Information System, National Institute on Alcohol Abuse and Alcoholism, www.alcoholpolicy.niaaa.nih.gov.

8 "Alcohol 101: Dram Shop Liability and Legislation," The Marin Institute, 2006, www.marininstitute.org.

9 McConnell, Alison L., "Underage Drinking Laws Take Aim at Parents," Pew Research Center, Oct. 13, 2004, www.stateline.org.

CHAPTER 3, HEALTHY DRINKING

1 "Heart Disease and Stroke Statistics—2009 Update," American Heart Association, Dallas, Texas, 2009, www.americanheart.org.

2 Ibid.

3 Mukamal, Kenneth J., Alberto Ascherio, Murray A. Mittleman, Katherine M. Conigrave, Carlos A. Camargo Jr., Ichiro Kawachi, Meir J. Stampfer, Walter C. Willett, and Eric B. Rimm, "Alcohol and Risk for Ischemic Stroke in Men: The Role of Drinking Patterns and Usual Behavior," *Annals of Internal Medicine,* Vol. 142:1, 11–19, Jan. 4, 2005.

4 Sacco, Ralph L., Mitchell Elkind, Bernadette Boden-Albala, I-Feng Lin, Douglas E. Kargman, W. Allen Hauser, Steven Shea, and Myunghee C. Paik, "The Protective Effect of Moderate Alcohol Consumption on Ischemic Stroke," *Journal of the American Medical Association,* Vol. 281:1, 53–60, January 6, 1999.

5 Reynolds, Kristi, L. Brian Lewis, John David L. Nolen, Gregory L. Kinney, Bhavani Sathya, and Jiang He, "Alcohol Consumption and Risk of Stroke," *Journal of the American Medical Association,* Vol. 289:5, 579–588, February 5, 2003.

6 Standridge, John B., Robert G. Zylstra, and Stephen M. Adams, "Alcohol Consumption: An Overview of Benefits and Risks," *Southern Medical Journal,* Vol. 97:7, 664–672, July 2004.

7 Ellison, R. Curtis, "Wine and a Healthy Lifestyle," *Wine Spectator,* Oct. 15, 2004, pp. 44–55.

8 Hu, Frank B., JoAnn E. Manson, Meir J. Stampfer, Graham A. Colditz, Simin Liu, Caren Solomon, and Walter C. Willett, "Diet, Lifestyle and the Risk of Type 2 Diabetes Mellitus in Women," *New England Journal of Medicine,* Vol. 345:11, 790–797, 2001.

9 Wannamethee, S. Goya, A. Gerald Shaper, I.J. Perry, and K.G.M.M. Alberti, "Alcohol Consumption and the Incidence of Type II Diabetes," *Journal of Epidemiology and Community Health,* 56:7, 542–548 (2002).

10 Ajani, Umed A. and Charles H. Hennekens, "Alcohol Consumption and Risk of Type 2 Diabetes Mellitus among U.S. Male Physicians," *Archives of Internal Medicine,* Vol. 160:7, 1025–1030, 2000.

11 "Alcoholism and Alcohol-related Problems," National Council on Alcoholism and Drug Dependence, June 2002, www.ncadd.org/facts/problems.html.

12 "Estimated economic costs of alcohol abuse in the United States, 1992 and 1998," National Institute on Alcohol Abuse and Alcoholism, U.S. Department of Health and Human Services, March 2004, www.niaaa.nih.gov.

13 National Council on Alcoholism and Drug Dependence, op.cit.

14 "Health Risks and Benefits of Alcohol Consumption," *Alcohol Research and Health,* Vol. 24:1, 5–11, 2000, National Institute on Alcohol Abuse and Alcoholism.

15 "Booze and Babies: How Much Danger?" *USA Today,* Nov. 20, 2002.

16 U.S. Department of Health and Human Services and U.S. Department of Agriculture, *Dietary Guidelines for Americans: 2005,* "Alcoholic Beverages," pp. 43–46.

Bibliography

"About Spirits," Association of Canadian Distillers, www.canadiandistillers.com.

"All About Distilled Spirits and Liquor," *Tastings*, Beverage Testing Institute, www.tastings.com/spirits/index.html.

Apstein, Michael, "To Your Health—or Is It?" *Boston Globe*, August 11, 1999, Sec. Food, F1.

Baldy, Marian W., *The University Wine Course*, South San Francisco: The Wine Appreciation Guild Ltd., 1995.

Bastianich, Joseph and David Lynch, *Vino Italiano: The Regional Wines of Italy*, New York: Clarkson Potter/Publishers, 2002.

Bazar, Emily, "Sunday Holidays Loosen Alcohol Laws," *USA Today*, December 20, 2006, Sec. News, 3a.

Belfrage, Nicolas, *Barolo to Valpolicella: The Wines of Northern Italy*, London: Faber and Faber Limited, 1999.

Belfrage, Nicolas, *Brunello to Zibibbo: The Wines of Tuscany, Central and Southern Italy*, London: Faber and Faber Limited, 2001.

"Benefits from Moderate Drinking Extended?" *Harvard Heart Letter*, July 2004, Harvard Health Publications, Harvard Medical School, www.health.harvard.edu.

"Beverage Service Training and Related Practices," Alcohol Policy Information System, September 19, 2007, National Institute on Alcohol Abuse and Alcoholism, www.alcoholpolicy.niaaa.nih.gov.

Bird, David, *Understanding Wine Technology*, South San Francisco: The Wine Appreciation Guild Ltd., 2002.

"Blood Alcohol Limits Worldwide," International Center of Alcohol Policies, Feb. 2007, www.icap.org.

Blue, Anthony Dias, *The Complete Book of Spirits: A Guide to Their History, Production, and Enjoyment*, New York: HarperCollins Publishers Inc., 2004.

Brandes, Richard and Robert Keane, "Spirits Training Guide: Distilled Spirits," *Stateways*, March/April 2001.

Bridenbaugh, Russ, "The 3-Tier System: Is Anyone Happy?" *Wines & Vines*, April 2002.

Brook, Stephen, ed., *A Century of Wine*, South San Francisco: The Wine Appreciation Guild Ltd., 2000.

Brook, Stephen, *Liquid Gold: Dessert Wines of the World*, London: Constable and Company Limited, 1987.

Bryce, J.H. and G.G. Stewart, eds., *Distilled Spirits: Tradition and Innovation*, Nottingham, England: Nottingham University Press, 2004.

Cass, Bruce ed., *The Oxford Companion to the Wines of North America*, Oxford, England: Oxford University Press, 2000.

Cernilli, Daniele and Gigi Piumatti, eds., *Italian Wines 2005: A Guide to the World of Italian Wine for Experts and Wine Lovers*, New York: Slow Food Editore, Gambero Rosso, 2004.

Clarke, Oz and Margaret Rand, *Oz Clarke's Encyclopedia of Grapes: A Comprehensive Guide to Varieties and Flavors*, New York: Harcourt, Inc., 2001.

Clarke, Oz, *Oz Clarke's New Wine Atlas*, London: Websters International Publishers, 2002.

Clarke, Oz, *Wine Atlas: Wines and Wine Regions of the World*, London: Pavilion Books, 2007.

Coates M.W., Clive, *The Wines of France*, South San Francisco: The Wine Appreciation Guild Ltd., 2000.

Comer, James, "Distilled Beverages." In *The Cambridge World History of Food, Vol. 1*, edited by Kiple, Kenneth F. and Kriemheld Conee Ornelas, Cambridge, England: The Press Syndicate of the University of Cambridge, 2000, 653–664.

DeCuir, Marissa, "More States Debate End to Blue Laws," *USA Today*, November 30, 2007, Sec. News, 3a.

Diel, Armin and Joel Payne, *Gault Millau Guide to German Wines*, London: Mitchell Beasley Publishers, 2005.

"Do Women Benefit from Light Drinking?" *Harvard Heart Letter*, September 1995, Harvard Health Publications, Harvard Medical School, www.health.harvard.edu.

Dornenburg, Andrew and Karen Page, *What to Drink with What You Eat: The Definitive Guide to Pairing Food with Wine, Beer, Spirits, Coffee, Tea—even Water—Based on Expert Advice from America's Best Sommeliers*, New York: Bullfinch Press, 2006.

Dufour, Mary C., "What Is Moderate Drinking? Defining Drinks and Drinking Levels," *Alcohol Research and Health*, Vol. 23:1, 5–14, 1999, National Institute on Alcohol Abuse and Alcoholism.

Ellison, R. Curtis, "Importance of Pattern of Alcohol Consumption," *Circulation*, Vol. 112:24, 3818–3819, Dec. 13, 2005.

Ellison, R. Curtis, "Wine and a Healthy Lifestyle," *Wine Spectator*, Oct. 15, 2004, 44–55.

"The Facts: Zero Tolerance," National Highway Transportation Safety Administration, www.nhtsa.dot.gov.

Fagan, Amy, "Bill Would Curb Online Alcohol Sales," *CQ Weekly*, Vol. 58:10, 480–481, March 4, 2000.

Forrestal, Peter, ed., *The Global Encyclopedia of Wine*, South San Francisco: The Wine Appreciation Guild Ltd., 2001.

Friedrich, Jacqueline, *A Wine and Food Guide to the Loire*, New York: Henry Holt and Company, Inc., 1996.

Friedrich, Jacqueline, *The Wines of France: The Essential Guide for Savvy Shoppers*, Berkeley, California: Ten Speed Press, 2006.

"Furnishing Alcohol to Minors," Alcohol Policy Information System, National Institute on Alcohol Abuse and Alcoholism, www.alcoholpolicy.niaaa.nih.gov.

Galet, Pierre, *A Practical Ampelography: Grapevine Identification*, Ithaca, New York: Cornell University Press, 1979.

Gasnier, Vincent, *A Taste for Wine*, New York: DK Publishing, 2006.

Gasnier, Vincent, *Drinks*, New York: DK Publishing, 2008.

Goldstein, Evan, *Perfect Pairings: A Master Sommelier's Practical Advice for Partnering Wine with Food*, Berkeley, California: University of California Press, 2006.

Goode, Jamie, *The Science of Wine: From Vine to Glass*, Berkeley, California: University of California Press, 2005.

Haraszthy, A., *Grape Culture, Wines, and Wine-Making*, New York: Harper and Brothers, 1862. Reprinted: Fairfield, Connecticut: James Stevenson Publisher, 2003.

Harrington, Robert J., *Food and Wine Pairing: A Sensory Experience*, Hoboken, New Jersey: John Wiley & Sons, Inc., 2008.

Henderson, J. Patrick and Dellie Rex, *About Wine*, New York: Thomson Delmar Learning, 2006.

Hunt, Maria C., "Softening Hard Liquor," *The San Diego Union-Tribune*, Sept. 14, 2005, www.signonsandiego.com.

Introducing Bordeaux Wines, Conseil Interprofessionnel du Vin de Bordeaux (CIVB), 1999.

Jeffs, Julian, *The Wines of Spain*, London: Mitchell Beasley Publishers, 2006.

Johnson, Hugh, *A Life Uncorked*, Berkeley, California: University of California Press, 2005.

Johnson, Hugh, *Hugh Johnson's How to Enjoy Wine*, London: Mitchell Beasley Publishers, 1985.

Julyan, Brian K., *Sales and Service for the Wine Professional*, 2nd ed., New York: Thomson Learning, 2003.

Kolpan, Steven, Brian H. Smith, and Michael A. Weiss, *Exploring Wine*, 3rd ed., Hoboken, New Jersey: John Wiley & Sons, Inc., 2010.

Kornaroff, Anthony L., "By the Way, Doctor," *Harvard Health Letter*, September 2006, Harvard Health Publications, Harvard Medical School, www.health.harvard.edu.

Lake, Max, *Food on the Plate, Wine in the Glass*, Manley, Australia: McMahon Publishing, 1994.

Laube, James, *Wine Spectator's California Wine*, New York: Wine Spectator Press, 1999.

Lea, A.G.H. and J.R. Piggott, eds., *Fermented Beverage Production*, Glasgow: Blackie Academic and Professional, 1995.

Lynch, Kermit, *Adventures on the Wine Route: A Wine Buyer's Tour of France*, New York: Farrar, Straus and Giroux, 1988.

MacNeil, Karen, *The Wine Bible*, New York: Workman Publishing Co., Inc., 2001.

Margalit, Dr. Yair, *Winery Technology and Operations: A Handbook for Small Wineries*, San Francisco: The Wine Appreciation Guild Ltd.,1996.

McGee, Harold, *On Food and Cooking: The Science and Lore of the Kitchen*, New York: Scribner, 2004, 761–771.

Meister, Kathleen, *Moderate Alcohol Consumption and Health*, American Council on Science and Health, 1999, www.acsh.org.

Peñín, José, *Peñín Guide to Spanish Wine*, 2007, Madrid: Grupo Peñín, 2007.

Peynaud, Emile, *The Taste of Wine: The Art and Science of Wine Appreciation*, South San Francisco: The Wine Appreciation Guild Ltd., 1987.

Platter, John, *South African Wine Guide*, 2007, South Africa: The John Platter SA Wine Guide Ltd., 2007.

"Quick Stats: Underage Drinking," Centers for Disease Control and Prevention, www.cdc.gov/alcohol/quickstats/underage_drinking.htm.

Radford, John, *The New Spain: A Guide to Contemporary Spanish Wine*, London: Mitchell Beasley Publishers, 2004.

Renaud, S. and M. de Lorgeril, "Wine, Alcohol, Platelets, and the French Paradox for Coronary Heart Disease." *Lancet*, Vol. 339:8808, 1523–1527, June 20, 1992.

Robinson, Jancis, ed., *The Oxford Companion to Wine*, 3rd ed., Oxford, England: Oxford University Press, 2006.

Robinson, Jancis, Vines, *Grapes and Wines: The Wine Drinker's Guide to Grape Varieties*, London: Mitchell Beazley Publishers, 1999.

Routh, Chuck, "Legal Issues in Importing Wine," American Bar Association, Spring 2006.

Sbrocco, Leslie, *Wine for Women: A Guide to Buying, Pairing, and Sharing Wine*, New York: HarperCollins Publishers, Inc., 2003.

Schreiner, John, *British Columbia Wine Country*, North Vancouver, Canada: Whitecap Books, 2007.

Schreiner, John, *Icewine: The Complete Story*, Toronto: Warwick Publishing Inc., 2001.

Simon, Joanna, *Discovering Wine*, London: Mitchell Beasley Publishers, 1994.

Simon, Joanna, *Wine with Food*, New York: Simon & Schuster, Inc., 1996.

"Social Host Liability," Alcohol Epidemiology Program, University of Minnesota, www.epi.umn.edu.

"Social Host Liability Laws and Teen Party Ordinances," Alcohol Policy, The Marin Institute, www.marininstitute.org.

"State of the Science Report on the Effects of Moderate Drinking," December 19, 2003, National Institute on Alcohol Abuse and Alcoholism, www.niaaa.nih.gov.

Stevenson, Tom, ed., *Wine Report 2009*, New York: DK Publishing, 2008.

Stevenson, Tom, ed., *Wine Report 2007*, New York: DK Publishing, 2006.

Stevenson, Tom, *World Encyclopedia of Champagne and Sparkling Wine*, South San Francisco: The Wine Appreciation Guild Ltd., 1999.

Vine, Richard P., *Wine Appreciation*, Hoboken, New Jersey: John Wiley & Sons, Inc., 1997.

Walzer, Janet, "To Your Health? Sorting through the Risks and Benefits of Drinking," *Tufts Nutrition*, Fall 2003, Friedman School of Nutrition Science and Policy, Tufts University.

The Wines and Spirits of France, Paris: Sopexa, 1989.

Wines from Spain Far from Ordinary Wine Guide, 2005–2006, New York: Wines from Spain, Trade Commission of Spain, 2005.

Zraly, Kevin, *Windows on the World: Complete Wine Course*, New York: Sterling Publishing Co., 2006.

Web Sites

www.abanet.org: American Bar Association

www.alcoholpolicy.niaaa.nih.gov: Alcohol Policy Information System, National Institute of Alcohol Abuse and Alcoholism

www.allaboutgreekwine.com: Experience the Greek Wine Renaissance

www.americanheart.org: American Heart Association

www.beaujolais.com: Les Vins du Beaujolais, official Beaujolais wines Web site

www.beerinstitute.org: The Beer Institute

www.bordeaux.com: Bordeaux, Everything about Wine: From the Vine to the Wine, Conseil Interprofessionnel du Vin de Bordeaux (CIVB)

www.brandydejerez.es: Consejo Regulador de Brandy de Jerez

www.burgundy-wines.fr: Bureau Interprofessionnel des Vins de Bourgogne (BIVB), official Burgundy wines Web site

www.canadiandistillers.com: Association of Canadian Distillers

www.cdc.gov: Centers for Disease Control and Prevention

www.champagne.com: Le Champagne, Comité Interprofessionnel du Vin de Champagne, official Champagne wines Web site

www.discus.org: Distilled Spirits Council of the United States

www.dorueda.com: Consejo Regulador Denominación de Origen Rueda, Vinos con DO Rueda (Wines from DO Rueda, Spain)

www.french-malbec.com: Cahors, Birthplace of Malbec

www.health.harvard.edu: Harvard Health Publications, Harvard Medical School

www.icap.org: International Center of Alcohol Policies

www.iihs.org: Insurance Institute for Highway Safety

www.jancisrobinson.com: Jancis Robinson, For People Who Love Wine

www.loirevalley.com: Loire Valley Wines, Loire Valley Wine Bureau

www.marininstitute.org: Alcohol Policy, The Marin Institute

www.matchingfoodandwine.com: Matching Food and Wine with Fiona Becket

www.napavintners.com: Napa Valley Vintners

www.nbwa.org: National Beer Wholesalers Association

www.ncadd.org: National Council on Alcoholism and Drug Dependence

www.newyorkwines.org: Uncork New York!, New York Wine and Grape Foundation

www.nhtsa.dot.gov: National Highway Transportation Safety Administration

www.niaaa.nih.gov: National Institute on Alcohol Abuse and Alcoholism

www.nih.gov: National Institutes of Health

http://nutrition.tufts.edu: Tufts Nutrition, Friedman School of Nutrition Science and Policy, Tufts University

www.psiloveyou.org: Petite Sirah Advocacy Organization

www.rhonerangers.org: The Rhône Rangers, Advancing Knowledge and Enjoyment of American Rhône-style Wine

www.servsafe.com: ServSafe, Educational Foundation, National Restaurant Association

www.tastings.com: Tastings, Web site of the Beverage Testing Institute

www.ttb.gov/alcohol/index.htm: Alcohol and Tobacco Tax and Trade Bureau

www.vibrantrioja.com: Vibrant Rioja, Consejo Regulador de la Denominación de Origen Calificada Rioja and Spanish Institute for Foreign Trade

www.vins-de-pays.info: Vin de Pays de France

www.vins-rhone.com: Vineyards of the Rhône Valley, official Rhône wines Web site

www.vinsalsace.com: Les Vins d'Alsace, Conseil Interprofessionnel des Vins d'Alsace

www.viticulture.hort.iastate.edu: Viticulture, University Extension, Department of Horticulture, Iowa State University

http://wine.appellationamerica.com: North American Wine Regions

www.wineinstitute.org: The Wine Institute

www.winepros.org: Professional Friends of Wine

www.winereviewonline.com: Wine Review Online

www.wines-france.com: Wines of France, Office National Interprofessionnel des Vins (ONIVINS)

www.winesfromspain.com: Wines from Spain, official Web site of the Spanish Institute for Foreign Trade

www.wswa.org: Wine and Spirits Wholesalers of America, Inc.

www.zinfandel.org: Zinfandel Advocates and Producers (ZAP)

Index

Page numbers in *italics* refer to photographs and illustrations.